Modern Aspects of Emulsion Science

Modern Aspects of Emulsion Science

Edited by
Bernard P. Binks
Department of Chemistry, University of Hull, UK

THE ROYAL
SOCIETY OF
CHEMISTRY
Information
Services

The cover picture shows emulsions of silicone oil-in-water ($\varphi = 0.7$) stabilised by nonyl-phenol heptaethoxylate. The upper left picture (scale bar $= 155\,\mu m$) is for the crude, polydisperse emulsion prepared in a mechanical mixer, whilst the lower right picture (scale bar $= 25\,\mu m$) is the truly monodisperse emulsion after controlled shearing at a rate of $10^3\,s^{-1}$. See Chapter 1 for further details.

(After T.G. Mason and J. Bibette, *Langmuir*, 1997, **13**, 4600; the author thanks Dr T.G. Mason for supplying the original photographs)

ISBN 0-85404-439-6

A catalogue record for this book is available from the British Library

Published by The Royal Society of Chemistry,
Thomas Graham House, Science Park, Milton Road,
Cambridge CB4 4WF, UK

For further information see our web site at www.rsc.org

Typeset by Keytec Typesetting Ltd, Bridport, Dorset.
Printed by Bookcraft (Bath) Ltd.

Preface

Emulsions occur either as end products or during the processing of products in a huge range of areas including the food, agrochemical, pharmaceutical, paint and oil industries. In some cases the desire is to stabilise an emulsion, *e.g.* in milk, in other cases to destabilise it, *e.g.* in water-in-crude oil emulsions. Despite over 100 years of research in emulsion science, a quantitative understanding of emulsions has been lacking.

The primary objective of this book is to describe the most important developments in understanding emulsions which have appeared within the last 10 years or so. Emphasis has been placed on relating the *chemistry* of the surfactant or protein adsorbed at the oil–water interface to the principles of the *physics* involved in the bulk emulsion property. Each chapter contains the relevant scientific background to the particular area, followed by an in-depth and up-to-date account of more recent research findings. Although this book is not a conference proceedings, the idea behind it came as a result of the three-day meeting entitled 'Emulsions' organised by the editor under the auspices of the Colloid and Interface Science Group (of the RSC) held at The Lawns, University of Hull, on 19–21 March 1997.

The first chapter, written by the editor, is a recent review of emulsions (including their formation, emulsion type, emulsion stability and interactions between adsorbed layers at oil–water interfaces) and introduces the necessary concepts and ideas for the subsequent chapters. Walstra and Smulders in Chapter 2 describe the processes involved during the preparation of emulsions by conventional means (homogenisers, colloid mills, *etc.*), whereas Vincent *et al.* discuss a novel preparation route involving nucleation and growth in Chapter 3.

The following six chapters deal in detail with the various mechanisms of instability occurring in emulsions. In Chapter 4, Robins and Hibberd discuss the use of ultrasound in following creaming in emulsions, in the presence or absence of flocculation. Dickinson relates the rheology of emulsions to their structure and stability to creaming and flocculation in Chapter 5. The important phenomenon of phase inversion in emulsions is dealt with by Brooks *et al.* in Chapter 6, who distinguish between transitional (brought about by changing factors which affect the hydrophile–lipophile balance of the system) and catastrophic (induced by increasing the fraction of dispersed phase) inversion. Emulsion coalescence forms the basis of the next two chapters. Kabalnov in Chapter 7 reviews the evidence relating the equilibrium properties of oil + water + surfactant systems to the emulsion type and stability, and then puts forward a new model for coalescence involving hole formation in thin liquid films. The agreement between the model and careful experiments is convincing. Deminière *et al.* in Chapter 8 describe the use of truly monodisperse emulsions in following the controlled coalescence of emulsions, and quantify the activation energies for the process in a variety of systems. Finally, it is shown how Ostwald ripening, covered in Chapter 9 by Weers, can lead to emulsion instability and a technique is described for following it.

The penultimate two chapters are concerned with concentrated emulsions. Petsev in Chapter 10 gives an extensive analysis of the various types of interactions that occur between emulsion drops on close approach and how these are affected by the presence of microemulsion droplets in the continuous phase. Chapter 11 by Solans *et al.* relates the structure and stability of so-called gel emulsions to the underlying phase behaviour of the surfactant system. The book rounds off with a chapter (12) by Förster and von Rybinski from Henkel on some of the more important current applications of emulsions in the world of today.

I would like to thank all the authors for their swift cooperation in preparing this book, and Mrs. Janet C. Freshwater of Information Services at the RSC, Cambridge, for guiding me through the various stages of its production.

Bernard P. Binks
March 1998

Contents

Chapter 12   Applications of Emulsions   395
Thomas Förster and Wolfgang von Rybinski

CHAPTER 1

Emulsions—Recent Advances in Understanding

BERNARD P. BINKS

Surfactant Science Group, Department of Chemistry, University of Hull, Hull HU6 7RX, UK

1.1 Introduction

This chapter reviews the progress in the understanding of emulsions over the last ten years or so. The emphasis is on the factors affecting the type and subsequent stability of emulsions, and on the associated properties of surface-active molecules adsorbed at the oil–water interface. The field of emulsions is a vast area and so the literature covered is selective rather than comprehensive.

An emulsion may be defined as an opaque, heterogeneous system of two immiscible liquid phases ('oil' and 'water') where one of the phases is dispersed in the other as drops of microscopic or colloidal size (typically around 1 μm). There are two kinds of simple emulsions, oil-in-water (O/W) and water-in-oil (W/O), depending on which phase comprises the drops.[1] Emulsions made by agitation of the pure immiscible liquids are very unstable and break rapidly to the bulk phases. Such emulsions may be stabilised by the addition of surface-active material which protects the newly formed drops from re-coalescence. An emulsifier is a surfactant which facilitates emulsion formation and aids in stabilisation through a combination of surface activity and possible structure formation at the interface.

This book is concerned with *macro*emulsions and not *micro*emulsions. The latter are thermodynamically stable dispersions of oil and water, which means that they form spontaneously and are stable indefinitely. Being optically clear, their characteristic size lies in the range 5–50 nm. Most (macro)emulsions require the input of considerable amounts of energy for their formation and can only be stable in a kinetic sense. However, many systems of oil + water + surfactant which form microemulsions may be emulsified to emulsions and there is a growing interest in relating the properties of these emulsions to the known equilibrium phase behaviour of the corresponding microemulsions. Several recent reviews on microemulsions exist[2–4] and an aim of this chapter is to discuss, as far as is possible, the behaviour of emulsions, stabilised by low molecular weight surfactants, in terms of the aggregation and adsorption in the micellar/microemulsion systems.

Several books on emulsions have appeared in the last decade.[5-13] The ones devoted solely to emulsions[10,12,13] include chapters on emulsion stability, food emulsions, crude oil emulsions, rheology, pharmaceutical emulsions and perfluoro-chemical emulsions as blood substitutes. Emulsions are so widely encountered in a huge variety of industries, *e.g.* agrochemical, food, pharmaceutical, paint, printing, petroleum, *etc.*, that two World Congresses on Emulsions have been held in 1993 and 1997. The proceedings have been published[14,15] and papers covered areas from manufacturing and stability to wetting and adhesion to applications. It would be impossible to review all aspects of emulsion science and technology in one article, and so the author refers to the many review articles which exist on the topics not to be discussed here, although some are the basis of subsequent chapters. These include emulsion formation,[16,17] rheology,[18,19] multiple emulsions,[20,21-23] solid-stabi-lised emulsions,[22,23] techniques for measurement,[24] parenteral (fluorochemical or phospholipid-stabilised) emulsions,[25-27] food emulsions,[28-30] crude oil emul-sions,[31,32] and applications.[33,34] This list is not exhaustive but it serves to illustrate the scope of the subject.

The chapter is organised into the following sections: Emulsion Type and the System Hydrophile–Lipophile Balance (HLB), Phase Inversion, Emulsion Stabi-lity, Gel Emulsions, and Forces between Oil–Water Interfaces.

1.2 Emulsion Type and the System Hydrophile–Lipophile Balance

1.2.1 Emulsions of Two Liquid Phases

Whether an emulsion is O/W or W/O depends on a number of variables like oil:water ratio, electrolyte concentration, temperature, *etc.* For most of this century, emulsion chemists have known that surfactants more 'soluble' in water tend to make O/W emulsions and surfactants more 'soluble' in oil tend to make W/O emulsions. This is the essence of Bancroft's rule, which states that the continuous phase of an emulsion tends to be the phase in which the emulsifier is preferentially soluble.[35] The word 'soluble' is misleading, however, for two reasons. Firstly, a surfactant may be more soluble in, say, oil than in water in a binary system, but in the ternary system of oil + water + surfactant it may partition more into water. A good example of this is with the anionic surfactant Aerosol OT (sodium bis-2-ethylhexylsulfosuccinate) which dissolves in heptane at 25 °C up to at least 0.5 M but has a solubility limit in water of only ~0.03 M.[36] An emulsion made from equal volumes of water and heptane at 25 °C is O/W, however.[37] Secondly, no distinction is made between the solubility of monomeric or aggregated surfactant in oil or water. We will see that this is an important omission.

The first quantitative measure of the balance between the hydrophilic and hydrophobic moieties within a particular surfactant came in 1949 when Griffin introduced the concept of the HLB, or hydrophile–lipophile balance, as a way of predicting emulsion type from surfactant molecular composition.[38] A major problem of the HLB concept is that the HLB numbers assigned to the neat surfactant take no account of the effective HLB of a surfactant *in situ* adsorbed at an oil–water interface. Thus, for example, a nonionic surfactant of low HLB number (and hence predicted to stabilise W/O emulsions) may form O/W emulsions at low enough temperatures. It therefore became clear that the prevailing conditions of temperature, electrolyte concentration, oil type and chain length and cosurfactant concentration can all modify the geometry of the surfactant at an interface[39] and thus change the curvature of the surfactant monolayer, which in some way affects the preferred emulsion type.[40,41]

There have been developments in understanding how the type of emulsion is related to the phase diagram of mixtures of oil + water + surfactant at equilibrium.[42–46] These have come from the studies with microemulsions and as an example we take the case of a nonionic surfactant of the polyoxyethylene glycol ether type, C_nE_m. Let us consider equilibrium systems of heptane and water (equal volumes) containing $C_{12}E_5$. At low [surfactant], monomer distributes between oil and water but heavily in favour of the oil.[47] The partition coefficient defined as (molar concentration in heptane)/(molar concentration in water) increases from ~130 at 10 °C to ~1500 at 50 °C. Above a critical surfactant concentration, reached in both phases and designated $C\mu C_{water}$ (typically 5×10^{-5} M) and $c\mu c_{oil}$ (typically $6-60 \times 10^{-3}$ M), all additional surfactant in excess of this concentration is present in the form of aggregates, hence the symbol μ for 'microemulsion'. At low temperatures (<28 °C) the aggregates are oil-in-water microemulsion droplets formed in the water phase, in equilibrium with excess oil containing monomeric surfactant (Winsor I system). The preferred curvature of the monolayer is around oil and may be termed positive. At higher temperatures (>30 °C), aggregation occurs in the oil phase in the form of water-in-oil microemulsion droplets which are in equilibrium with monomeric surfactant in the aqueous phase (Winsor II system). The monolayer curvature is now negative. At intermediate temperatures (28–30 °C), three phases are formed (Winsor III system) consisting of a surfactant-rich phase and both excess oil and aqueous phases.[48] Here, the monolayer has, on average, zero net curvature.

It is frequently observed that the type of emulsion (O/W, W/O or intermediate) formed by homogenisation of the Winsor system is the same as that of the equilibrium microemulsion,[49–51] *e.g.* emulsification of an O/W microemulsion plus excess oil generally gives an O/W emulsion, the continuous phase of which is itself an O/W microemulsion. It is not obvious why this should be so, since in the case of microemulsions their behaviour is determined in part by the spontaneous curvature of the surfactant monolayer stabilising the nanometre-sized droplets. For emulsions, the radii of curvature of micrometre-sized drops are of the order of 1000 times the molecular dimensions and so it is difficult to see how curvature effects are implicated. In order to test and understand Bancroft's rule, the type of emulsion at different temperatures must be determined at various fixed surfactant concentrations. This is readily done *via* conductivity measurements (Figure 1.1). At high [$C_{12}E_5$], sufficient to form microemulsion aggregates in equilibrium systems, emulsions invert from O/W to W/O at temperatures around the Winsor III region.[52] Thus, Bancroft's rule holds but it is impossible to say whether preferred emulsion

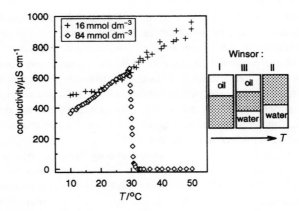

Figure 1.1 *Conductivity of emulsions ($\phi = 0.5$) vs. temperature for two initial [surfactant] in the system ($C_{12}E_5$ in heptane)–0.01 M NaCl. In the Winsor equilibria shown on the right, the shaded areas represent microemulsions*
(Redrawn with permission from ref. 52)

type is determined by the distribution of monomeric or aggregated surfactant. At low [$C_{12}E_5$], where only monomer surfactant exists, emulsions are O/W at low temperatures and remain O/W even at temperatures where Winsor II systems would form in systems of higher surfactant concentration. Thus, we see an apparent violation of Bancroft's rule since despite the surfactant monomer distribution being in favour of oil, the stable emulsions are O/W. Similar findings were reported by Harusawa *et al.*[53,54] using nonylphenol ethoxylated surfactants.

A more useful concept than that of surfactant HLB numbers is that of the system HLB, related to the locus of aggregate formation. Thus 'high' HLB systems are those in which the aggregates (micelles or O/W microemulsions) form in the water phase, whereas 'low' HLB systems are those in which aggregation occurs in the oil phase, either as reverse micelles or W/O microemulsion droplets. It is now more correct to say that the phase containing the surfactant aggregates becomes the continuous phase of an emulsion. From the foregoing discussion, it is clear that the continuous phase of an emulsion is not necessarily the phase containing the highest concentration of surfactant. Bancroft's rule and the system HLB apply to surfactant *aggregates* rather than to *total* surfactant present in the system.

The results quoted above have allowed testing of theories put forward to explain preferred emulsion type. In the kinetic theory of Davies,[55] it was assumed that the type of emulsion was determined by the relative rates of coalescence of oil and water drops after emulsification. Coalescence can be modelled as (i) the approach of the drops and the formation of a plane-parallel film and (ii) thinning of the film to a critical thickness followed by film rupture. The kinetics of thinning of such emulsion films was predicted theoretically using a hydrodynamic model.[56–58] This velocity of thinning is dependent upon the balance of forces acting at the interface of the approaching drops. As the two drops come close to each other, liquid flows out of the film to its thicker parts resulting in the convective flux of surfactant at the surface, thus perturbing its equilibrium distribution. This generates reverse fluxes tending to restore equilibrium, including surface diffusive flux and bulk fluxes from the film and drop. The difference in surface concentration of surfactant results in a variation of the interfacial tension which produces a surface force (tension gradient) opposite to the liquid flow. The rise in tension depends on the Gibbs elasticity G of the film, equal to $2\varepsilon_D$ if sufficiently thick, where ε_D is the surface dilational modulus. G is a measure of the film's resistance to thinning. With this model, it is shown that the velocity of thinning depends on the location of non-adsorbed surfactant. If surfactant is present mainly in the continuous (film) phase, it has to diffuse a long way from the film perimeter in order to reduce G. Since the driving force for this process is the gradient of surfactant concentration along the surface, this diffusion cannot eliminate the tension gradient which opposes thinning. If, however, the surfactant is more soluble in the dispersed (drop) phase, it must diffuse a much shorter distance, and since this flux is driven by the normal gradient of the concentration, it can counterbalance the convective flux and so relieve the elasticity and increase the thinning velocity. Summarising, theory predicts that emulsions containing most of the surfactant in the *dispersed* phase will coalesce faster than those where surfactant is mainly in the *continuous* phase, and so will not be the preferred emulsion type, in apparent accordance with Bancroft's rule. The results in Figure 1.1, alongside the known partitioning of monomers and aggregates, demonstrate that arguments of emulsion type based on film elasticity are not applicable in cases where significant partitioning of monomer to an oil phase occurs.

Recently, Petsev *et al.*[59] have advanced an argument based on interfacial bending in order to explain the correspondence between emulsion (sub-micron in size) and microemulsion types. They assume that drop surfaces can deform on approach and, in addition to the energy of interfacial stretching upon collision, the energy of interfacial bending is also taken into account. This is the energy contribution caused by the variation of the interfacial curvature, $H = -1/r$, r being the drop radius. The corresponding contribution to the drop deformation energy is

$$W_b = -2\pi a^2 B_0 H \qquad (1.1)$$

provided $(a/r)^2 \ll 1$, where a is the film radius and B_0 is the interfacial bending moment of a flat interface, equal to $-4KH_0$, with K being the bending elasticity constant[60] and H_0 the spontaneous curvature. K is a measure of the energy required to bend unit area of a spherical interface. Assuming $a \approx r/50$ and using equation (1.1), one obtains $|W_b| = r(\pi/1250)|B_0|$. For oil–water interfaces saturated with surfactant, K is of the order of the thermal energy kT,[61] and $|B_0|$ is of the order of 5×10^{-11} N.[62] With $r = 5 \times 10^{-7}$ m, then $W_b \approx 15kT$. In other words, bending effects can be significant for the interaction between sub-micron emulsion drops. Although the bending energy is known to be important for the interfaces of very high curvature as in microemulsion droplets,[63] the above result seems surprising for emulsion drops. It is due to the fact that when the drop radius r increases, the bent area increases faster ($a^2 \propto r^2$) than the bending energy per unit area decreases ($H \propto 1/r$). For positive B_0, the interface is bent around the oil phase (O/W) and the bending moment facilitates the formation of a flat film between two aqueous drops in oil ($W_b < 0$) but opposes the formation of a flat film between two oil drops in water ($W_b > 0$). In emulsions where the interfacial tension is low (say < 0.1 mN m^{-1}), simultaneous formation of microemulsion droplets and emulsion drops can occur (Figure 1.2). One can expect that when $B_0 H < 0$, the formation of stable drops is facilitated since $W_b > 0$, opposing deformation. This is the case for both macro- and microdrops depicted in Figure 1.2a, where microdroplets are in the continuous phase. On the other hand, H has opposite signs for macro- and microdrops in the configuration depicted in Figure 1.2b, where microdroplets are in the dispersed phase. $B_0 H < 0$ for microemulsion droplets but is positive for the macroemulsion drops. Consequently, the bending moment will favour the formation of a plane film between approaching macrodrops. This will facilitate substantial deformation, leading to flocculation and coalescence of the emulsion drops containing microemulsion droplets inside. The emulsion containing microemulsion droplets in the continuous phase will be more stable and survive, as observed in practice.

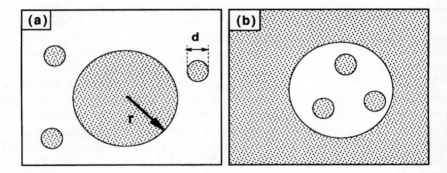

Figure 1.2 *Sketch of a miniemulsion drop coexisting with smaller microemulsion droplets which could be in either the continuous phase (a) or in the dispersed phase (b). State (a) is more stable and survives*
(Redrawn with permission from ref. 59)

A similar theoretical argument, proposed by Kabalnov and Wennerström,[64] explains the correspondence between the equilibrium phase behaviour and the emulsion type and stability by considering the final stage of drop coalescence in which two drops which have made contact are connected by a narrow neck or hole, as shown schematically in Figure 1.3. The surfactant monolayer is highly curved in the region of the neck whereas it is virtually planar everywhere else. As the neck grows there is a change in both the area of the intervening liquid film (ΔA) and in its curvature. On one hand, emulsion film rupture is driven by reducing the interfacial area of the planar part of the film. On the other, the edge of the neck creates extra interfacial area and therefore a free energy penalty. The energy barrier to hole nucleation thus comes from an interplay between the free energy penalty at the edge of the hole and the free energy gain at the planar part.

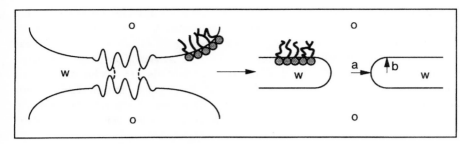

Figure 1.3 *Schematic representation of hole nucleation in a liquid water film originally separating two oil emulsion drops. Following the situation depicted on the left in which thermal corrugations exist, the drops come into contact, the monolayers fuse (dotted lines) and a narrow neck (hole) forms between the drops of radius a and water film thickness 2b. The radius of the oil drops is assumed to be much larger than 2b, so that the water film is considered as being flat. The two pictures are not to scale with respect to each other*

The total free energy of the hole, W_T, is the sum of three terms:

$$W_T = W_1 + W_2 + W_3 \qquad (1.2)$$

W_1 is the interfacial free energy

$$W_1 = \gamma \Delta A = 2\pi\gamma(\pi - 3)b^2 + (\pi - 2)ab - a^2 \qquad (1.3)$$

where γ is the oil–water interfacial tension, a is the radius of the hole and $2b$ is the liquid film thickness. W_2 is the mean curvature energy and accounts for the 'extra' interfacial tension of a curved interface relative to a planar one

$$W_2 = 2K \int (H^2 - 2H_0 H)\, dA \qquad (1.4)$$

and W_3 is the Gaussian curvature contribution accounting for the fact that fusion of two drops changes the topology of the film

$$W_3 = -4\pi\overline{K} \qquad (1.5)$$

where $\overline{K}$ is the saddle-splay elasticity constant.[65] Of the parameters appearing in eqns. (1.2)–(1.5), K and H_0 are known from microemulsion studies whilst $\overline{K}$ does not change as the neck grows. Expressing the tension γ as a function of the spontaneous curvature of the film, the total free energy of the hole (W_T) can be calculated as a function of a and b. For large values of the neck radius a, the term W_1 is large and negative, but for values such that a and b are of similar magnitude, W_1 and W_2 are of similar magnitude. W_1 is insensitive to the sign of the spontaneous curvature H_0 whereas W_2 depends on it, since the neck is a region of the surfactant monolayer which is strongly curved towards the continuous phase.

Calculations reveal that it is this term which has a profound influence on the energetics of neck growth. For large, positive values of H_0, oil–water–oil (O/W/O) films are stable with a coalescence barrier equal to W_T of around $40kT$, whereas water–oil–water (W/O/W) films break without a barrier. Conversely, for large negative values of H_0, W/O/W films are stable whilst O/W/O films break without a barrier. In the vicinity of phase inversion, a very steep change in film stability with H_0 is predicted with the barrier going to zero, and in careful experiments Kabalnov and Weers[66] have demonstrated that this prediction is quantitatively accurate for the $C_{12}E_5$/octane/aqueous NaCl system.

1.2.2   Emulsions of Three Liquid Phases

Little work has been done on three-phase emulsions formed from excess oil and water phases plus the third phase in Winsor III systems, probably due to their low stability. It is more difficult to determine the morphology of a three-phase emulsion completely from conductivity measurements because many more possibilities exist than for two-phase emulsions, *e.g.* third-in-oil-in water, water-in-third-in oil, and because the theory of dispersion conductivities has not been developed extensively. Nonetheless, in a series of papers,[67–70] Smith and co-workers describe experiments aimed at determining emulsion type in stirred, steady-state emulsions prepared from equilibrium phases. They test two hypotheses in the older literature; one says that the continuous phase is always the third phase regardless of all other parameters,[71] whilst the other, based on wettability arguments from experiments of single drops of two fluids in a large volume of a third, predicts that the 'choice' of continuous phase depends solely on the wettability among the three fluids.[72] Thus, for a three-component, isobaric system the continuous phase must depend solely on temperature, as do the interfacial tensions. Smith *et al.* not only show that three-phase emulsions may be either water, oil or third phase continuous but also that morphology changes at constant temperature can be induced by changes in the volume fraction of one of the phases. Some of their results are shown in Figure 1.4 for the system C_4E_1 + decane + 10 mM aqueous NaCl. At equilibrium, a three-phase region exists between 23.8 and 46.6 °C and the third phase wets the interface between oil and water at all temperatures. For dispersions of isotropic phases in which the differences among the phase conductivities are not too large, the conductivity of the dispersion, K_{dis}, is given by the approximate equation[73]

$$K_{dis} = \left[\sum_i \phi_i(K_i)^{1/3} \right]^3 \qquad (1.6)$$

where the sum is over all phases, i, of phase conductivity K_i and phase volume fraction ϕ_i. Similar equations to this are used to deduce which phase is continuous. Starting from two-phase emulsions (comprised of either third phase plus oil or third phase plus water), the other phase is progressively added and the conductivity of the resulting emulsion determined. Figure 1.4 shows that, depending on the relative phase volume fractions, three-phase emulsions can be water continuous (morph-

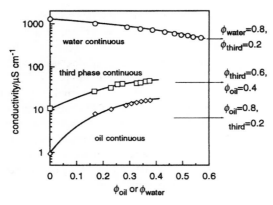

Figure 1.4 *Conductivities of (initially) two-phase emulsions, of composition given, on progressive addition of the other phase in three-phase systems of C_4E_1 + decane + 0.01 M NaCl at 35 °C. Third phase wets oil–water interface*
(Redrawn with permission from ref. 69)

ology oil/third/water), oil continuous (morphology water/third/oil) or third-phase continuous (morphology (water and (third phase/oil))/third phase). Using a longer chain nonionic surfactant, conditions are chosen such that the third phase does not wet the oil–water interface. For C_6E_2 + tetradecane + 10 mM NaCl, three phases form from 9 to 49 °C, where the third phase is non-wetting below 37 °C and wetting above it. Measurements conducted at 25 °C in which the oil phase is successively added to an initial third phase-in-water emulsion (Figure 1.5) now show that the conductivities of the three-phase emulsions fall into three ranges of oil volume fraction. Within each range the conductivity changes smoothly whereas the boundaries of the ranges are defined by large discontinuous changes in

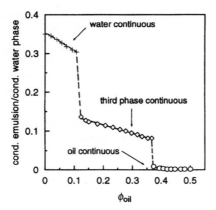

Figure 1.5 *Normalised conductivities of initial third phase-in-water emulsions (ϕ_{third} = 0.6, ϕ_{water} = 0.4) upon addition of oil phase in systems containing C_6E_2 + tetradecane + 0.01 M NaCl at 25 °C. Third phase is non-wetting*
(Redrawn with permission from *Langmuir*, 1993, **9**, 2089)

conductivity. The three ranges correspond to water continuous, third phase continuous and oil continuous emulsions, respectively. Smith and Johnson[74] have also
questioned the phase inversion temperature (PIT) idea[71] that, below the PIT,
emulsions (two phase or three phase) have a water-rich continuous phase, and
above it the emulsions have an oil-rich continuous phase. They provide evidence
that proves the exact opposite of this idea, namely by forming three-phase
emulsions which are oil continuous below the PIT and water continuous above it.
Their conclusions are that the number of phases, phase conductivities, phase
volume fractions, wettabilities and emulsion morphologies are all functions of
temperature, and that variable temperature, constant composition experiments
which are frequently carried out are difficult to interpret for three-phase emulsions
unless requisite constant temperature data are available.

1.3   Phase Inversion

Phase inversion in emulsions can be one of two types. Transitional inversion is
induced by changing factors which affect the HLB of the system, *e.g.* temperature,
electrolyte concentration. Catastrophic inversion is induced by increasing the
fraction of the dispersed phase and has the characteristics of a catastrophe, meaning
a sudden change in behaviour of the system as a result of gradually changing
conditions. Catastrophe theory, a mathematical framework to describe catastrophes,
has been successfully applied by Vaessen and Stein[75] to reproduce the qualitative
features of catastrophic phase inversion following the suggestion first by
Dickinson.[76] It has been shown[77] that catastrophic inversions are not reversible—
the value of the water:oil ratio at the transition when oil is added to water is not the
same as that when water is added to oil—and that the inversion point depends on
the intensity of agitation and on the rate of liquid addition to the emulsion. A
detailed study has been made of the dynamics of both transitional and catastrophic
phase inversion in emulsions stabilised by nonionic surfactants.[78] In the former, the
inversion is caused by changing the system HLB at constant temperature using
surfactant mixtures. In Figure 1.6 it can be seen that the average emulsion drop size
decreases and the emulsification rate (defined as the time required to achieve a
stable drop size) increases as inversion is approached. Both trends are consistent
with the oil–water interfacial tension passing through a minimum within the HLB
range where three phases form. It was also noted that emulsions formed by
transitional inversion were finer and required less input of energy than those
produced by direct emulsification. Optical microscopy was used to show that, at
inversion, multiple emulsions of various types were formed for both inversion
mechanisms.

   It is now well established that the choice of emulsification conditions is an
important consideration in determining the ultimate drop size (and hence stability)
of an emulsion. Using nonionic surfactants, Shinoda and Saito[79] demonstrated that
emulsification at the phase inversion temperature (PIT) followed by cooling led to
the formation of stable O/W emulsions of small drop size. Emulsification at
temperatures higher than the PIT, initially producing W/O emulsions, resulted in
very stable emulsions on subsequent cooling. The inversion process, forming a

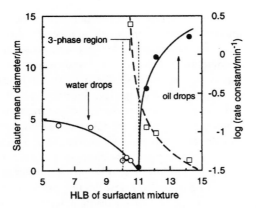

Figure 1.6 *Emulsion drop diameters (circles) and rate constants for attaining steady size (squares) as a function of surfactant HLB in cyclohexane + 0.067 M KCl systems containing nonylphenol ethoxylates at 25 °C*
(Redrawn with permission from ref. 78)

third surfactant-rich phase, is implicated in this phenomenon. Since then, various authors have confirmed these findings in a variety of systems. Mitsui *et al.*[80] applied the PIT method to the emulsification of commercial polyoxyethylene oleyl ether surfactants used in cosmetics. Enever[81] correlated the rate of emulsion drop coalescence with the PIT in systems containing $C_{16}E_4$. On increasing the electrolyte concentration in water, the PIT was lowered and emulsions prepared and stored at temperatures close to the PIT were very unstable. Saito *et al.*[82] have shown how the PIT and O/W emulsion stability depend on the distribution of both ethylene oxide chain length and alkyl chain length with samples of $C_{12}E_6$. Their conclusion is that samples possessing a wide distribution of chain lengths had improved emulsion stability. For a series of polyoxyethylene oleyl ether surfactants of different head group size, Pasternacki-Surian *et al.*[83] concluded that the type of surfactant-rich phase as controlled by the emulsification temperature plays a significant role in the efficiency of the emulsification process. Emulsions prepared at temperatures where the surfactant + water phase contained a liquid crystal (or L_α) dispersion were the most stable. Phase inversion in emulsions is accompanied by sharp changes in their physical properties. This can be used to detect, say, the PIT, by different methods such as measurements of electrical conductivity, viscosity and dielectric constant.[84]

In a series of papers, Förster and co-workers have described very clearly the production of finely dispersed O/W emulsions with long term stability using the PIT method by careful selection of the emulsification route.[85-89] At temperatures around phase inversion, three phases can form at equilibrium. Upon increasing surfactant concentration, the amount of third phase increases by solubilising the excess water and oil phases. Eventually, a single surfactant-rich phase exists, being either a lamellar liquid crystal, L_α, a bicontinuous microemulsion or a so-called L_3 phase of reverse bilayer structure.[90] For the system cetearyl isononanoate (a polar oil) + $C_{16/18}E_{12}$/glycerol monostearate + water, this occurs at around 6 wt% at an oil:water ratio of 3:7. At this constant oil:surfactant mixture ratio of 4.5:1, the water content in the system is varied and the phase behaviour with respect to temperature is determined (Figure 1.7). In the phase inversion region, in addition to an L_α phase at high and low water content, a bicontinuous microemulsion also appears at low water content (<30%). For the preparation of finely dispersed emulsions, different emulsification routes, which utilise the phase behaviour in different ways, are represented in Figure 1.7. In route 1, all components are emulsified at 80 °C where an L_α phase is present. On cooling to room temperature, a finely dispersed oil phase with a monomodal size distribution at ~100 nm in diameter is obtained. In the two-step process in route 2 the bicontinuous micro-emulsion is used. A pre-concentrate of oil and surfactant mixture is emulsified with a water content of 15% at 85 °C, followed by dilution with water at 40 °C. Here only part of the total water must be heated (*cf.* low energy emulsification of Lin[91]) and the oil drops are also finely dispersed of mean diameter ~110 nm. In route 3 the same pre-concentrate as that used in route 2 is emulsified at 65 °C. This temperature lies within the phase inversion range but does not result in the appearance of a bicontinuous phase. The phase formed is probably an L_3 phase. On subsequent dilution, a coarse emulsion with a broad size distribution is obtained (mean diameter ~1400 nm). The results show that for this reasoning to be successful, it is not just sufficient to enter a single phase region near inversion, but that the interfacial tension must become low enough for either an L_α or bicontinuous microemulsion phase to form.

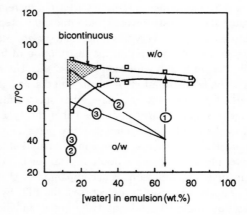

Figure 1.7 *Partial phase diagram for the system $C_{16/18}E_{12}$/glycerol monostearate/cetearyl isononanoate (0.66:0.33:4.5) versus water content. Numbers refer to the emulsification routes*
(Redrawn with permission from ref. 87)

Sagitani[92] has described an alternative method for producing fine O/W emulsions called 'surfactant-phase emulsification'. Here, the addition of polyols like 1,3-butanediol to the usual emulsion components oil, water and surfactant is necessary. In the method, oil is added dropwise to an isotropic solution of concentrated surfactant in a water/polyol mixture. A clear gel eventually forms which is an oil-in-surfactant phase emulsion. Dilution of this gel with water results in the final O/W emulsion. Thus fine emulsions can be prepared without the need for adjusting the system HLB. A final link between phase behaviour and emulsion inversion is provided by the work of Kahlweit *et al.*,[90,93] who correlate the shape of the emulsion conductivity/temperature curves with the amphiphilic strength of the surfactant. In systems of alkane + water + C_nE_m surfactants, they observe a transition between those surfactants of short n and m lengths like C_4E_1 which are weak amphiphiles forming molecular solutions, and those of longer chain lengths like $C_{12}E_5$ which are strong amphiphiles and form structured microemulsions.[94] For emulsions made with equal mass fractions of octane and water, the change of conductivity with temperature is very shallow for weakly structured mixtures but becomes sharper and more pronounced as one moves to long chain surfactants. The conductivity in the O/W emulsions at low T rises only slightly, whereas that in the W/O emulsions decreases steeply with increasing stability of the structure formed at equilibrium.

1.4   Emulsion Stability

Emulsion stability is a kinetic concept. In a stable emulsion, there is no discernible change in the number, size distribution and spatial arrangement of drops within the experimental timescale. This timescale can vary from a few seconds to years, implying that stability is also a relative concept. Emulsion instability may involve a number of processes which take place simultaneously or consecutively, depending on conditions. The four main ways in which an emulsion may become unstable are creaming (or sedimentation), flocculation, coalescence and Ostwald ripening. Curt[95] has reviewed the techniques used for the evaluation of emulsion stability, including the principle of each method and their main advantages and drawbacks.

1.4.1   Creaming

In an O/W emulsion, creaming is the movement of oil drops under gravity or in a centrifuge to form a concentrated layer at the top of the sample, but with no accompanying change in the drop size distribution. Initially, a concentration gradient of drops develops in the vertical direction, often followed by the appearance of a distinct boundary between an upper cream layer and a lower depleted emulsion layer. In a W/O emulsion, the equivalent phenomenon is called sedimentation. Creaming arises from the action of gravity on drops of lower density than that of the continuous phase and is reversible in that gentle agitation can re-establish the original uniform distribution of drops. In very dilute emulsions, the creaming speed v_S of an isolated, spherical drop of radius r moving through a fluid medium of density ρ_0 and Newtonian shear viscosity η_0 is given by Stokes' Law[96]

$$v_S = \frac{2r^2(\rho_0 - \rho)g}{9\eta_0} \qquad (1.7)$$

where g is the acceleration due to gravity and ρ is the density of the dispersed phase. In the absence of flocculation, eqn. (1.7) indicates that creaming in a dilute emulsion can be inhibited by reducing the average drop size, reducing the density difference of the phases or by increasing the viscosity of the continuous phase. Producing emulsion drops less than a micron in diameter is now possible using either high-pressure valve homogenisers[97] or the jet homogeniser.[98] Matching the densities of dispersed and continuous phases by, say, adding sugar or ethanol to water or by using heavy water in the case of triglyceride oils has been used to lower $\Delta\rho$. Adjusting the viscosity of the dispersion medium is normally achieved by thickening the aqueous phase with gums or starches, although this often leads to non-Newtonian character of the continuous phase and hence a complication in applying eqn. (1.7).

Equation (1.7) is strictly only applicable to unaggregated spheres at infinite dilution. Real emulsions are far from being monodisperse and a modification to this equation is to define an average speed with r^2 replaced by

$$\langle r^2 \rangle = \sum_i n_i r_i^5 \bigg/ \sum_i n_i r_i^3 \tag{1.8}$$

where n_i is the number of drops of radius r_i. The problems of *finite drop concentration* and *flocculation* are not so easily dealt with. Up to volume fractions of about 0.05, the average creaming speed $\langle v \rangle$ is less than the Stokes value by an amount proportional to ϕ. Batchelor's result[99] for $\langle v \rangle / v_s = 1 - 6.55\phi$ follows from the fact that, in a container with walls, each movement of a drop upwards must be accompanied by simultaneous movement downwards of an equivalent volume of the continuous phase. At ϕ above a few per cent, several theories of creaming/sedimentation have been developed with varying degrees of success,[100-104] but Barnea and Mizrahi[105] give a semi-empirical equation that fits experimental data reasonably well:

$$\langle v \rangle = v_s(1 - \phi)/(1 + \phi^{1/3}) \exp\left[5\phi/3(1 - \phi)\right] \tag{1.9}$$

Recently, a computer model which simulates creaming in a concentrated, non-flocculated polydisperse emulsion has been forwarded by Pinfield *et al.*[106] The model, making use of eqn. (1.9), includes the effects of hydrodynamic hindrance and thermal diffusion and considers the system as a set of fixed horizontal layers. Between each pair of layers there is a flux of drops during any small time interval. Polydispersity is accounted for by dividing the size distribution into discrete intervals of radius, each with the appropriate contribution to the total volume fraction. The time dependence is modelled by calculating, in each time interval, the volume contribution of each size of drop which crosses between adjacent layers. Then the creaming velocity in a given layer can be calculated from the local volume fraction within the layer. The numerical results are given for idealised emulsions (20% alkane-in-water and 10% sunflower oil-in-water) in the form of concentration profiles for the total volume fraction and for an individual particle size, along with the size distribution. In summary, a region of increasing concentration with height appears for the largest drops. These drops accumulate fastest at the top of the emulsion and occupy a greater proportion of the volume fraction in the upper regions. The smallest drops move downward (entrained with continuous phase fluid) from the top of the emulsion and increase in concentration at a lower level. All drops except the largest have a region of decreasing concentration with increasing height, caused by the slowing down of the descent of the cream interface with time. It is hoped that such predictions could be tested by experiment.

The traditional method to study creaming in dilute emulsions has been to follow the heights of cream or serum layers *versus* time either visually or using light scattering. In concentrated, opaque emulsions the most direct way has been to remove samples periodically at various heights and examine the volume fraction and drop size distribution.[107] This technique is invasive, if not destructive, and time consuming. Novel, non-invasive methods of studying creaming have become available recently. These include the use of conductivity,[108,109] magnetic resonance imaging (MRI)[110-112] and ultrasound.[113-116] Bury *et al.*[109] describe a method based on the measurement of the conductivity of an emulsion simultaneously at different heights by employing a multi-electrode conductivity cell. Using dispersion dielectric theory, data conversion of the qualitative values of conductivity into the quantitative values of dispersed phase volume fraction is performed. The approach, used in O/W emulsions stabilised by a range of ionic and nonionic surfactants, is claimed to be capable of detecting creaming processes with good accuracy in a reproducible manner. Using MRI, images of O/W emulsions can be acquired from the nuclear magnetic resonance signal generated by a spin-echo pulse sequence,[110] providing qualitative visualisation of creaming. Determination of the spin–lattice relaxation time T_1 quantitatively allows the migrating oil phase to be followed in a vertical plane through the emulsion and advantage is taken of the fact that the T_1 of protons in oil is about an order of magnitude lower than that of protons in water. Creaming of vegetable oil-in-water emulsions with Tween 80 surfactant has been studied as a function of oil volume fraction and surfactant concentration.[111]

The use of ultrasound to measure changes in the oil volume fraction profile as a function of height in concentrated emulsions has been investigated by a number of workers over the past few years. The power of the technique is that it can detect creaming in an emulsion long before it becomes visible to the eye in the form of a distinct cream layer. Ultrasonics is also a sensitive method of monitoring melting and crystallisation in emulsions.[117] So long as individual drops are smaller than the wavelength of the ultrasound, a simple relationship exists between the ultrasonic velocity u and the properties of the two phases:

$$u = (\beta\rho)^{-1/2} \tag{1.10}$$

$$\beta = \phi\beta_1 + (1 - \phi)\beta_0 \tag{1.11}$$

$$\rho = \phi\rho_1 + (1 - \phi)\rho_0 \tag{1.12}$$

Equations (1.10)–(1.12) together form what is known as the Urick equation, in which the quantities ρ and β are the average emulsion density and adiabatic compressibility, and the subscripts 1 and 0 refer to the dispersed and continuous phases, respectively. The ultrasound velocity is therefore directly related to the total volume fraction of dispersed phase and is independent of drop size. This is valid when scattering of ultrasound by the drops is not significant. A good review of the various experimental techniques, relevant theories and applications involving ultrasonics in emulsions (and suspensions), including the effects of scattering, has been given by McClements.[118]

Robins and co-workers have studied in detail the creaming of 20 vol% alkane-in-water emulsions stabilised by the nonionic surfactant Brij 35 ($\sim C_{12}E_{23}$) in the presence and absence of a non-adsorbing, nonionic hydroxyethylcellulose polymer (HEC) in the continuous phase.[113,114,119–123] Using a combination of ultrasound measurements, visual observation and drop sizing, they conclude that, up to 0.02 wt% HEC in the continuous phase, creaming is consistent with the movement of individual polydisperse drops. At 0.04%, creaming is much more rapid and the oil volume fraction profiles (Figure 1.8) show that the drops cream as a sharp boundary which rises rapidly from the base of the emulsion, indicative of a highly flocculated system, eventually leading to separation into a drop-rich phase and a drop-poor phase. The behaviour is consistent with a depletion (as opposed to bridging) mechanism[124] whereby exclusion of polymer from the immediate vicinity of emulsion drops produces an attraction between them, allowing flocculation in a shallow secondary minimum. The creaming kinetics and rheology of mineral oil-in-water emulsions stabilised by either Tween 20 (polyoxyethylene (20) sorbitan monolaurate) or SDS (sodium dodecyl sulfate) has also been investigated by Dickinson *et al.*[125–127] in systems containing the polysaccharides rhamsan, xanthan or dextran, all of which are non-adsorbing. Creaming is enhanced at low added polymer content due to depletion flocculation, but it is inhibited at higher concentrations due to thickening or gelation behaviour. In addition, the apparent shear viscosity of the emulsions increases from low values independent of applied shear stress (Newtonian) without polymer to high, shear-dependent values with

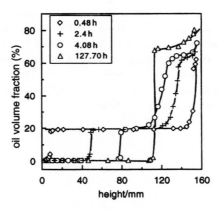

Figure 1.8 *Creaming profiles of 20 vol% heptane–hexadecane (9:1)-in-water emulsions stabilised by 0.28 wt% Brij 35 at 20°C in the presence of 0.04 wt% HEC in the aqueous phase. The times in hours are after emulsion formation* (Redrawn with permission from ref. 120)

polymer. An example of these effects is shown in Figure 1.9 for 18 vol% mineral oil-in-water emulsions of initial diameter 0.65 μm stabilised by 2 wt% Tween 20, where the creaming profiles are compared after ~19 h. Without added xanthan, the oil volume fraction is uniformly equal to the average value over the entire height of the sample, corresponding to an emulsion with good creaming stability. At the lowest xanthan content in the aqueous phase (0.017 wt%), a phase boundary has developed at ~30 mm up from the bottom of the tube. At 0.035%, an increase in the rate and extent of creaming is evident. The distinct boundary is now located at ~70 mm and both a serum layer at the bottom and a cream layer at the top of the sample have formed. A further increase in the rate of destabilisation is observed at 0.069% xanthan, so that within one day a cream layer ($\phi \approx 65\%$) is present in the top quarter of the tube and a serum layer ($\phi \sim 3\%$) occupies the rest of the space. At 0.173%, however, the emulsion is much more stable again, although it still creams faster than the original polymer-free sample. The results demonstrate the importance of surfactant–polymer interactions in controlling the stability of emulsions.

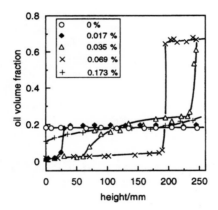

Figure 1.9 *Effect of [xanthan] in the aqueous phase on the creaming profiles after 19 hours of 18 vol% mineral oil-in-water emulsions stabilised by 2 wt% Tween 20 at 30 °C*

(Reproduced with permission from ref. 126)

1.4.2 Flocculation

Flocculation is the process in which emulsion drops aggregate, without rupture of the stabilising layer at the interface, if the pair interaction free energy becomes appreciably negative at a certain separation. It may be weak (reversible) or strong (not easily reversible) depending on the strength of the interdrop forces. Flocculation usually leads to enhanced creaming because flocs rise faster than individual drops due to their larger effective radius. Exceptions occur in concentrated emulsions where the formation of a gel-like network structure can have a stabilising influence. Flocculation is enhanced by polydispersity since the differential creaming speeds of small and large drops cause them to come into close proximity (and hence possibly aggregate) more often than would occur in a monodisperse system. The cream layer formed towards the end of the creaming process is actually a concentrated floc. The rate of flocculation can be estimated from the product of a frequency factor (how often drops encounter each other) and a probability factor (how long they stay in contact). The former can be calculated for the case of Brownian motion (perikinetic flocculation) or under shear flow (orthokinetic flocculation) whilst the latter depends on the interaction energy, *i.e.* the free energy required to bring drops from infinity to a specified distance apart. In calculating the interaction energy as a function of interdrop distance, three terms are normally considered: van der Waals attraction (depends on the drop diameter and the Hamaker constant), electrostatic repulsion (depends mainly on the surface potential of the drops, drop diameter and the ionic strength of the continuous phase) and steric repulsion due to protruding, flexible chains (depends on the surface density and length of these chains and on the solvent quality). We will see below that an additional attractive term may be added if depletion (mentioned above) occurs.

Experimental studies of flocculation have been reported for several different types of emulsion system.[128-133] Techniques commonly used to follow the kinetics are photon correlation spectroscopy, turbidimetry and drop-counting methods such as the Coulter counter. In dilute O/W emulsions stabilised by ionic surfactants, the flocculation rate, calculated from the change in the number of drops with time and applying the theory of von Smoluchowski, decreased whilst the zeta potential of the drops increased as the surfactant concentration was increased up to the critical micelle concentration (CMC).[128] Flocculation was also reduced by addition of either anionic[129] or cationic[130] surfactants to nonionic emulsions of Tween 80. The results were discussed in terms of the effect on the zeta potential causing increased repulsion between drops and on the changes in the adsorbed amount of nonionic surfactant at the oil–water interface. Jansson *et al.*[131] found that organic electrolytes containing tetraalkylammonium ions induced flocculation of sodium dodecanoate-stabilised emulsions at 10 times lower concentrations compared to divalent inorganic salts. The destabilisation could not be rationalised by electrostatic considerations alone, and arguments based on the bulky counterion inhibiting liquid crystal formation and on the adsorption of the ions modifying the drop surfaces were advanced. Perfluorochemical oil-in-water emulsions, developed as artificial blood substitutes, containing phospholipids as the main stabiliser can be prevented from flocculating by the addition of charged surfactant or by using a saccharide solution as the continuous phase.[132] Both electrostatic and hydration (steric) forces play a role in this effect. Washington[133] has reviewed the stability to flocculation of triglyceride oil-in-water emulsions used in total parenteral nutrition mixtures.

Perhaps the most interesting findings experimentally have arisen from investigating the effect of surfactant concentration on emulsion stability. Various surfactants, when present at a critical concentration above their CMC, cause emulsions to flocculate leading to rapid creaming/sedimentation. The presence therefore of 'free' (*i.e.* not emulsified) surfactant is implicated in this process. The systems include O/W emulsions stabilised by either nonionic[134-137] or ionic surfactants,[138-141] and flocculated by aqueous micelles, vesicles[142] or water-soluble polymers[143] and W/O emulsions flocculated by W/O microemulsion droplets[144] or reverse micelles.[145] In the case of nonionic surfactants, the critical surfactant concentration must be above the $c\mu c$ in oil. The results have been interpreted by Aronson[135] and Bibette *et al.*[138] in terms of depletion flocculation caused by excess surfactant, similar in some respects to the flocculation of sterically stabilised colloids at a critical concentration of free, non-adsorbed polymer[146] or surfactant.[147] Although the exact mechanism is still a matter of debate, it may arise from volume restrictions imposed on the free polymer by the presence of the larger colloid particle. This process depends upon particle size, molecular weight of the polymers and the adsorbed layer thickness.[148]

An example of the effect of free surfactant on creaming is given in Figure 1.10 for alkane-in-water emulsions stabilised by pure nonionic surfactants.[136] Samples of a stock emulsion containing 35 wt% oil were diluted to 25 wt% oil with varying concentrations of surfactant in water. Stability was assessed by allowing the emulsions to cream for a fixed time and then analysing gravimetrically a sample from the bottom of each vial for oil content. The results are expressed as the percentage of the total oil that is lost from this part of the emulsion. The extensive creaming above 1 wt% surfactant, attributed to flocculation without change in the drop-size distribution, has been confirmed by direct microscopic observation. Below 1 wt%, the drops are discrete and move independently while at higher concentrations tight flocs are developed which flow as a single unit. The onset of flocculation occurs at lower surfactant concentrations on increasing the average emulsion drop size. Bulk rheological measurements[135,149] showed that the emulsions are almost Newtonian at low [surfactant] and become shear thinning above the critical [surfactant] due to the gradual breakup of the flocs under shear. Even in emulsions stabilised by food protein surfactants, the effect of free protein on the creaming behaviour can be attributed to depletion flocculation. Using a combination of the ultrasound velocity scanning technique and light microscopy, the

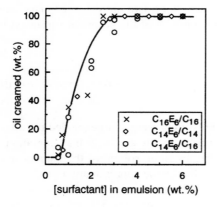

Figure 1.10 *Effect of surfactant concentration on the stability to creaming of 25 wt% alkane (C_n)-in-water emulsions stabilised by pure nonionic surfactants (C_nE_m) at 25 °C*
(Redrawn with permission from ref. 136)

influence of protein content on the stability of concentrated O/W emulsions (35 or
45 vol% tetradecane) containing sodium caseinate as the sole emulsifier has been
investigated.[150] At low overall caseinate content (1 wt%), the surface concentration
of protein is around half that required for a saturated monolayer and bridging
flocculation occurs accompanied by coalescence. For 2 wt% caseinate, the emulsions formed are the most stable with respect to creaming since there is sufficient
protein present for complete surface coverage and electrostatic stabilisation and
since the concentration of unbound protein is so low that no depletion flocculation
occurs. Enhanced creaming due to depletion flocculation is observed in emulsions
containing between 3 and 5 wt% caseinate. The flocculation is reversible and the
creaming profiles show evidence of a phase coexistence between a flocculated
(network) phase and a dilute phase containing discrete drops. At high caseinate
contents ($\geq$6 wt%), extensive depletion flocculation occurs which produces a very
strong emulsion drop network which reorganises slowly, and is therefore much
more stable to creaming and serum separation. The entities responsible for
inducing flocculation are thought to be casein sub-micelles of between 10 and
20 nm in size.

The basic model[135] for describing these systems consists of large (~ 1 µm), non-
deformable emulsion drops in a continuous phase containing small ($\sim$5 nm),
spherical surfactant aggregates (micelles or microemulsion droplets). When emulsion drops approach sufficiently close that the thickness of the film of continuous
phase between them is less than the diameter of the micelle, micelles are then
excluded ('depleted') from that thin film. Since this region is devoid of micelles,
the activity of the solvent exceeds that in the surrounding medium, resulting in a
driving force for dilution and an osmotic pressure difference between the micelle-
free zone and the bulk continuous phase containing micelles. This gradient results
in attraction and serves to force emulsion drops closer together. Aronson's analysis,
based on earlier treatments by Asakura and Oosawa[151] and Sperry,[152] estimates the
contribution of micellar exclusion to the total interaction free energy of emulsion
drops. Assuming drops can flocculate into a secondary minimum at a drop
separation of twice the thickness δ of the adsorbed surfactant layer, this contribution is given by

$$\Delta G_{osm} = \int_{\infty}^{2\delta} \Pi[(r + r_m + \delta)^2 - (r + 0.5h)^2]\,\mathrm{d}h \qquad (1.13)$$

where Π is the osmotic pressure arising from excluded micelles, r is the emulsion
drop radius, r_m is the micelle radius and h is the separation between the outer
surfaces of emulsion drops. In dilute solutions, the osmotic pressure is related to
the concentration C of micelles by the approximation

$$\Pi = (CRT/M)(1 + 2C/\rho) \qquad (1.14)$$

where M is the micelle molecular weight and ρ is the micelle density. Integration
of eqn. (1.13) allows the calculation of ΔG_{osm} for various values of r and C. The

numbers range from -1 to $-10kT$, becoming more negative (attraction) with increasing C and decreasing r,[135,137] consistent with experimental observations. Similar calculations by Binks et al.[144] reveal that ΔG_{osm} is more or less independent of microemulsion droplet radius (r_m), using an alternative formula for the osmotic pressure in the case of hard spheres. The depletion force profile has been measured directly between mica surfaces coated with a bilayer of cationic surfactant immersed in micellar solutions, using the surface force apparatus.[153]

Bibette et al.[138–140] see similar effects in silicone oil-in-water emulsions stabilised by SDS. They describe the enhanced creaming effect in terms of a reversible fluid–solid phase transition caused by depletion attraction. A typical polydisperse emulsion ($\phi = 10\%$, drop diameters between 0.2 and 5 µm) separates into a dense phase made of aggregated oil drops and a dilute phase of free drops on addition of SDS in powder form. The upper cream is removed and further surfactant (of increased concentration) is added to the dilute phase, promoting flocculation. The cream is removed again and the procedure continued. The cream samples ($\phi = 50\%$) are diluted to about 10% in oil and flocculation is repeated as before. This original method, called fractionated crystallisation,[139] leads to a set of monodisperse emulsions of various sizes. It relies on the fact that larger drops are trapped in the flocs which becomes the cream layer, leaving smaller drops behind. These monodisperse emulsions can then be flocculated themselves by excess surfactant in order to study the physics of the phase separation. Using phase-contrast Normarsky microscopy, the phase diagram in the surfactant concentration–oil volume fraction plane has been determined for different emulsion sizes (Figure 1.11a). The experimental points separate the fluid phase, where only free drops are seen, from the solid phase, where aggregated and free drops coexist. The dense phase possesses long-range ordering of oil drops and light diffraction experiments are consistent with a face-centred cubic crystalline structure (Figure 1.11b). The dilute phase exhibits a correlation peak indicating hard sphere-like interactions between emulsion drops at low [surfactant], replaced by intense small angle scattering indicative of attractive interactions at higher [surfactant]. The phase transition is analysed as a first order one between an ideal gas and a harmonic solid and the equation describing the phase boundary is[140]

$$\phi_m = \frac{\sigma_m}{9\sigma}(-\ln \phi + \mu_s^\circ - \mu_g^\circ - 6w) \qquad (1.15)$$

where ϕ_m and ϕ are the micelle and oil volume fractions, respectively, σ_m and σ are the micelle and emulsion drop diameters, μ° is the reference chemical potential for the gas (g) and solid (s) and w is the van der Waals contribution to the lattice energy. The lines in Figure 1.11a represent the fit of the data (ϕ_m versus ϕ) to eqn. (1.15), leading to sensible values for the terms before and within parentheses. Monodisperse SDS emulsions were shown to flocculate at all [surfactant] on the addition of >0.3 M NaCl to the aqueous phase.[140] The strongly adhesive nature of the drops leads to a reversible gelation transition, in which a stable, rigid gel network of deformed drops forms on lowering the temperature.[154] Unlike most gels, these possess a well-defined characteristic length scale, and if the adhesion between

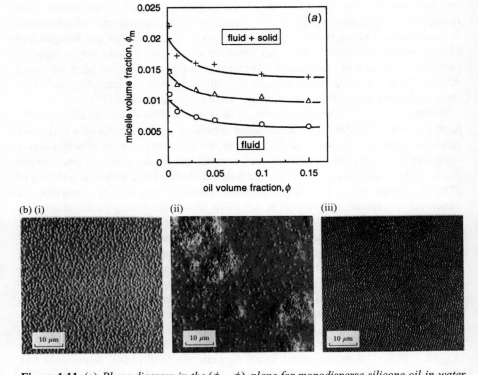

Figure 1.11 (a) *Phase diagram in the* (ϕ_m, ϕ) *plane for monodisperse silicone oil-in-water emulsions of SDS of diameter 0.46 µm (crosses), 0.60 µm (triangles) and 0.93 µm (circles). Lines are fits to eqn. (1.15)*
(b) *Microscope pictures of 0.7 µm diameter monodisperse emulsions of* $\phi = 0.1$ *in* (i) *the fluid region,* $\phi_m = 0$ *and* (ii) *the fluid + solid region,* $\phi_m = 0.0092$. *In* (iii), *the long range ordering of the solid phase is visible for a diameter of 0.93 µm*
(Part (a) redrawn with permission from ref. 138)

drops is increased further (by lowering T), the gel fractures and well separated, compact flocs are formed. The authors stress that all these results could only have been obtained with monodisperse drops. Other methods of preparing monodisperse emulsions have been described. Kandori *et al.*[155] make use of the shirasu-porous glass filter technique for W/O systems and Obey and co-workers[156,157] prepare silicone oil (PDMS)-in-water emulsions using a controlled nucleation and growth route involving the base-catalysed hydrolysis and polymerisation of dimethyl-diethoxysilane.

A novel approach to emulsification by Mason and Bibette[158] considers the shear-induced rupturing of viscous drops in viscoelastic complex fluids, in which a pre-mixed emulsion of large, polydisperse drops can be ruptured into a monodisperse emulsion of uniform drops. The method is based on the physics involved in emulsification in which work is done by the applied shear stresses against interfacial tension (γ) to elongate and rupture larger drops into smaller ones. In order to deform a *single* drop of radius r, the viscous shear stress of the continuous phase, $\eta_c \dot{\gamma}$, must overcome the Laplace pressure γ/r, where η_c is the continuous phase viscosity and $\dot{\gamma}$ is the shear rate. For rupturing to occur, the capillary number, C_a, must exceed a critical value of the order of unity where

$$C_a = \frac{\eta_c \dot{\gamma} r}{\gamma} \qquad (1.16)$$

In certain cases seen experimentally, however, a drop is stretched under shear into an elongated liquid thread that undergoes a capillary instability and breaks into a chain of many drops. The critical value of C_a is much greater than unity and the elongated drops resemble liquid cylinders whose radius becomes the radius of the drops. In a real emulsion of *many* interacting drops, the viscosity η_c in eqn. (1.16) is replaced by an elevated effective viscosity η_{eff} which takes into account the effect of volume fraction, ϕ. Thus, the viscosity of the emulsion and not just that of the continuous phase alone determines the shear stress governing the rupturing and hence the drop radius r. To eliminate inhomogeneous flow which leads to fracturing (non-uniform rupturing) and high polydispersity, a shearing geometry with a very narrow gap is employed and the pre-mixed emulsion is subject to two types of controlled shear. In *oscillatory shear*, a rheometer fitted with circular parallel plates separated by around 200 μm is used, in which both the shear strain amplitude, γ_s, and the frequency, ω, can be controlled. Since γ_s increases linearly from zero at the centre of the plate to the perimeter, the emulsion drop radius varies with radial distance from the centre. Figure 1.12a shows this dependence for a silicone oil-in-water emulsion stabilised by a nonylphenol ethoxylate surfactant at three frequencies. For small deformations below a critical value of $\gamma_s \approx 6$ (vertical dotted line), the emulsion remains polydisperse at all frequencies. At higher deformations, it is possible to obtain truly monodisperse emulsions in which the drop radius decreases with an increase in both γ_s and ω. In *steady shear*, a concentric cylinder couette geometry is used where the steady shear rate $\dot{\gamma}$ can be controlled. At low shear rates (Figure 1.12b), the drops are not ruptured, but above $\dot{\gamma} \approx 100$ s^{-1}, rupturing occurs and the monodisperse drop radius falls sharply with $\dot{\gamma}$. It is important to note that in the example given, the viscosity of the continuous surfactant + water phase η_c is Newtonian in behaviour but that of the emulsion is non-Newtonian or shear thinning exhibiting a smaller η_{eff} at higher $\dot{\gamma}$. By controlling the rheology of the emulsion, either by manipulating η_c through the surfactant concentration or η_{eff} by changes in the drop volume fraction, precise control of the monodisperse emulsion drop size is possible.

Theoretically, flocculation has been described using a kinetic model[159,160] and using statistical thermodynamics.[161] When two drops of millimeter size collide, one usually observes the formation of a planar film between them.[57] For micron-size emulsions it is not clear whether such a deformation will occur, since on the one hand smaller drops possess higher capillary pressure which opposes their deformation, whereas on the other hand they undergo intensive Brownian motion giving rise to an additional force enhancing the deformation.[162] Only a few studies[163,164] were published where emulsion stability to flocculation was explained taking into

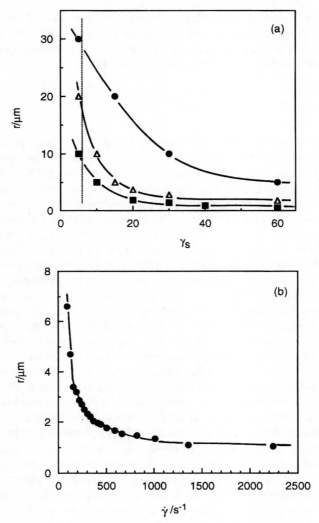

Figure 1.12 (a) *Oscillatory shear. Dependence of the drop radius,* r, *on the shear strain amplitude,* γ_s, *at fixed frequencies* ω *(in rad s^{-1}) of 1* (●), *10* (△) *and 100* (■). *To the left of the vertical dotted line the emulsion remains polydisperse. The O/W emulsions are composed of nonylphenol heptaethoxylate (mass fraction in water = 0.4)/water/PDMS (350 cP) with* $\phi = 0.7$. (b) *Steady shear. Dependence of the drop radius on the steady shear rate,* $\dot{\gamma}$, *for the same emulsion as in* (a)
(Redrawn with permission from ref. 158)

account drop deformability. The role of the deformability-of drops during their approach is considered in detail by Petsev *et al.*[59,165–168] The total interaction energy (being the sum of van der Waals, electrostatic, steric and surface extension energies) between sub-micron emulsion drops is calculated and it is shown that the attractive interactions may lead to the formation of a planar film between two drops. This deformation, leading to an increase of the interfacial area, is particularly important at high electrolyte concentration and low interfacial tension. The presence of a low volume fraction of micelles (depletion effect) is shown to cause stronger attraction and more pronounced drop deformation, whereas at higher concentrations the energy barrier produced by the oscillatory structural forces[169] actually stabilises the liquid films. Expressions derived for the energy contributions were also used to calculate the equilibrium film thickness and film radius.[59,166]

A very recent study makes use of particle simulations in modelling the combined creaming and flocculation in emulsions.[170] The model consists of the formation and breakage of bonds between particles (drops) and the variation of the creaming rate and diffusion coefficient with the size of the floc formed. In the 3-D model, particles are placed at random on a cubic lattice; they can move between lattice sites according to creaming and diffusional rules of motion. The spacing between lattice sites is equal to the diameter of the particles so that particles on adjacent sites are just touching. At any time, particles occupying neighbouring sites are allowed to stick together permitting the growth of flocs, and then all the particles in a floc move together as a single unit. Bonds between particles are also allowed to break, representing reversible flocculation with finite bond energies. The evolution of the sample is modelled by a sequence of time intervals. A particle is selected and either the particle itself or the floc to which it is joined is given the chance to move to a new site. All neighbouring particle pairs within the floc or those which touch the particle or floc are allowed to stick together or break apart. This is repeated for another particle until all particles or flocs have been given the chance to move and flocculate. This marks the end of a time step, with the whole cycle starting again as a new time interval. With these considerations in mind, simulations have been made for a total of 98 000 particles in a lattice of dimensions $70 \times 70 \times 200$ for a particle diameter of 1.2 μm. Flocculation is controlled by the bond formation probability, P_{bond}, and the bond breakage probability, P_{break}.

Various cases have been described. For *irreversibly flocculated* emulsions ($P_{break} = 0$), Figure 1.13 shows the effect of increasing P_{bond} for a 'long' time of 20 minutes. In the concentration profiles (plots of the lattice occupation, defined as the proportion of the lattice sites which are occupied in any horizontal plane, *versus* height up the cube), the general features of creaming are seen: a serum (of low particle concentration) forms at the bottom and a cream (high concentration) appears at the top, separated by a region which is almost uniform. At $P_{bond} = 0$, corresponding to a non-flocculated system, there is no sharp step change in concentration because of the diffusion of the particles. As P_{bond} increases, flocculation is more rapid and particle gel formation occurs in a shorter time (not shown). A number of features are visible as P_{bond} increases: (i) the concentration of the cream is reduced since extensive cross-linking occurs before a concentrated region forms, (ii) the spatial extent of the cream is also reduced, (iii) creaming ceases at a much earlier time as demonstrated by the lack of movement of the serum interface, (iv) the position of the serum interface occurs much closer to the bottom of the sample. For *reversibly flocculated* emulsions ($P_{break} > 0$), the breakup of flocs delays the onset of a particle network structure which would reduce the creaming

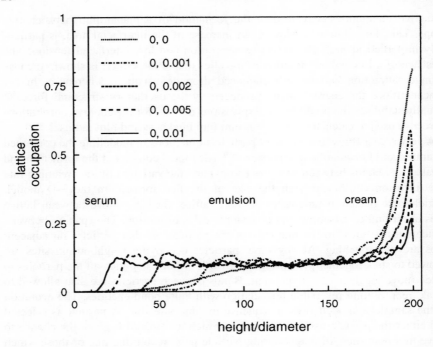

Figure 1.13 *Proportion of lattice sites occupied as a function of height after 20 minutes, simulating an emulsion undergoing irreversible flocculation. The numbers in the key refer to the break-up probability, P_{break}, and the bonding probability, P_{bond}, respectively*
(Data taken with permission from ref. 170)

rate. As P_{break} increases, a larger cream region develops and the serum interface position moves higher up the sample. The concentration in the cream is also greater due partly to the longer time before particle movement is prevented by cross-linking, and it indicates that the cream is formed by smaller or more compact flocs which have a greater packing efficiency. Most of these predictions are borne out with measurements on real emulsions, although quantitative comparisons are difficult due to the small scale of the simulations.

1.4.3 Coalescence

Coalescence is the process in which two or more emulsion drops fuse together to form a single larger drop, and is irreversible. For coalescence to occur, the forces between the drop surfaces must be such that the film of continuous phase separating them can become sufficiently thin that film rupture becomes a likely possibility. The film thinning stage depends on the hydrodynamics of film flow and on the forces acting across the film, whilst film rupture depends on fluctuations in film thickness and on the mechanical properties of the film. For two large (around millimetre) emulsion drops approaching each other, when the distance between them is less than ~ 1 μm, their interfaces interact and begin to deform.[57] The various stages of the deformation have been discussed by Ivanov and Dimitrov.[171] In the final stage, a plane-parallel thin film is formed. The rate of thinning of this film and its stability against rupture are among the main factors determining the overall stability of the emulsion. The stability of thin plane-parallel films has been analysed extensively and reviewed. The models incorporate measurable bulk and surface properties such as shear and dilational viscosity, surfactant diffusion coefficients, interfacial tension, film radius and thickness, *etc.*, to predict the drainage and lifetime of the films. The agreement of the models with experimental data was found to be reasonably good. More recent treatments of film behaviour have been given,[172–180] but until 1993 neither a theory relating the lifetime of single films to emulsion stability,[177] nor a theory describing thin film stability in the case of micron-sized drops where the effect of Brownian motion becomes significant,[178] was available. Lobo *et al.*[177] developed a model which can be used to estimate the change (due to coalescence) in the average drop size of a creamed emulsion with time, knowing the lifetime of a single film. It applies to large drops (20–50 μm) and assumes that no drop deformation occurs and that the film surfaces are immobile.

Danov *et al.*[178] have extended the Smoluchowski treatment for particle coagulation to include the case of coalescing emulsion drops (<1 μm) which can deform. In all stages of the process from approach to rupture of the plane film, the intermolecular and hydrodynamic interactions are considered separately. They calculate the distance at which drop deformation sets in (h_d), the total force acting between drops and the coalescence rate. The influence of different factors, *e.g.* surface potential, drop size and Hamaker constant, is investigated. It is shown that attraction between the drops increases h_d, whereas repulsion decreases it, sometimes to such an extent that deformation is unlikely and the drops behave as non-deformable charged spheres. In the absence of electrostatic repulsion, the rate-determining stage is the approach of the drops before their deformation. In an alternative view, Sharma and Ruckenstein[172] have investigated the effect of surface waves on the thinning and rupture process. Their reason for doing this is summarised here. Reynolds[181] derived an expression for the rate of approach of two, rigid parallel discs under the action of an external pressure. His law provided a convenient model for the drainage of tangentially immobile emulsion films. Investigations[182] of thin films have used the law to obtain the critical thickness and lifetime of the films. However, the law assumes the interfaces to be non-deformable and plane-parallel, whereas experiments[183] demonstrate that the film surfaces have a wavy appearance and that the amplitude of these surface waves increases with the film radius. Furthermore, the velocities of thinning calculated using the Reynolds equation were up to several orders of magnitude smaller than those determined experimentally. In view of this, Sharma and Ruckenstein have formulated a theory of film thinning based on the fact that the film interfaces do not remain plane-parallel during thinning, but assume this wavy profile especially for film radii >0.05 mm. The surface waves originate from an unidentified hydrodynamic instability and are distinct from the very small amplitude waves arising from the thermal motion of the molecules. The latter grow due to dispersion interactions when the film becomes very thin ($< 10^3$ Å); their growth destroys the film at some critical thickness. The surface waves travel away from the film centre and the 'pumping' action generated by these moving waves was shown to be responsible for enhancing the velocity of thinning. The case of immobile surfaces and the effects of interfacial mobility on the wave-induced drainage were both studied. The velocities of thinning, critical thicknesses and film lifetimes calculated from theory are in quantitative agreement with experimental results.

Direct measurement of coalescence between two unsupported liquid drops is difficult to perform, although the use of acoustic levitation and acoustic cavitation to study oil drop coalescence in water has been described.[184] Instead, the kinetic studies involve either measuring the rest times of large drops with planar interfaces[185–189] or forming liquid films on suitable supports.[189–192] For a drop of one phase immersed in the other and coalescing with its homophase, the lifetime distribution curve normally consists of two regions; one is associated with the drainage of the continuous phase between drop and interface, and the other relates to the rupture of the thin lamellae. Palermo[185] has reviewed various experimental and theoretical approaches in this kind of study. Aveyard et al.[186] have shown a correlation between the coalescence time for oil or water drops and emulsion type in systems containing the ionic surfactant Aerosol OT. For alkane oil phases, it is found that oil drops are more stable than water drops at low [electrolyte] whereas the reverse is true at high [electrolyte]. Interestingly, emulsions invert from O/W to W/O on increasing the electrolyte concentration, possibly indicating a link between film stability and the type of dispersion. Nakache et al.[187,188] studied single drops of heptane stabilised by varying concentrations of nonionic surfactant, and correlated the coalescence time with the stability of the emulsions. In studies of this kind, varying one parameter in the system invariably affects many others, e.g. interfacial tension hence drop size and therefore film area, and so understanding the results unambiguously often proves difficult. Investigations of thin emulsion films are more scarce than those on foam films. Mass transfer of material across the film surfaces has been shown to have drastic effects on the behaviour of such films.[190–193] For aqueous films between oil phases, Velev et al.[191] report a fascinating phenomenon of cyclic dimpling caused by the re-distribution of surfactant between phases. The nonionic surfactants (Tween 20 and 80) are initially dissolved in the water phase (at concentrations $\gg$CMC) and aqueous films ~300 μm in diameter are formed. After the films thin down to their equilibrium thickness (~150 nm), a lens-shaped dimple forms spontaneously by sucking in water, and starts to grow around the film centre. On reaching a certain size, the dimple forms a channel to the periphery (meniscus) and liquid flows out leaving a plane-parallel film behind. A new dimple forms immediately and the process is repeated. It has been shown that the driving force is the transfer of surfactant (probably as monomer) from water to oil, since films formed from pre-equilibrated phases showed no dimpling. Initially, surfactant diffuses from the film interior and the film surfaces to the oil phase (Figure 1.14a). The lowering of the surface concentration of surfactant at the film surfaces causes a tangential movement of surfactant from the surfaces in the menisci (Marangoni effect) which is coupled with an influx of water into the film center, forming the dimple. The complementary case in which

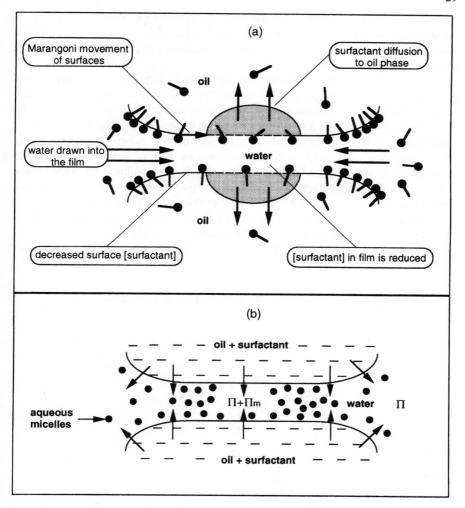

Figure 1.14. *Schematic representation of* (a) *cyclic dimpling caused by transfer of surfactant from an aqueous film to the adjacent oil phases, and* (b) *osmotic swelling of an aqueous film caused by transfer of surfactant from the oil phases to the film containing micelles. The excess osmotic pressure,* Πm, *in the film counterbalances the outer capillary pressure and arrests further thinning of the film* (Redrawn with permission from *Colloids Surf. A*, 1997, **128**, 155)

surfactant is initially dissolved in the oil phase and diffuses into the aqueous film has also been described.[192] Instead of thinning to ~10 nm and rupturing, as is the case with equilibrated phases, these films remain thick and very stable during the mass transfer process of over 48 h. Channels of larger thickness also form connecting two points of the film periphery. The observations are rationalised by assuming that the stabilisation of these films originates from the higher osmotic pressure of surfactant micelles (once formed) in the film, with a consequent influx of liquid, compared to that in the surrounding meniscus (Figure 1.14b). This is because the film becomes enriched with micelles due to their low diffusion into the Plateau borders. The findings could have consequences for the stability of freshly formed emulsions.

In contrast to the above findings for films formed from non-pre-equilibrated phases, other phenomena appear when the two phases have been in contact and surfactant distribution is complete. The effect of surfactant concentration on the stability of oil-water-oil films of diameter 330 μm shows the importance of selecting the right type of emulsifier for stabilisation.[194] Aqueous films formed in glass capillaries thin down with time in various stages; some reach a critical thickness and either rupture or form a stable black film. Defining the lifetime of a film as the time elapsed between film formation and either rupture or black film appearance, it can be seen in Figure 1.15 that this lifetime increases linearly with the logarithm of the (predominantly) water-soluble Tween 20 surfactant concentration both below and above the CMC. The increase above the CMC may be a result of the measured increase in the adsorption of surfactant at the oil–water interface.

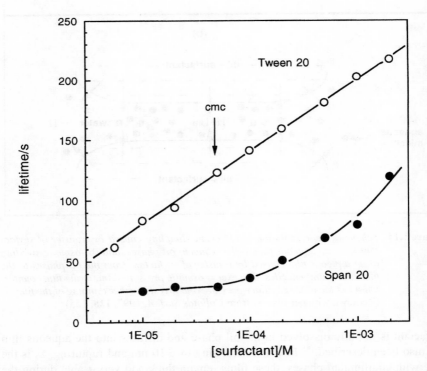

Figure 1.15 *Average lifetime of aqueous emulsion films from pre-equilibrated phases as a function of initial surfactant concentration in water (for Tween 20) or in oil (for Span 20). The oil is xylene, the aqueous phase is 0.1 M NaCl and the temperature is 27.5 °C*
(Redrawn with permission from ref. 194)

For the mainly oil-soluble Span 20 surfactant, however, the lifetimes are much less and films rupture prematurely, in line with predictions based on Bancroft's rule. At concentrations well above the CMC where the effective volume fraction of micelles is significant (>5 vol%), thin liquid films may drain in a stepwise fashion by stratification.[195] This phenomenon, seen initially with foam films, was explained by the formation of periodic colloidal structures inside the film that results in layering of the micelles. At a step-transition, a layer of micelles leaves the film and the film thickness decreases by approximately the effective micellar diameter. It can also occur in emulsion films shown recently for hexadecane–aqueous sodium caseinate–hexadecane systems.[196] The step-height seen of around 20 nm is very close to the measured diameter of the casein micelles of between 20 and 25 nm. The layering ultimately increases the lifetime of a film, but a critical film area exists below which step transitions are inhibited; such thick films containing layers of micelles are even more stable.

In concentrated emulsions, the forces between drops are particularly important in relation to coalescence. Drops in creamed layers become distorted polyhedra separated by thin films of continuous phase, and ultracentrifuge studies exist in which the volume of bulk phase produced by coalescence is used as a measure of emulsion stability.[197] More recently, Bibette and co-workers[198] describe the use of an osmotic stress technique for observing the coalescence of oil drops in mono-disperse emulsions. Silicone oil-in-water emulsions of SDS of oil volume fraction >0.5 are prepared and placed in a dialysis bag, which is immersed in a reservoir containing water, surfactant and dextran polymer. The bags are permeable to water and surfactant but impermeable to oil drops and polymer. Dextran is chosen because the osmotic pressure Π and the surfactant chemical potential μ_s can be controlled independently by varying the polymer concentration and surfactant concentration, respectively. The effect of the applied osmotic pressure is to squeeze water out from the emulsion, resulting in the drops becoming deformed and separated by thin flat films. Coarsening of the emulsion due to coalescence is then followed microscopically. The stability of a set of emulsions of fixed drop diameter equal to 1.6 μm and varying Π and μ_s is summarised in Figure 1.16. An emulsion is defined as stable if no coalescence is visible 15 days after placing in the reservoir. The rate of coalescence does not change gradually but exhibits a sharp threshold. Emulsions are always unstable in the reservoir below a critical [SDS] which is approximately one third of the CMC. Above this concentration, an emulsion only becomes unstable if Π is raised above a critical value Π^*; Π^* is independent of [SDS] above the CMC. The low concentration instability is due to insufficient surfactant adsorbed at oil–water interfaces around the drops, whereas above the CMC the interfaces are saturated and the stability is controlled by the osmotic pressure. Using monodisperse emulsions of different diameters, it is found that Π^* increases from ~0.04 atm for 0.3 μm drops to ~0.9 atm at 3.3 μm, and the corresponding volume fractions of oil at Π^* are 0.6 and 0.96. Coalescence is seen to proceed by the growth of a small number of randomly distributed drops. These larger drops coalesce with small neighbouring drops until large drops meet each other. Then, surprisingly, no further coalescence occurs. The reason for this is that with time the increase in the average drop size leaves the (new) emulsion below the

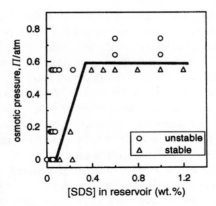

Figure 1.16. *Osmotic pressure of monodisperse silicone oil-in-water emulsions* versus *[SDS] in the reservoir for $\phi \approx 0.5$ and diameter = 1.6 μm. The line represents the stability (to coalescence) threshold*
(Redrawn with permission from ref. 198)

threshold pressure where it is stable again. Coalescence can then be initiated by a further increase in Π. The authors suggest that their results can be understood if the critical osmotic pressure they measure is equal to the disjoining pressure[199] of the film. The disjoining pressure Π_d is the net force per unit area acting in the direction normal to the film surfaces and can be regarded as a kind of osmotic pressure difference between liquid in the film and in the bulk continuous phase. A positive value of Π_d implies that the colloidal forces are in opposition to the thinning of the film. Π_d is generally not a monotonic function of film thickness h, but on decreasing h may intially rise due to double-layer repulsion, reach a maximum and then decrease due to van der Waals attraction. A metastable thickness only exists at pressures less than Π_{max}. Above Π_{max}, the film drains continuously, leading to black film formation or rupture. Petsev and Bibette[200] have calculated Π_{max} (or Π^*) explicitly from thin film and DLVO theories in the case of large drops. The disjoining pressure is given by the expression[201]

$$\Pi_d = -\frac{A_H}{6\pi h^3} + 64 \times 10^3 c_{el} N_A kT \tan h^2 \left(\frac{e\psi_0}{4kT}\right) \exp\left(-\kappa h\right) \qquad (1.17)$$

where A_H is the Hamaker constant, c_{el} is the molar concentration of electrolyte, N_A is Avogadro's number, kT is the thermal energy, ψ_0 is the surface potential of a drop and κ is the Debye length, proportional to the elementary charge e and to $(c_{el})^{0.5}$. Π_{max} can be deduced from eqn. (1.17) by setting $\partial\Pi_d/\partial h = 0$, corrresponding to the maximum in the disjoining pressure isotherm. Using known values for the parameters appearing in eqn. (1.17), Π_{max} is calculated to be 1.206 atm, very close to the maximum Π^* measured experimentally. Π_{max} decreases with a decrease in the magnitude of ψ_0, reaching zero at a surface potential of ~18 mV, suggesting that replacing some of the ionic with nonionic surfactant may induce coalescence at zero applied pressure.

1.4.4 Ostwald Ripening

Ostwald ripening, referred to as isothermal distillation or molecular diffusion in some texts, is the result of the solubility differences of, say, oil contained within drops of differing sizes. According to the Kelvin equation,[202] the solubility of a substance in the form of spherical particles increases with decreasing size:

$$c(r) = c(\infty) \exp \left(\frac{2\gamma V_m}{rRT} \right) \tag{1.18}$$

where $c(r)$ is the aqueous phase solubility (m^3 m^{-3}) of oil contained within a drop of radius r, $c(\infty)$ is the solubility in a system with only a planar interface, γ is the interfacial tension between the two phases and V_m is the molar volume of the oil (m^3 mol^{-1}). As a consequence of its increased solubility, material contained within the smaller drops tends to dissolve and diffuse through the aqueous phase, recondensing onto the larger drops. This results in an overall increase in the size of the emulsion drops and an accompanying decrease in the interfacial area. This provides the driving force for the growth of the drops. Theoretically, the process of Ostwald ripening should be completed by merging all the drops into one, but this does not occur in practice owing to a considerable decrease in the rate of the process as the average size of the drops increases. Although the theory of Ostwald ripening is well developed, there have been few experimental studies of the process in emulsions until relatively recently. The theory in its contemporary form was initially formulated by Lifshitz and Slezov (LS)[203] and independently by Wagner (W).[204] The LSW theory assumes that the particles of the dispersed phase are spherical and separated from each other by distances which are much larger than the particle sizes, and that the mass transport of, say, oil is limited by the molecular diffusion in the continuous phase. Under these conditions, their treatment gives the rate of Ostwald ripening, ω, as

$$\omega = \frac{dr_c^3}{dt} = \frac{8c(\infty)\gamma V_m D}{9RT} \tag{1.19}$$

where D is the diffusion coefficient (m^2 s^{-1}) of the dissolved species in the aqueous phase and r_c is the critical radius of the system at a given time. This is the radius of a drop which at a given time is neither growing or diminishing. Drops with $r > r_c$ will grow at the expense of the smaller ones. It can be shown that the critical radius may be reasonably approximated to the number average radius $\bar{r}_n$ of the drops. The right-hand side of eqn. (1.19) can be multiplied by a correction coefficient $f(\phi)$ to take into account the diffusional interaction of particles at finite values of the dispersed phase volume fraction ϕ. It takes the values[205] from 1 as ϕ tends to zero up to ~2.5 at $\phi = 0.3$. The main results of the LSW theory can be summarised as follows. (i) The cube of the critical radius increases linearly with time. Since the number of drops in unit volume of the system (n) is linked to the volume fraction ϕ and the critical radius by $n = 3\phi/4\pi r_c^3$, eqn (1.19) corresponds to the following timewise change in the number of particles in unit volume

$$-\frac{dn}{dt} = \frac{4\pi\omega}{3\phi} n^2 \tag{1.20}$$

i.e. dn/dt is second order with respect to drop concentration n. (ii) The size distribution of the drops assumes a time independent form, $g(u)$, given by

$$g(u) = cu^2(3 + u)^{-7/3}(1.5 - u)^{-11/3} \exp\left(\frac{-u}{1.5 - u}\right) \qquad 0 < u \leq 1.5 \tag{1.21}$$

and

$$g(u) = 0 \qquad u > 1.5$$

where u is the normalised drop radius, $u = r/r_c$.

1.4.4.1   One-component Dispersed Phase

There have been several studies aimed at verifying equations (1.19) and (1.21) in O/W emulsion systems. Kabalnov and co-workers[206–211] have followed Ostwald ripening in systems stabilised by SDS. At a volume fraction of oil equal to 0.1 and at 0.1 M surfactant in water (drop diameters 50–200 nm), the experimental ripening rates fall from 6.8×10^{-25} to 8.7×10^{-29} m³ s⁻¹ in going progressively from nonane to hexadecane[208] (*i.e.* the rates decrease by an order of magnitude for an increase of two CH₂ groups, *cf.* the decrease of the solubility of the oil in water), in good agreement with theory if the *molecular* (as opposed to the *micellar*) solubility of the hydrocarbon in water is used. Similar agreement was seen for emulsions of perfluoro compounds[206,209,210] and 1,2-dichloroethane.[207] Recently, Kabalnov has studied the effect of the micelle concentration on the ripening rate in SDS emulsions of undecane.[211] Potentially, micelles could solubilise oil from emulsion drops and act as carriers, enhancing the rate. Figure 1.17 shows that ω is virtually independent of the micelle concentration, even though there is a 30-fold rise in [SDS] and a 100-fold increase in oil solubilisation. It is argued that due to electrostatic repulsion, SDS micelles cannot absorb oil directly from emulsion drops. This may be possible, however, with nonionic surfactants where, in addition, the solubilisation capacity of micelles can be much larger, leading to the formation of O/W microemulsion droplets.

Taylor and Ottewill[212] investigated Ostwald ripening in 'miniemulsions' (radius ~100 nm, volume fraction of oil ≤ 0.004) prepared by dilution of O/W microemulsions. The system chosen by them contained SDS, pentanol and dodecane. The growth rate of such miniemulsions was followed by measuring the turbidity spectra and applying Mie theory to the data. The values of ω fall linearly with [SDS] up to the CMC (solely due to the decrease in γ), and then increase very slightly above it. Again, agreement between experiment and theory is good if the aqueous phase solubility is used. They conclude that although dodecane may be transferred through the aqueous phase within a micelle, the rate determining step is one involving the transfer of dodecane through the aqueous phase in unsolubilised form. In subsequent

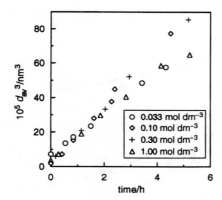

Figure 1.17 *Variation of the average drop diameter with time for SDS undecane (10 vol%)-
in-water emulsions undergoing Ostwald ripening at 20°C for different [SDS] in
water (given)*
(Redrawn with permission from ref. 211)

work, Taylor[213] has measured ω for emulsions containing SDS and decane prepared
by sonication. The ripening rate increases by a factor of 3 above the CMC and then
levels out at ~0.1 M SDS. As implied above, this increase is not as large as might be
expected if the transfer mechanism of decane was simply micellar transport from
one drop to another. The weak influence of micelles on the ripening rate has been
observed in emulsions stabilised by other anionic surfactants.[214,215]

The mass transport of oil molecules in emulsions containing aqueous micelles,
in which there is a competition between solubilisation of oil into micelles and
transfer of oil from small emulsion drops to larger ones, has been studied by
McClements *et al.*[216–219] The experiments have involved preparing stock emulsions
of oil-in-water stabilised by the nonionic surfactant Tween 20 (polyoxyethylene
sorbitan monolaurate) and diluting them with either pure water or an aqueous phase
containing Tween 20 above the CMC. Using a combination of static and dynamic
light scattering and turbidity measurements, the time dependence of the emulsion
drop concentration and size was determined. No change in the size or concentration
of the drops with time was observed using pure water as diluent. However, in the
case of dilution with micellar solutions, a decrease in the drop concentration and an
increase in the drop size with time occurred (Figure 1.18). The rate of these
changes was found to be very dependent on the type and chain length of the
hydrocarbon oil. The decrease in drop concentration with time is a result of the
transfer of oil from emulsion drops to (empty) micelles by solubilisation. This
would be expected to result in a decrease in the emulsion drop size, however, and
not an increase as observed. The explanation for this apparent anomaly is that
alongside the uptake of oil into micelles, Ostwald ripening of emulsion drops
occurs which is enhanced by the presence of surfactant micelles. Clearly, the
ripening process leading to a larger average emulsion drop size is dominant.

As Figure 1.18 shows, the solubilisation rate and the micellar solubilisation
capacity increase as the chain length of the hydrocarbon decreases, and both are

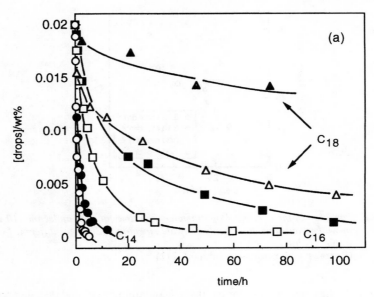

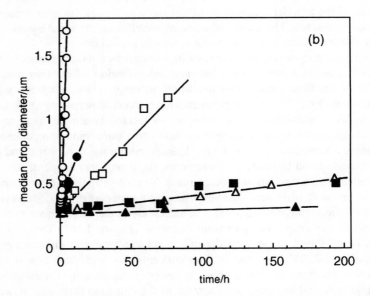

Figure 1.18 (a) *Time dependence of the drop concentration for O/W emulsions containing 0.02 wt% hydrocarbon in a 2 wt% Tween 20 aqueous phase at 30°C. Filled symbols are for alkanes, unfilled symbols are for alkenes, of chain length given.* (b) *Time dependence of the median drop diameter for the same emulsions as in* (a). *Circles—C_{14}, squares—C_{16}, triangles—C_{18}; filled—alkanes, unfilled—alkenes*
(Redrawn with permission from ref. 218)

higher for alkenes compared with alkanes. Mixing of oil with the chain region of a surfactant monolayer is favoured on entropic grounds for short alkanes of low molar volume, and this enhanced penetration compared with that for long chain alkanes manifests itself in both microemulsion properties[61] and in the behaviour of mixed monolayers of oil + surfactant at the air–water surface.[220] The larger uptake of the unsaturated 1-alkene compared with the saturated alkane is interesting and may be a result of the greater polarisability of the C=C double bond (2.80 nm^3, *cf.* 0.97) promoting further penetration towards the head group region of the monolayer. This effect would presumably be modified if the double bond was positioned further down the linear chain. Precisely how oil is transported from emulsion drops to micelles has been the subject of recent molecular dynamics simulations.[221,222] Three possible mechanisms are: (i) dissolution of oil molecules into water followed by uptake into micelles, (ii) collision of (dry) micelles with emulsion drop interfaces and (iii) spontaneous 'budding-off' of oil and surfactant molecules from the surface of an emulsion drop to form a swollen micelle. For the systems discussed above, it is believed that the collision mechanism is important since it has been shown that oil molecules from one emulsion drop can be solubilised in micelles, transported through an aqueous phase and transferred into a different emulsion drop.

1.4.4.2 *Two-component Dispersed Phase*

Higuchi and Misra[223] were the first to show that if one of the components of a dispersed phase is completely insoluble in the continuous phase, then even small amounts of such a substance may stop the Ostwald ripening in the system. The reason for this is as follows. In a two-component dispersed phase system, the mass transfer of the more soluble component from small to larger drops caused by the difference in the Laplace pressures changes the composition of the drops. Namely, it increases the concentration of the poorly soluble component in the small drops and decreases it in the larger ones. According to Raoult's law, this results in a compensation of the difference in chemical potentials of the more soluble component caused by the difference in capillary pressures. When the capillary and concentration effects completely compensate, the mass transfer terminates and the drops come to 'equilibrium'. This equilibrium implies the equality of the chemical potentials of the major component in all of the drops of the polydisperse emulsion. Such an equality is unattainable for the second component if its solubility in the continuous phase is truly zero. Kabalnov *et al.*[224-226] have considered two cases as follows.

(i) Zero solubility of the second component in the continuous phase. If the initial mole fraction of a second component in the dispersed phase is high, the initial distribution function undergoes only slight deformation when passing to the 'equilibrium' state. If the initial mole fraction of a second component is low, the drops cannot attain equilibrium. The size distribution becomes represented by a bimodal function, in which the coarse fraction changes in accordance with the LSW theory, whilst the fine fraction grows at the expense of the coarse drops.

(ii) Non-zero solubility of the second component in the continuous phase. In this case a peculiar kinetic regime may be realised when the rate of Ostwald ripening is limited by diffusion of the barely soluble component 2 even if its content in the dispersed phase is small. If the difference in the molar volumes and interfacial tensions (with continuous phase) of the two components is neglected compared with the difference in their solubilities, then the overall ripening rate is given by

$$\omega = \left(\frac{\phi_1}{\omega_1} + \frac{\phi_2}{\omega_2} \right)^{-1} \tag{1.22}$$

where ω_1 and ω_2 refer to the ripening rate from drops of pure component 1 and 2, respectively. If ω_1 and ω_2 are markedly different, then an asymmetric dependence develops. Here, minor amounts of the second component retard the Ostwald ripening of the first component, whereas substantial quantities of the first only slightly affect the ripening of the second. When $\phi_2 \ll \omega_2/\omega_1$, mass transfer of the major component determines the rate and $\omega \approx \omega_1$. If $\phi_2 \gg \omega_2/\omega_1$, mass transfer of the second component governs the process and $\omega = \omega_2/\phi_2$.

Experiments confirm the decrease in the ripening rate on addition of a poorly soluble component to the dispersed phase. Buscall *et al.*[227] examined emulsions of SDS after adding long chain alkanes. Kabal'nov *et al.*[228] showed that eqn. (1.22) was well obeyed for emulsions of SDS and hexane to which small quantities of the alkanes octane to hexadecane were added. Recently, Weers and Arlauskas[229] have provided the first experimental evidence that molecular diffusion in two-component mixtures leads to an increase in the mole fraction of the less soluble component in the smaller drops. Using the technique of sedimentation field-flow fractionation coupled with gas chromatography, the disperse phase composition can be determined for monosized drop fractions. Evidence is also provided for the development of bimodal distributions at low volume fractions of the less soluble component, in excellent agreement with the above predictions. Their system was perfluorooctyl bromide (plus perfluorododecyl bromide) as oil stabilised by egg yolk phospholipid.[230]

1.5 Gel Emulsions

As mentioned earlier, certain emulsions may be phase inverted on increasing the dispersed phase volume fraction (ϕ). This usually occurs at ϕ values between 0.5 and 0.8. However, other systems do not phase invert even up to ϕ values as high as 0.99. In these concentrated emulsions, referred to as biliquid foams or aphrons,[231] the dispersed phase content is so high that close packing of spherical drops ($\phi_{max} = 0.74$ for a face centred cubic arrangement of monodisperse drops) is no longer possible. Consequently, the drops are distorted in shape from spheres to polyhedra in order to fill the available space. The high internal phase emulsions produced tend to be viscous and clear and have been called gel emulsions, finding applications in the cosmetics and explosives industries. One of the aims of the recent interest in gel emulsions has been to correlate their behaviour with the solution properties of the mixtures in pure surfactant systems. Investigations on

W/O gel emulsions, stabilised by hydrogenated surfactants with hydrocarbon oils,[232–245] and fluorinated surfactants with fluorocarbon oils,[246–249] and on O/W gel emulsions[250–257] have been conducted. It is convenient to summarise their preparation,[232–235,257] macro- and microstructure,[236–239,246–248] stability,[240–244,251,252] and rheological properties[245,250,254–256] separately. Molecular modelling of such emulsions, studying the simultaneous effects of polydispersity and drop distortion on ϕ_{max} for various packing types, has been carried out by Das *et al.*[258]

There are at least three methods for preparing gel emulsions. For nonionic surfactants, W/O gels form above the PIT from a Winsor II system[235] and O/W gels form below the PIT from Winsor I systems.[257] One route for W/O gels involves taking an isotropic solution (either an O/W or bicontinuous microemulsion which exist below the PIT) of overall composition equal to that of the final gel, and heating it rapidly to temperatures above the PIT, during which phase separation and emulsification take place.[234] Gel emulsions prepared this way have small drop sizes (<1 μm) with a narrow distribution, but the maximum amount of water that can be incorporated is 95 wt%. Similarly, O/W gels form spontaneously by rapid cooling of a W/O microemulsion so long as the reverse L_3 phase is passed in the vicinity of the PIT. By contrast, W/O emulsions made by mixing all the components above the PIT can contain >99 wt% dispersed phase. Addition of glass beads or fabric material facilitates emulsification. Here, an ordinary O/W emulsion forms initially after shaking the components gently. With further agitation, white flocs form which separate from a turbid solution below; these increase in volume with time until the gel forms. It is shown that W/O emulsification proceeds inside the (unstable) oil emulsion drops to give a multiple W/O/W emulsion. The external water phase is then taken up as the internal phase until a W/O emulsion is formed.[233,234] The role of the solid surfaces is to promote coalescence of the original oil drops. A third method, described by Lissant,[259] involves adding the dispersed phase slowly with constant mixing to the continuous phase containing surfactant. The critical stage in processing occurs at an internal phase volume of $\sim75\%$, since the emulsion has a tendency either to invert or to form multiple emulsions in which the 'wrong' phase becomes continuous. This has been used to prepare both O/W gels with ionic surfactants like SDS and alkylpyridinium halides[250,251] and W/O gels.[243]

Gel emulsion structure has been determined using a variety of techniques including optical microscopy,[233,243] video-enhanced microscopy,[237] small angle X-ray (SAXS)[236,248] and neutron (SANS)[246–248] scattering, and NMR[237] and ESR[238,239] spectroscopy. Since, in the case of nonionic surfactants, O/W gels form in the Winsor I régime where an O/W microemulsion coexists with an oil phase, and W/O gels form in the Winsor II régime (W/O microemulsion plus an excess water phase), it is reasonable to assume that the gels consist of large drops of one phase separated by thin films of continuous phase, which itself contains small microemulsion droplets. Ravey *et al.*,[248] employing SAXS and SANS on W/O gels, show that the scattering patterns are consistent with adsorbed surfactant monolayers, as opposed to multilayers, around large (>1 μm) water drops. By superposing the spectra of made-up microemulsion samples on those of gel emulsions, they conclude that the high q region of the curves is due to scattering from W/O microemulsions (radius ~ 50 Å) in the continuous phase of the gels. However, the water content is higher than the solubilisation capacity in made-up microemulsions at the same temperature, suggesting that these aggregates are not at equilibrium within the gel. Table 1.1 indicates how both the emulsion and microemulsion sizes change with water volume fraction, oil-to-surfactant ratio and temperature.[248] These changes have been rationalised in terms of the interfacial tensions and the availability of surfactant in the systems.[236]

Table 1.1 *Structural data* * *from SANS for W/O gels in the system water–perfluoro-decalin–$C_6F_{13}C_2H_4SC_2H_4(OC_2H_4)_2OH$*

$T/°C$	R	ϕ	$r/\mu m$	$r_m/\text{Å}$	$W_c/w/w$
29	3.2	0.75	2.7	47	8.3
		0.85	4.5	47.5	8.0
		0.91	5.1	47	8.5
29	3.9	0.70	4.0	48	7.0
		0.78	4.0	48	7.0
		0.89	4.5	47.5	7.0
39	3.9	0.60	4.4	42.5	6.0
		0.70	7.0	41	6.0
		0.78	5.5	43	6.0
		0.89	9.0	43.5	6.0
29	4.8	0.76	3.3	45.4	6.5
		0.90	3.1	44	6.8

* R = oil/surfactant (w/w), ϕ = wt. fraction of water, r = average emulsion drop radius, r_m = microemulsion droplet radius, W_c = water content in continuous phase of gel.

Pulsed gradient spin echo NMR measurements have also been useful in elucidating the nature of the continuous phase in W/O gels. Solans *et al.*[237] show that concentrated gel emulsions form when the continuous phase contains spherical microemulsion droplets. On increasing the water content or decreasing the temperature towards the PIT, the measured self-diffusion coefficients of the components indicate that this structure transforms to a bicontinuous one and the emulsion gel is either very unstable or cannot be formed. Gel emulsions are not thermodynamically stable and they separate into distinct phases at equilibrium. Coarsening occurs mainly as a result of coalescence, with the emulsions exhibiting a significant decrease in viscosity and a loss in 'whiteness'.[243] With nonionic surfactants, the stability decreases as the temperature is changed towards the PIT, *i.e.* on increasing T for O/W gels[253] and on decreasing T for W/O gels.[242] This is consistent with a lowering of the interfacial tension facilitating coalescence, already observed in ordinary emulsions.[79] Addition of salt to nonionic W/O gels invariably leads to an increase in stability.[240,243] The effect is dependent on the type and concentration of electrolyte, so that salts like Na_2SO_4 are more effective than, say, NaCl. Figure 1.19 shows how the volume average diameter of the emulsion drops varies with time in the presence and absence of K_2SO_4 for various surfactants. In all cases, the incorporation of electrolyte reduces the initial drop size and retards the coarsening of the emulsion. Various explanations have been put forward to explain the stabi-

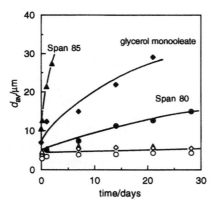

Figure 1.19 *Change in volume average diameter with time at 25 °C of 91 vol% water-in-mineral oil emulsions stabilised by nonionic surfactants with (unfilled) and without (filled) 0.1 M K_2SO_4 in the aqueous phase*
(Redrawn with permission from ref. 243)

lising influence of salt. Kizling and Kronberg[241] argue that salt reduces the attractive forces between aqueous drops across thin oil films by lowering the refractive index difference between oil and water to zero, although this has been contradicted in a later paper.[237] Aronson and Petko,[243] using commercial surfactants, state that salt affects the surface concentration of surfactant by increasing the cohesion of the adsorbed molecules. This is not convincing since, although their measured (post-CMC) oil–water tension decreases on adding salt, correct conclusions concerning the adsorbed amount can only be made by careful measurements below the CMC. The most plausible reasoning is by Solans *et al.*,[237] who correctly state that the PIT is reduced with salt, which results in the system being further from the PIT (and hence more stable) compared to the pure water case. Consistent with this is an increase in the interfacial tension and drop size in the salt-containing emulsions.[236] An electron spin spectroscopy study[238] found that adding salt increased the apparent order parameter of the spin label, suggesting that the packing of surfactant hydrocarbon chains increases. Salt addition may also lead to dehydration of ethyleneoxy head groups, increasing lateral interactions within the monolayer.

The viscoelastic properties of gel emulsions can be investigated by measuring time-dependent rheological properties. In *transient* measurements, a step function shear strain is applied to the gel and the shear stress is measured as a function of time. The simplest mechanical model describing a viscoelastic system is the Maxwell material, consisting of a spring (elastic element) and a dashpot (viscous element) connected in series. The spring corresponds to a shear modulus G_0 and the dashpot to a viscosity η_0. In *dynamic* measurements, a sinusoidal deformation is applied and, if the sample behaves as a viscoelastic liquid, the stress is out of phase with the strain and the storage modulus G' (characterises elastic properties) and the loss modulus G'' ($\propto$ energy dissipated during flow) can be calculated. At low values of deformation the gels behave as an elastic solid, but at high values viscous flow processes occur. The threshold value between the two is called the yield value. According to the theory of Princen and Kiss[260] for hypothetical monodisperse gels, the static shear modulus G_0 is given by

$$G_0 = 1.769 \frac{\gamma}{r_{32}} \phi^{1/3}(\phi - 0.712) \tag{1.23}$$

where r_{32} is the surface-volume mean radius and γ is the interfacial tension. Experimentally, values of G_0 show more or less the right dependence as γ, r_{32} and ϕ are varied in a variety of systems.[245,248,250,256] To a first approximation, gel emulsion behaviour can be adequately described by a Maxwell element, where the elastic component is related to the foam-like structure, and the viscous element is related to the intrinsic viscosity of the continuous phase. Addition of salt to nonionic W/O gels increases their yield value and slows down considerably the loss in yield stress with time.[243]

1.6   Forces between Oil–Water Interfaces

The measurement of forces between surfaces at small separation is of great importance in gaining a fundamental understanding of the complex behaviour of colloidal systems. Interactions of solid surfaces (coated or uncoated) across a fluid medium have been made through development of both the surface forces apparatus (SFA)[261] and the atomic force microscope.[262] Interactions between liquid surfaces are generally discussed in terms of the variation of the disjoining pressure with surface separation, and the majority of studies relate to foam (*i.e.* vapour–liquid–vapour) films.[263] Investigations of various aspects of emulsion (*i.e.* liquid–liquid–liquid) films do exist[264,265] but invariably the thin film radius is of the order of 100 µm, far larger than the contact film radius likely to be formed when two micron-sized emulsion drops approach. Two recent techniques have been developed capable of determining the force–distance curves between either sub-micron-sized emulsion drops or between a drop and a planar oil–water interface where the emulsion film radius is only a few microns.

1.6.1   Forces between Emulsion Drops

Oil-in-water emulsions, in which the dispersed phase is a ferrofluid of kerosene containing ~6 vol% iron oxide (γ-Fe_2O_3) grains of diameter 90 Å, can be prepared in a monodisperse state.[266] In the absence of an external magnetic field, these emulsions behave like non-magnetic emulsions where the drops have no permanent dipole moment since the magnetic grains within each drop are randomly oriented. An external magnetic field induces a dipole moment in each drop, the magnitude of which increases with the field strength. As a result, the drops interact through dipole forces which can be controlled. Ferrofluid emulsions are essentially brown, and although strong scattering occurs as in the case of a white, milky emulsion, light absorption also arises by the iron oxide particles enclosed within oil drops. In a magnetic field, the brown colour changes instantaneously to either yellow, green, red or blue depending on the field strength and drop size. The colours are only visible in the back-scattering direction which is also parallel to the applied field. The behaviour of these emulsions has been studied by optical microscopy, light transmission and static light scattering techniques.[266] Samples are loaded into flat wedge cells varying in thickness between 5 and 800 μm. A rich phase behaviour is seen in such confined colloidal dispersions. The appearance of different phases depends on a number of parameters including the magnetic field strength, H, the drop volume fraction and the cell thickness. At low H, the emulsion drops are randomly dispersed. With increasing H, drops merge initially forming small chains, which themselves combine to give long chains. The chains can either stay separated or join to form columns. At even higher H, the chains and columns are ordered until they become so rigid that they are locked in position. For small cell thicknesses (5 μm), the chains, equal in thickness to the drop diameter (0.51 μm), are around 10 times the drop diameter in length and are separated by about 5 μm. Increasing the cell thickness allows longer chains to form and columns comprised of around six chains appear. The columns are stable because each chain is shifted by one drop radius from neighbouring ones, resulting in chains attracting each other and 'zippering' together. By varying the rate at which the magnetic field is applied, the authors conclude that the columnar structure is the equilibrium phase, equilibrium being reached at low field rates. Light scattering from these systems gives rise to a number of concentric rings, indicating that a partially ordered structure is formed.[266,267] The average separation between the columns, d_c, can be determined precisely from the spectra. Surprisingly, d_c is independent of volume fraction although the degree of ordering improves. Instead of forming more columns with lower spacing between them, the column width increases as $\phi^{0.5}$ so as to maintain constant spacing. By contrast, the column spacing increases with increasing cell thickness, L, a reasonable fit leading to the relation $d_c = 1.33 L^{0.37}$.[268] A new theory which models the column shape as cylinders accounts for these results.[269]

The most exciting exploitation of the properties of magnetic emulsions has been to provide a novel approach for determining directly the force–distance relationships between small, non-deformable emulsion drops.[270–272] Leal Calderon et al.[270] make use of the fact that the magnitude of the induced dipole can be controlled by the strength of the applied field, allowing the force to be determined. Since the drops are uniform in size, light is Bragg scattered from the chains, enabling the separation to be measured. Octane-in-water emulsions stabilised by SDS in which the oil phase contains 10 vol% iron oxide grains change colour upon increasing the intensity of the magnetic field. The centre-to-centre spacing (d) between drops within chains is obtained from the spectral distribution of the scattered light. Using drops of known radius (r), the minimum separation between interfaces of neighbouring drops, $h = d - 2r$, can be determined. The repulsive force between charged drops must balance exactly the attractive force caused by the induced dipoles. This attractive force, augmented by van der Waals attraction at short distances, can be calculated exactly knowing the induced magnetic moment of each drop (which varies with H) and d. The repulsive force between emulsion drops, deduced in this way, is plotted as a function of the interfacial separation h in Figure 1.20 for three concentrations of SDS in the continuous phase. It can be seen that the method is able to resolve separations of about 1.5 nm and to measure forces

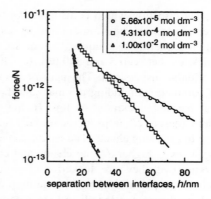

Figure 1.20 *Repulsive force vs. separation between drop interfaces for monodisperse octane ferrofluid-in-water emulsions (radius = 0.094 μm, φ = 0.001) stabilised by SDS of varying concentrations in the continuous phase (given)*
(Redrawn with permission from ref. 270)

down to 10^{-13} N. The minimum force that can be measured is that needed to hold drops in a chain so that the interaction energy just overcomes the thermal energy. Increasing the [SDS] from well below the CMC to just above it reduces the range of the Coulombic interaction by changing the surface potential (ψ_0) of the drops and the Debye length. Calculations of the repulsive force based on DLVO theory, in which the only unknown parameter is ψ_0, show excellent agreement with experimental findings. The value of ψ_0 obtained from such fits becomes more negative with increasing bulk surfactant concentration as a result of increasing adsorption onto emulsion drop surfaces. Although this leads to a larger repulsion, it is compensated by a decrease in the Debye screening length. Similar force–distance profiles have been given for ferrofluid emulsions stabilised by a cationic surfactant in the presence of varying concentrations of micelles in the continuous phase.[271] Here, the repulsive force profile contains two contributions, one of which is repulsive and is due to the double layers around drops, whilst the other is attractive and originates from the depletion of micelles between drops. The electrostatic repulsion between micelles and emulsion drops enhances the depletion force by effectively increasing the emulsion drop diameter.

1.6.2   Forces between a Drop and a Planar Oil–Water Interface

A novel liquid surface forces apparatus (LSFA) developed at Hull has been used to study oil–water–oil films.[273–275] An oil drop of radius a few microns is supported in an aqueous phase on the tip of a fine glass micropipette that is held initially below an oil–water interface. The other thick end of the micropipette is connected to a manometer containing the same oil. The radius of the oil drop is controlled by the balance of the Laplace pressure inside the drop and the applied pressure head. When the oil drop is moved up to the oil–water interface, a thin water film is formed between the apex of the drop and the interface; the disjoining pressure in the film turns out to be simply one half of the pressure head since there are two oil–water interfaces. The thickness of the emulsion film is determined by analysis of the optical interference pattern collected with a microscope objective lens, video camera and digitizing electronics. Moving the drop upwards into the oil–water interface at constant hydrostatic pressure results in an increase in the film radius at constant film thickness. Therefore, measurements of the film thickness at different applied hydrostatic pressures yields the disjoining pressure isotherm. In addition, the total force exerted on the oil drop as it is pushed up to the interface can be determined from the extent of the vertical deflection of the micropipette. The LSFA is thus capable of measuring disjoining pressures in the range 50–2000 Pa and thicknesses in the range 5–100 nm. It has been used so far to examine emulsion films for which the net interaction between the oil–water interfaces is either repulsive at all separations (ionic surfactant AOT without electrolyte[273]) or mutually attractive (ionic surfactant SDS in the presence of high concentrations of salt[274]). The effects of electrolyte in aqueous films stabilised by the nonionic surfactants $C_{12}E_5$ or decyl-β-D-glucopyranoside (sugar) and AOT were reported recently[275] and some results are shown in Figure 1.21. The isotherms for AOT fall into two series;

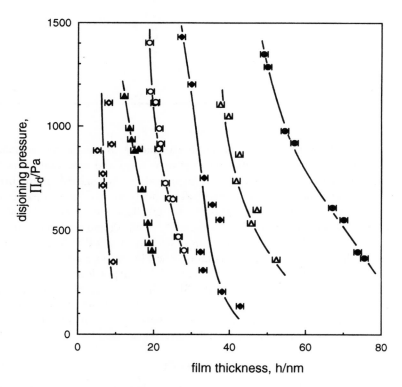

Figure 1.21 *Variation of the disjoining pressure with film thickness for dodecane–water–dodecane emulsion films stabilised by AOT at 20 °C determined with the LSFA. The [AOT]/mM and [NaCl]/mM pairs are:* ●*(0.3, 0);* △*(1.0, 0);* ◆*(3.0, 0);* ○*(0.11, 8.55);* ▲*(0.031, 51.3);* ◇*(0.0135, 136.7)*
(Redrawn with permission from ref. 275)

in one no salt is present and the AOT concentration is increased from below the CMC (2.2 mM) to above it, whereas in the other containing salt the AOT concentration is adjusted below the CMC so as to keep the oil–water tension constant at 18 mN m^{-1}. Assuming that the total disjoining pressure is made up only of a repulsive electrostatic component and an attractive van der Waals term, the calculated curves have been compared with the experimental ones using the surface potential ψ_0 as the only adjustable parameter. Table 1.2 lists the best fit values for ψ_0 and the calculated surface charge densities (σ_0) from them. As seen, ψ_0 is roughly constant at around −80 to −90 mV and close to the zeta potentials of O/W emulsion drops stabilised by the same surfactant.

Table 1.2 *Derived values of the surface potential and surface charge density for AOT monolayers at the dodecane–water interface at 20°C obtained by fitting the disjoining pressure-film thickness isotherms for oil–water–oil films determined with the LSFA to DLVO theory*

[AOT] in water/mM	[NaCl]/mM	ψ_0/mV	σ_0/charges nm^{-2}
0.3	0	−90	0.036
1.0	0	−85	0.059
3.0	0	−90	0.114
0.11	8.55	−80	0.16
0.031	51.3	−80	0.38
0.0135	136.7	−80	0.62

This overview of recent advances in the understanding of emulsions concludes with a short description of the similarities between foam and emulsion films. A vertical soap (air–water–air) film may be formed by drawing a suitable frame from a surfactant solution. The film, composed of two monolayers and intervening liquid, may thin under the action of gravitational and capillary forces and ultimately become black when observed in reflected light.[276–278] With ionic surfactants in the presence of certain salts, such films thin down to one of two equilibrium thicknesses. At low [salt] and high temperature, the film is more than 10 nm thick and is termed a common black film, whereas at high [salt] or low temperature its thickness is only a few nm (Newton black film). In the latter films, a non-zero contact angle θ develops between the film and surrounding meniscus where the adhesion between the two surfaces arises from the presence of attractive interactions. The energy of adhesion ε between the surfaces is given by

$$\varepsilon = 2\gamma(1 - \cos\theta) \qquad (1.24)$$

and measurements reveal that ε increases (*i.e.* θ increases since γ is roughly constant) with a decrease in temperature. From X-ray reflectivity analysis, SDS Newton black films consist of a surfactant bilayer containing hydration water of total thickness equal to 32 Å.[279] If air is replaced by oil, similar films may be

investigated with pairs of large O/W emulsion drops using optical microscopy.[280]
The temperature at which the drops begin to adhere, T^*, is found to be linearly
related to the salt concentration, as seen in Figure 1.22. The inset shows how ε
varies with temperature at 0.7 M NaCl; for a given temperature, ε is larger the
higher the [salt]. In order to gain insight into the structure of these thin films and
whether it changes as ε is varied, neutron scattering was performed on bulk
emulsion samples. Using monodisperse O/W drops of 0.4 µm in diameter and a
drop volume fraction equal to 70%, the scattering results mainly from the thin
films. Without adhesion, the drops remain essentially spherical and the system
may be considered as a collection of randomly distributed curved monolayers.
With adhesion, the surfactant films are randomly distributed flat bilayers, and an
increase in ε causes an increase in both the lateral extension of the flat films and
an increase in the contact angle. For adhesive emulsions, the fitted spectra yield a
value for the total thickness of the thin film of 29 Å, very close to that in foam
films. Interestingly, this thickness is constant with respect to temperature even

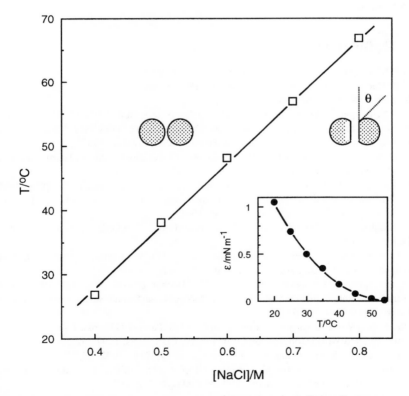

Figure 1.22 *Effect of salt concentration on the temperature at which adhesion occurs (non-
zero contact angle θ) for 50 µm diameter dodecane-in-water emulsion drops
stabilised by 2 mM SDS. The inset shows the variation of the energy of
adhesion, ε, with temperature for the 0.7 M NaCl system*
(Redrawn with permission from ref. 280)

though the adhesion energy changes by ~ 1 mN m^{-1}. It is also independent of salt concentration even though ε changes. The constant thickness over all adhesive conditions suggests the existence of a short range repulsion which is unaffected by the prevailing conditions. Its specific origin, however, remains unclear.

Adhesion between emulsion drops also manifests itself in the morphology of the bulk emulsion. In exactly the same system as described above for emulsion films, monodisperse emulsions have been observed at different temperatures and salt concentrations.[281] At high temperatures the drops are dispersed. On lowering the temperature a phase separation takes place between a dense phase of aggregated drops and a dilute phase of discrete drops. At low temperatures, no separated drops remain, *i.e.* the attractive interactions between the drops are strong and only aggregates exist. As with large drops reported in Figure 1.22, the temperature at which phase separation occurs in emulsions of small drops (radius <0.5 µm), T_s, increases on adding salt, and is lower for smaller drops. The proximity of T^* and T_s shows that the aggregation of the drops in emulsions is most likely due to the adhesion phenomena mentioned above occurring between two drop surfaces.

1.7   References

1. B.P. Binks, *Annu. Rep. Prog. Chem., Sect. C*, 1996, **92**, 97.
2. G.D. Rees and B.H. Robinson, *Adv. Mater.*, 1993, **5**, 608.
3. R. Zana, *Heterog. Chem. Rev.*, 1994, **1**, 145.
4. M.J. Schwuger, K. Stickdorn and R. Schömacker, *Chem. Rev.*, 1995, **95**, 849.
5. K. Shinoda and S. Friberg, 'Emulsions and Solubilisation', Wiley, New York, 1986.
6. E. Dickinson, ed., 'Food Emulsions and Foams', RSC, London, 1987.
7. E. Dickinson and G. Stainsby, eds., 'Advances in Food Emulsions and Foams', Elsevier Applied Science, London, 1988.
8. K. Larsson and S.E. Friberg, eds., 'Food Emulsions', 2nd edn., Dekker, New York, 1990.
9. M. El-Nokaly and D. Cornell, eds., 'Microemulsions and Emulsions in Foods', ACS Symposium Series 448, ACS, Washington, 1991.
10. J. Sjöblom, ed., 'Emulsions—A Fundamental and Practical Approach', Kluwer, Dordrecht, 1992.
11. L.L. Schramm, ed., 'Emulsions: Fundamentals and Applications in the Petroleum Industry', Adv. Chem. Series No. 231, ACS, Washington, 1992.
12. 'Encyclopedia of Emulsion Technology', ed. P. Becher, Dekker, New York, 1996, vol. 4.
13. 'Emulsions and Emulsion Stability', ed. J. Sjöblom, Surfactant Science Series, vol. 61, Dekker, New York, 1996.
14. 'Proc. First World Congress on Emulsion(s)', EDS, Paris, 1993, vols. 1–4.
15. 'Proc. Second World Congress on Emulsion(s)', EDS, Paris, 1997, vols. 1–4.
16. E.H. Lucassen-Reynders and K.A. Kuijpers, *Colloids Surf.*, 1992, **65**, 175.
17. E. Dickinson, in 'Controlled Particle, Droplet and Bubble Formation', ed. D.J. Wedlock, Butterworth-Heinemann, Oxford, 1994, ch. 7, p. 191.
18. H.A. Barnes, *Colloids Surf. A*, 1994, **91**, 89.
19. R. Pal, *Chem. Eng. J.*, 1997, **67**, 37; *Chem. Eng. Sci.*, 1997, **52**, 1177; *J. Rheol.*, 1997, **41**, 141.
20. N. Garti and A. Aserin, *Adv. Colloid Interface Sci.*, 1996, **65**, 37; N. Garti, *Colloids Surf. A*, 1997, **123–124**, 233; *Food Sci. Technol.*, 1997, **30**, 222.

21. V.B. Menon and D.T. Wasan, *Sep. Sci. Technol.*, 1988, **23**, 2131; *Colloids Surf.*, 1988, **29**, 7.
22. D.E. Tambe and M.M. Sharma, *Adv. Colloid Interface Sci.*, 1994, **52**, 1.
23. N.X. Yan and J.H. Masliyah, *J. Colloid Interface Sci.*, 1996, **181**, 20; *Colloids Surf. A*, 1996, **117**, 15; *Ind. Eng. Chem. Res.*, 1997, **36**, 1122.
24. R. Pal, *Colloids Surf. A*, 1994, **84**, 141.
25. J.G. Riess, *Colloids Surf. A*, 1994, **84**, 33; *New J. Chem.*, 1995, **19**, 891.
26. C. Washington, *Adv. Drug Delivery Rev.*, 1996, **20**, 131.
27. J.G. Riess and J.G. Weers, *Curr. Opin. Colloid Interface Sci.*, 1996, **1**, 652.
28. E. Dickinson, *J. Food Eng.*, 1994, **22**, 59.
29. S. Matsumoto, *J. Jpn. Soc. Food Sci. Technol.*, 1995, **42**, 69.
30. D.S. Horne, *Curr. Opin. Colloid Interface Sci.*, 1996, **1**, 752.
31. O. Urdahl and J. Sjöblom, *J. Disp. Sci. Technol.*, 1995, **16**, 557.
32. Y.H. Kim, D.T. Wasan and P.J. Breen, *Colloids Surf. A*, 1995, **95**, 235.
33. Th.F. Tadros, *Adv. Colloid Interface Sci.*, 1993, **46**, 1.
34. M. Chappat, *Colloids Surf. A*, 1994, **91**, 57.
35. W.D. Bancroft, *J. Phys. Chem.*, 1913, **17**, 501.
36. O. Ghosh and C.A. Miller, *J. Phys. Chem.*, 1987, **91**, 4528.
37. B.P. Binks, *Colloids Surf. A*, 1993, **71**, 167.
38. W.C. Griffin, *J. Soc. Cosmet. Chem.*, 1949, **1**, 311.
39. R. Aveyard, B.P. Binks and P.D.I. Fletcher, in 'The Structure, Dynamics and Equilibrium Properties of Colloidal Systems', eds. D.M. Bloor and E. Wyn-Jones, Kluwer, Dordrecht, 1990, p. 557.
40. A. Graciaa, J. Lachaise, G. Marion and R.S. Schechter, *Langmuir*, 1989, **5**, 1315.
41. H.T. Davis, *Colloids Surf. A*, 1994, **91**, 9.
42. R. Aveyard, B.P. Binks, T.A. Lawless and J. Mead, *J. Chem. Soc., Faraday Trans. 1*, 1985, **81**, 2155.
43. R.D. Hazlett and R.S. Schechter, *Colloids Surf.*, 1988, **29**, 53.
44. D.H. Smith and K-H. Lee, *J. Phys. Chem.*, 1990, **94**, 3746.
45. B.P. Binks, *Chem. Ind. (London)*, 1993 (14), 537.
46. D.H. Smith, G.C. Covatch and K-H. Lim, *J. Phys. Chem.*, 1991, **95**, 1463.
47. R. Aveyard, B.P. Binks, S. Clark and P.D.I. Fletcher, *J. Chem. Soc., Faraday Trans.*, 1990, **86**, 3111.
48. R. Aveyard, B.P. Binks and P.D.I. Fletcher, *Langmuir*, 1989, **5**, 1210.
49. H. Kunieda and K. Shinoda, *J. Colloid Interface Sci.*, 1985, **106**, 107.
50. J.L. Salager, G. Lopéz-Castellanos, M. Miñana-Pérez, C. Parra, C. Cucuphat, A. Graciaa and J. Lachaise, *J. Disp. Sci. Technol.*, 1991, **12**, 59.
51. A. Kabalnov, T. Tarara, R. Arlauskas and J. Weers, *J. Colloid Interface Sci.*, 1996, **184**, 227.
52. B.P. Binks, *Langmuir*, 1993, **9**, 25.
53. F. Harusawa, T. Saito, H. Nakajima and S. Fukushima, *J. Colloid Interface Sci.*, 1980, **74**, 435.
54. F. Harusawa, H. Nakajima and M. Tanaka, *J. Soc. Cosmet. Chem.*, 1982, **33**, 115.
55. J.T. Davies, in 'Proceedings of the 2nd Int. Congr. Surf. Activity', ed. J.H. Schulman, Butterworths, London, 1957, vol. 1, p. 426.
56. T.T. Traykov and I.B. Ivanov, *Int. J. Multiphase Flow*, 1977, **3**, 471.
57. I.B. Ivanov, *Pure Appl. Chem.*, 1980, **52**, 1241.
58. Z. Zapryanov, A.K. Malhotra, N. Aderangi and D.T. Wasan, *Int. J. Multiphase Flow*, 1983, **9**, 105.
59. D.N. Petsev, N.D. Denkov and P.A. Kralchevsky, *J. Colloid Interface Sci.*, 1995, **176**, 201.

60. W. Helfrich, *Z. Naturforsch. C*, 1973, **28**, 693.
61. H. Kellay, B.P. Binks, Y. Hendrikx, L.T. Lee and J. Meunier, *Adv. Colloid Interface Sci.*, 1994, **49**, 85.
62. T.D. Gurkov, P.A. Kralchevsky and I.B. Ivanov, *Colloids Surf.*, 1991, **56**, 119.
63. B.P. Binks, H. Kellay and J. Meunier, *Europhys. Lett.*, 1991, **16**, 53.
64. A. Kabalnov and H. Wennerström, *Langmuir*, 1996, **12**, 276.
65. H. Kellay, J. Meunier and B.P. Binks, *Phys. Rev. Lett.*, 1993, **70**, 1485.
66. A. Kabalnov and J. Weers, *Langmuir*, 1996, **12**, 1931.
67. D.H. Smith and Y-H.C. Wang, *J. Phys. Chem.*, 1994, **98**, 7214.
68. G.K. Johnson, D.B. Dadyburjor and D.H. Smith, *J. Phys. Chem.*, 1994, **98**, 12097.
69. D.H. Smith, G.K. Johnson, Y-H.C. Wang and K-H. Lim, *Langmuir*, 1994, **10**, 2516.
70. G.K. Johnson, D.B. Dadyburjor and D.H. Smith, *Langmuir*, 1994, **10**, 2523.
71. K. Shinoda and H. Kunieda, in 'Encyclopedia of Emulsion Technology', ed. P. Becher, Dekker, New York, 1983, vol. 1, p. 337.
72. S. Torza and S.G. Mason, *Science*, 1969, **163**, 813.
73. L.D. Landau and E.M. Lifshitz, 'Electrodynamics of Continuous Media', Pergamon Press, New York, 1960.
74. D.H. Smith and G.K. Johnson, *J. Phys. Chem.*, 1995, **99**, 10853.
75. G.E. Vaessen and H.N. Stein, *J. Colloid Interface Sci.*, 1995, **176**, 378.
76. E. Dickinson, *J. Colloid Interface Sci.*, 1981, **84**, 284.
77. B.W. Brooks and H.N. Richmond, *Colloids Surf.*, 1991, **58**, 131.
78. B.W. Brooks and H.N. Richmond, *Chem. Eng. Sci.*, 1994, **49**, 1053, 1065.
79. K. Shinoda and H. Saito, *J. Colloid Interface Sci.*, 1969, **30**, 258.
80. T. Mitsui, Y. Machida and F. Harusawa, *Am. Cosmet. Perfum.*, 1972, **87**, 33.
81. R.P. Enever, *J. Pharm. Sci.*, 1976, **65**, 517.
82. Y. Saito, T. Sato and I. Anazawa, *J. Am. Oil Chem. Soc.*, 1990, **67**, 145.
83. J.M. Pasternacki-Surian, R.L. Schnaare and E.T. Sugita, *Pharm. Res.*, 1992, **9**, 406.
84. S. Lehnert, H. Tarabishi and H. Leuenberger, *Colloids Surf. A*, 1994, **91**, 227.
85. Th. Förster, F. Schambil and H. Tesmann, *Int. J. Cosmet. Sci.*, 1990, **12**, 217.
86. Th. Förster, F. Schambil and W. von Rybinski, *J. Disp. Sci. Technol.*, 1992, **13**, 183.
87. A. Wadle, Th. Förster and W. von Rybinski, *Colloids Surf. A*, 1993, **76**, 51.
88. T. Engels, T. Förster and W. von Rybinski, *Colloids Surf. A*, 1995, **99**, 141.
89. T. Förster, W. von Rybinski and A. Wadle, *Adv. Colloid Interface Sci.*, 1995, **58**, 119.
90. M. Kahlweit, G. Busse and J. Winkler, *J. Chem. Phys.*, 1993, **99**, 5605.
91. T.J. Lin, *J. Soc. Cosmet. Chem.*, 1978, **29**, 117.
92. H. Sagitani, *J. Disp. Sci. Technol.*, 1988, **9**, 115.
93. M. Kahlweit, *Ber. Bunsenges. Phys. Chem.*, 1994, **98**, 490.
94. M. Kahlweit, R. Strey and G. Busse, *Phys. Rev. E*, 1993, **47**, 4197.
95. C. Curt, *Sci. Aliments*, 1994, **14**, 699.
96. G.G. Stokes, *Philos. Mag.*, 1851, **1**, 337.
97. L.W. Phipps, *J. Phys. D*, 1975, **8**, 448.
98. I. Burgaud, E. Dickinson and P.V. Nelson, *Int. J. Food Sci. Technol.*, 1990, **25**, 39.
99. G.K. Batchelor, *J. Fluid Mech.*, 1972, **52**, 245.
100. Y.T. Shih, D. Gidaspow and D.T. Wasan, *Powder Technol.*, 1987, **50**, 201.
101. R.A. Williams and W.P.K. Amarasinghe, *Trans. Inst. Min. Metall.*, 1989, **C98**, 68.
102. G.C. Barker and M.J. Grimson, *Colloids Surf.*, 1990, **43**, 55.
103. Y. Zimmels, *Powder Technol.*, 1992, **70**, 109.
104. K. Stamatakis and C. Tien, *Powder Technol.*, 1992, **72**, 227.
105. E. Barnea and J. Mizrahi, *Chem. Eng. J.*, 1973, **5**, 171.
106. V.J. Pinfield, E. Dickinson and M.J.W. Povey, *J. Colloid Interface Sci.*, 1994, **166**, 363.

107. P. Walstra and H. Oortwijn, *J. Colloid Interface Sci.*, 1969, **29**, 424.
108. M. Bury, J. Gerhards and W. Erni, *Int. J. Pharm.*, 1991, **76**, 207.
109. M. Bury, J. Gerhards, W. Erni and A. Stamm, *Int. J. Pharm.*, 1995, **124**, 183.
110. R. Kauten, J. Maneval and M.J. McCarthy, *J. Food Sci.*, 1990, **56**, 799.
111. G.M. Pilhofer, M.J. McCarthy, R.J. Kauten and J.B. German, *J. Food Eng.*, 1993, **20**, 369.
112. B. Newling, P.M. Glover, J.L. Keddie, D.M. Lane and P.J. McDonald, *Langmuir*, 1997, **13**, 3621.
113. A.M. Howe, A.R. Mackie and M.M. Robins, *J. Disp. Sci. Technol.*, 1986, **7**, 231.
114. C. Carter, D.J. Hibberd, A.M. Howe, A.R. Mackie and M.M. Robins, *Prog. Colloid Polym. Sci.*, 1988, **76**, 37.
115. D.J. McClements and M.J.W. Povey, *Adv. Colloid Interface Sci.*, 1987, **27**, 285.
116. D.J. Wedlock, I.J. Fabris and J. Grimsey, *Colloids Surf.*, 1990, **43**, 67.
117. D.J. McClements, E. Dickinson, S.R. Dungan, J.E. Kinsella, J.G. Ma and M.J.W. Povey, *J. Colloid Interface Sci.*, 1993, **160**, 293.
118. D.J. McClements, *Adv. Colloid Interface Sci.*, 1991, **37**, 33.
119. P.A. Gunning, M.S.R. Hennock, A.M. Howe, A.R. Mackie, P. Richmond and M.M. Robins, *Colloids Surf.*, 1986, **20**, 65.
120. D.J. Hibberd, A.M. Howe and M.M. Robins, *Colloids Surf.*, 1988, **31**, 347.
121. A.M. Howe and M.M. Robins, *Colloids Surf.*, 1990, **43**, 83.
122. A.J. Fillery-Travis, P.A. Gunning, D.J. Hibberd and M.M. Robins, *J. Colloid Interface Sci.*, 1993, **159**, 189.
123. D. Hibberd, A. Holmes, M. Garrood, A. Fillery-Travis, M. Robins and R. Challis, *J. Colloid Interface Sci.*, 1997, **193**, 77.
124. D.H. Napper, 'Polymeric Stabilisation of Colloidal Dispersions', Academic Press, London, 1983.
125. E. Dickinson, M.I. Goller and D.J. Wedlock, *Colloids Surf. A*, 1993, **75**, 195.
126. E. Dickinson, J. Ma and M.J.W. Povey, *Food Hydrocolloids*, 1994, **8**, 481.
127. E. Dickinson, M.I. Goller and D.J. Wedlock, *J. Colloid Interface Sci.*, 1995, **172**, 192.
128. A. Avranas, G. Stalidis and G. Ritzoulis, *Colloid Polym. Sci.*, 1988, **266**, 937.
129. G. Stalidis, A. Avranas and D. Jannakoudakis, *J. Colloid Interface Sci.*, 1990, **135**, 313.
130. A. Avranas and G. Stalidis, *J. Colloid Interface Sci.*, 1991, **143**, 180.
131. M. Jansson, L. Eriksson and P. Skagerlind, *Colloids Surf.*, 1991, **53**, 157; M. Jansson and M.A. Pés, *J. Colloid Interface Sci.*, 1994, **163**, 512.
132. C.B. Oleksiak, S.S. Habif and H.L. Rosano, *Colloids Surf. A*, 1994, **84**, 71.
133. C. Washington, *Int. J. Pharm.*, 1990, **66**, 1.
134. D. Fairhurst, M.P. Aronson, M.L. Gum and E.D. Goddard, *Colloids Surf.*, 1983, **7**, 153.
135. M.P. Aronson, *Langmuir*, 1989, **5**, 494.
136. M.P. Aronson, *Colloids Surf.*, 1991, **58**, 195.
137. D.J. McClements, *Colloids Surf. A*, 1994, **90**, 25.
138. J. Bibette, D. Roux and F. Nallet, *Phys. Rev. Lett.*, 1990, **65**, 2470.
139. J. Bibette, *J. Colloid Interface Sci.*, 1991, **147**, 474; Fr. Pat. 90/01990.
140. J. Bibette, D. Roux and B. Pouligny, *J. Phys. II*, 1992, **2**, 401.
141. F. Leal Calderon, J. Biais and J. Bibette, *Colloids Surf. A*, 1993, **74**, 303.
142. M-P. Krafft, J-P. Rolland and J.G. Riess, *J. Phys. Chem.*, 1991, **95**, 5673.
143. A. Meller and J. Stavans, *Langmuir*, 1996, **12**, 301.
144. B.P. Binks, P.D.I. Fletcher and D.I. Horsup, *Colloids Surf.*, 1991, **61**, 291.
145. F. Leal-Calderon, B. Gerhardi, A. Espert, F. Brossard, V. Alard, J.F. Tranchant, T. Stora and J. Bibette, *Langmuir*, 1996, **12**, 872.

146. B. Vincent, P.F. Luckham and F.A. Waite, *J. Colloid Interface Sci.*, 1980, **73**, 508.
147. C. Ma, *Colloids Surf.*, 1987, **28**, 1.
148. A. Vrij, *Pure Appl. Chem.*, 1976, **48**, 471.
149. R. Pal, *Colloids Surf.*, 1992, **64**, 207; *Colloids Surf. A*, 1993, **71**, 173.
150. E. Dickinson, M. Golding and M.J.W. Povey, *J. Colloid Interface Sci.*, 1997, **185**, 515;
     E. Dickinson and M. Golding, *Food Hydrocolloids*, 1997, **11**, 13.
151. S. Asakura and F. Oosawa, *J. Chem. Phys.*, 1954, **22**, 1255.
152. P.R. Sperry, *J. Colloid Interface Sci.*, 1982, **87**, 375.
153. P. Kékicheff, F. Nallet and P. Richetti, *J. Phys. II*, 1994, **4**, 735.
154. J. Bibette, T.G. Mason, H. Gang and D.A. Weitz, *Phys. Rev. Lett.*, 1992, **69**, 981;
     J. Bibette, T.G. Mason, H. Gang, D.A. Weitz and P. Poulin, *Langmuir*, 1993, **9**, 3352.
155. K. Kandori, K. Kishi and T. Ishikawa, *Colloids Surf.*, 1991, **55**, 73.
156. T.M. Obey and B. Vincent, *J. Colloid Interface Sci.*, 1994, **163**, 454; K.R. Anderson,
     T.M. Obey and B. Vincent, *Langmuir*, 1994, **10**, 2493.
157. M.I. Goller, T.M. Obey, D.O.H. Teare, B. Vincent and M.R. Wegener, *Colloids Surf. A*,
     1997, **123**, 183.
158. T.G. Mason and J. Bibette, *Phys. Rev. Lett.*, 1996, **77**, 3481; *Langmuir*, 1997, **13**, 4600;
     J. Bibette and T.G. Mason, *Fr. Pat.* 96/04736.
159. R.P. Borwankar, L.A. Lobo and D.T. Wasan, *Colloids Surf.*, 1992, **69**, 135.
160. K.D. Danov, I.B. Ivanov, T.D. Gurkov and R.P. Borwankar, *J. Colloid Interface Sci.*,
     1994, **167**, 8.
161. F.A.M. Leermakers, Y.S. Sdranis, J. Lyklema and R.D. Groot, *Colloids Surf. A*, 1994,
     85, 135.
162. G.K. Batchelor, *J. Fluid Mech.*, 1976, **74**, 1.
163. M.P. Aronson and H.M. Princen, *Nature (London)*, 1980, **286**, 370.
164. J.A.M.H. Hofman and H.N. Stein, *J. Colloid Interface Sci.*, 1991, **147**, 508.
165. N.D. Denkov, D.N. Petsev and K.D. Danov, *Phys. Rev. Lett.*, 1993, **71**, 3226.
166. K.D. Danov, D.N. Petsev, N.D. Denkov and R. Borwankar, *J. Chem. Phys.*, 1993, **99**,
     7179.
167. N.D. Denkov, D.N. Petsev and K.D. Danov, *J. Colloid Interface Sci.*, 1995, **176**, 189.
168. D.N. Petsev and P. Linse, *Phys. Rev. E*, 1997, **55**, 586.
169. N.D. Denkov and P.A. Kralchevsky, *Prog. Colloid Polym. Sci.*, 1995, **98**, 18.
170. V.J. Pinfield, E. Dickinson and M.J.W. Povey, *J. Colloid Interface Sci.*, 1997, **186**, 80.
171. I.B. Ivanov and D.S. Dimitrov, in 'Thin Liquid Films', ed. I.B. Ivanov, Dekker, New
     York, 1988, p. 379.
172. A. Sharma and E. Ruckenstein, *Langmuir*, 1987, **3**, 760; *Colloid Polym. Sci.*, 1988,
     266, 60.
173. V.G. Babak, *Colloids Surf.*, 1988, **30**, 307; *Colloids Surf. A*, 1994, **85**, 279.
174. Ya.I. Rabinovich and A.A. Baran, *Colloids Surf.*, 1991, **59**, 47.
175. A.K. Chesters, *Trans. Inst. Chem. Eng. A*, 1991, **69**, 259.
176. J.K. Klahn, W.G.M. Agterof, F. van Voorst Vader, R.D. Groot and F. Groeneweg,
     Colloids Surf., 1992, **65**, 151.
177. L. Lobo, I. Ivanov and D. Wasan, *AIChE J.*, 1993, **39**, 322.
178. K.D. Danov, N.D. Denkov, D.N. Petsev, I.B. Ivanov and R. Borwankar, *Langmuir*, 1993,
     9, 1731.
179. Ph.T. Jaeger, J.J.M. Janssen, F. Groeneweg and W.G.M. Agterof, *Colloids Surf. A*,
     1994, **85**, 255.
180. A. Bhakta and E. Ruckenstein, *Langmuir*, 1995, **11**, 4642.
181. O. Reynolds, *Philos. Trans. Roy. Soc. (London)*, 1886, **A177**, 57.
182. For example, A. Scheludko, *Adv. Colloid Interface Sci.*, 1967, **1**, 391.

183. B. Radoev, A. Scheludko and E. Manev, *J. Colloid Interface Sci.*, 1983, **95**, 254.
184. E.A. Gardner and R.E. Apfel, *J. Colloid Interface Sci.*, 1993, **159**, 226.
185. T. Palermo, *Rev. Inst. Fr. Petrole*, 1991, **46**, 325.
186. R. Aveyard, B.P. Binks, P.D.I. Fletcher and X. Ye, *Prog. Colloid Polym. Sci.*, 1992, **89**, 114.
187. Sv.K. Chakarova, M. Dupeyrat, E. Nakache, C.D. Dushkin and I.B. Ivanov, *J. Surf. Sci. Technol.*, 1990, **6**, 201.
188. E. Nakache, P-Y. Longaive and S. Aiello, *Colloids Surf. A*, 1995, **96**, 69.
189. O.D. Velev, T.D. Gurkov, Sv.K. Chakarova, B.I. Dimitrova, I.B. Ivanov and R.P. Borwankar, *Colloids Surf. A*, 1994, **83**, 43.
190. I.B. Ivanov, Sv.K. Chakarova and B.I. Dimitrova, *Colloids Surf.*, 1987, **22**, 311.
191. O.D. Velev, T.D. Gurkov and R.P. Borwankar, *J. Colloid Interface Sci.*, 1993, **159**, 497.
192. O.D. Velev, T.D. Gurkov, I.B. Ivanov and R.P. Borwankar, *Phys. Rev. Lett.*, 1995, **75**, 264.
193. K.D. Danov, T.D. Gurkov, T. Dimitrova, I.B. Ivanov and D. Smith, *J. Colloid Interface Sci.*, 1997, **188**, 313.
194. K.P. Velikov, O.D. Velev, K.G. Marinova and G.N. Constantinides, *J. Chem. Soc., Faraday Trans.*, 1997, **93**, 2069.
195. A.D. Nikolov and D.T. Wasan, *J. Colloid Interface Sci.*, 1989, **133**, 1.
196. K. Koczo, A.D. Nikolov, D.T. Wasan, R.P. Borwankar and A. Gonsalves, *J. Colloid Interface Sci.*, 1996, **178**, 694.
197. M.J. Vold, *Langmuir*, 1985, **1**, 74.
198. J. Bibette, *Langmuir*, 1992, **8**, 3178; J. Bibette, D.C. Morse, T.A. Witten and D.A. Weitz, *Phys. Rev. Lett.*, 1992, **69**, 2439.
199. B.V. Derjaguin, *Acta Physicochim. URSS*, 1939, **10**, 25.
200. D.N. Petsev and J. Bibette, *Langmuir*, 1995, **11**, 1075.
201. B.V. Derjaguin, 'Theory of Stability of Colloids and Thin Liquid Films', Plenum Press, New York, 1989.
202. W. Thomson (Lord Kelvin), *Proc. R. Soc. Edinburgh*, 1871, **7**, 63.
203. I.M. Lifshitz and V.V. Slezov, *J. Phys. Chem. Solids*, 1961, **19**, 35.
204. C. Wagner, *Z. Elektrochem.*, 1961, **65**, 581.
205. P.W. Voorhees, *J. Stat. Phys.*, 1985, **38**, 231.
206. A.S. Kabal'nov, Yu.D. Aprosin, O.B. Pavlova-Verevkina, A.V. Pertsov and E.D. Shchukin, *Colloid J. USSR*, 1986, **48**, 20.
207. A.S. Kabalnov, A.V. Pertzov and E.D. Shchukin, *J. Colloid Interface Sci.*, 1987, **118**, 590.
208. A.S. Kabalnov, K.N. Makarov, A.V. Pertzov and E.D. Shchukin, *J. Colloid Interface Sci.*, 1990, **138**, 98; A.S. Kabalnov and K.N. Makarov, *Colloid J. USSR*, 1990, **52**, 589.
209. A.S. Kabalnov, K.N. Makarov and O.V. Shcherbakova, *J. Fluorine Chem.*, 1990, **50**, 271; E.A. Amelina, E.Z. Kumacheva, A.V. Pertsov and E.D. Shchukin, *Colloid J. USSR*, 1990, **52**, 185; A.S. Kabalnov, L.D. Gervits and K.N. Makarov, *ibid.*, 1990, **52**, 915.
210. A.S. Kabalnov, K.N. Makarov and E.D. Shchukin, *Colloids Surf.*, 1992, **62**, 101.
211. A.S. Kabalnov, *Langmuir*, 1994, **10**, 680.
212. P. Taylor and R.H. Ottewill, *Colloids Surf. A*, 1994, **88**, 303; *Prog. Colloid Polym. Sci.*, 1994, **97**, 199.
213. P. Taylor, *Colloids Surf. A*, 1995, **99**, 175.
214. Y. de Smet, J. Malfait, C. de Vos, L. Deriemaeker and R. Finsy, *Bull. Soc. Chim. Belg.*, 1996, **105**, 789; L. Bremer, B. de Nijs, L. Deriemaeker, R. Finsy, E. Geladé and J. Joosten, *Part. Part. Syst. Charact.*, 1996, **13**, 350.

215. J. Soma and K.D. Papadopoulos, *J. Colloid Interface Sci.*, 1996, **181**, 225.

216. D.J. McClements and S.R. Dungan, *Colloids Surf. A*, 1995, **104**, 127.

217. J. Weiss, J.N. Coupland and D.J. McClements, *J. Phys. Chem.*, 1996, **100**, 1066.

218. J. Weiss, J.N. Coupland, D. Brathwaite and D.J. McClements, *Colloids Surf. A*, 1997, **121**, 53; J.N. Coupland, J. Weiss, A. Lovy and D.J. McClements, *J. Food Sci.*, 1996, **61**, 1114.

219. J.N. Coupland, D. Brathwaite, P. Fairley and D.J. McClements, *J. Colloid Interface Sci.*, 1997, **190**, 71.

220. R. Aveyard, B.P. Binks, P. Cooper and P.D.I. Fletcher, *Adv. Colloid Interface Sci.*, 1990, **33**, 59.

221. S. Karaborni, N.M. van Os, K. Esselink and P.A.J. Hilbers, *Langmuir*, 1993, **9**, 1175.

222. K. Esselink, P.A.J. Hilbers, N.M. van Os, B. Smit and S. Karaborni, *Colloids Surf. A*, 1994, **91**, 155.

223. W.I. Higuchi and J. Misra, *J. Pharm. Sci.*, 1962, **51**, 459.

224. A.V. Pertsov, A.S. Kabalnov and E.D. Shchukin, *Kolloidn. Zh.*, 1984, **46**, 1172.

225. A.S. Kabal'nov, A.V. Pertzov and E.D. Shchukin, *Colloids Surf.*, 1987, **24**, 19.

226. A.S. Kabalnov and E.D. Shchukin, *Adv. Colloid Interface Sci.*, 1992, **38**, 69.

227. R. Buscall, S.S. Davis and D.C. Potts, *Colloid Polym. Sci.*, 1979, **257**, 636.

228. A.S. Kabal'nov, A.V. Pertsov, Yu.D. Aprosin and E.D. Shchukin, *Colloid J. USSR*, 1985, **47**, 898.

229. J.G. Weers and R.A. Arlauskas, *Langmuir*, 1995, **11**, 474.

230. J.G. Weers, Y. Ni, T.E. Tarara, T.J. Pelura and R.A. Arlauskas, *Colloids Surf. A*, 1994, **84**, 81; A. Kabalnov, J. Weers, R. Arlauskas and T. Tarara, *Langmuir*, 1995, **11**, 2966.

231. N.R. Cameron and D.C. Sherrington, *Adv. Polymer Sci.*, 1996, **126**, 163.

232. C. Solans, F. Comelles, N. Azemar, J. Sánchez Leal and J.L. Parra, *J. Com. Esp. Deterg.*, 1986, **17**, 109; C. Solans, N. Azemar and J.L. Parra, *Prog. Colloid Polym. Sci.*, 1988, **76**, 224.

233. H. Kunieda, C. Solans, N. Shida and J.L. Parra, *Colloids Surf.*, 1987, **24**, 225.

234. R. Pons, I. Carrera, P. Erra, H. Kunieda and C. Solans, *Colloids Surf. A*, 1994, **91**, 259.

235. H. Kunieda, Y. Fukui, H. Uchiyama and C. Solans, *Langmuir*, 1996, **12**, 2136.

236. R. Pons, J.C. Ravey, S. Sauvage, M.J. Stebe, P. Erra and C. Solans, *Colloids Surf. A*, 1993, **76**, 171.

237. C. Solans, R. Pons, S. Zhu, H.T. Davis, D.F. Evans, K. Nakamura and H. Kunieda, *Langmuir*, 1993, **9**, 1479.

238. V. Rajagopalan, C. Solans and H. Kunieda, *Colloid Polym. Sci.*, 1994, **272**, 1166.

239. H. Kunieda, V. Rajagopalan, E. Kimura and C. Solans, *Langmuir*, 1994, **10**, 2570.

240. H. Kunieda, N. Yano and C. Solans, *Colloids Surf.*, 1989, **36**, 313.

241. J. Kizling and B. Kronberg, *Colloids Surf.*, 1990, **50**, 131.

242. R. Pons, C. Solans, M.J. Stébé, P. Erra and J.C. Ravey, *Prog. Colloid Polym. Sci.*, 1992, **89**, 110.

243. M.P. Aronson and M.F. Petko, *J. Colloid Interface Sci.*, 1993, **159**, 134.

244. M.P. Aronson, K. Ananthapadmanabhan, M.F. Petko and D.J. Palatini, *Colloids Surf. A*, 1994, **85**, 199.

245. R. Pons, P. Erra, C. Solans, J-C. Ravey and M-J. Stébé, *J. Phys. Chem.*, 1993, **97**, 12320.

246. J.C. Ravey and M.J. Stébé, *Physica B*, 1989, **156/157**, 394.

247. J.C. Ravey and M.J. Stébé, *Prog. Colloid Polym. Sci.*, 1990, **82**, 218.

248. J.C. Ravey, M.J. Stébé and S. Sauvage, *Colloids Surf. A*, 1994, **91**, 237; *J. Chim. Phys.*, 1994, **91**, 259.

249. J.M. Williams, *Langmuir*, 1988, **4**, 44; J.M. Williams and D.A. Wrobleski, *ibid.*, 1988, **4**, 656; J.M. Williams, A.J. Gray and M.H. Wilkerson, *ibid.*, 1990, **6**, 437; J.M. Williams, *ibid.*, 1991, **7**, 1370.

250. G. Ebert, G. Platz and H. Rehage, *Ber. Bunsenges. Phys. Chem.*, 1988, **92**, 1158.
251. E. Ruckenstein, G. Ebert and G. Platz, *J. Colloid Interface Sci.*, 1989, **133**, 432.
252. H.H. Chen and E. Ruckenstein, *J. Colloid Interface Sci.*, 1990, **138**, 473.
253. H. Kunieda, D.F. Evans, C. Solans and M. Yoshida, *Colloids Surf.*, 1990, **47**, 35.
254. D.M.A. Buzza and M.E. Cates, *Langmuir*, 1994, **10**, 4503.
255. T.G. Mason, J. Bibette and D.A. Weitz, *Phys. Rev. Lett.*, 1995, **75**, 2051.
256. R. Pons, C. Solans and Th.F. Tadros, *Langmuir*, 1995, **11**, 1966.
257. K. Ozawa, C. Solans and H. Kunieda, *J. Colloid Interface Sci.*, 1997, **188**, 275.
258. A.K. Das and P.K. Ghosh, *Langmuir*, 1990, **6**, 1668; D. Mukesh, A.K. Das and P.K. Ghosh, *ibid.*, 1992, **8**, 807; A.K. Das, D. Mukesh, V. Swayambunathan, D.D. Kotkar and P.K. Ghosh, *ibid.*, 1992, **8**, 2427.
259. K.J. Lissant, *J. Soc. Cosmet. Chem.*, 1970, **21**, 141.
260. H.M. Princen and A.D. Kiss, *J. Colloid Interface Sci.*, 1986, **112**, 427.
261. J.N. Israelachvili and G.E. Adams, *J. Chem. Soc., Faraday Trans. 1*, 1978, **74**, 975.
262. H.-J. Butt, *J. Colloid Interface Sci.*, 1994, **166**, 109.
263. 'Thin Liquid Films—Fundamentals and Applications', ed. I.B. Ivanov, Surfactant Science Series, vol. 29, Dekker, New York, 1988.
264. D. Platikanov and E. Manev, 'Proc. 4th Int. Congr. Surf. Act. Subst. (Brussels, 1964)', ed. J.Th.G. Overbeek, Gordon and Breach, London, 1967, vol. 2, p. 1189.
265. T.M. Herrington, B.R. Midmore and S.S. Sahi, *J. Chem. Soc., Faraday Trans. 1*, 1982, **78**, 2711.
266. J. Bibette, *J. Magn. Magn. Mater.*, 1993, **122**, 37; E.M. Lawrence, M.L. Ivey, G.A. Flores, J. Liu, J. Bibette and J. Richard, *Int. J. Mod. Phys. B*, 1994, **8**, 2765.
267. M. Hagenbuchle, P. Sheaffer, Y. Zhu and J. Liu, *Int. J. Mod. Phys. B*, 1996, **10**, 3057.
268. T. Mou, G.A. Flores, J. Liu, J. Bibette and J. Richard, *Int. J. Mod. Phys. B*, 1994, **8**, 2779.
269. J. Liu, E.M. Lawrence, A. Wu, M.L. Ivey, G.A. Flores, K. Javier, J. Bibette and J. Richard, *Phys. Rev. Lett.*, 1995, **74**, 2828.
270. F. Leal Calderon, T. Stora, O. Mondain Monval, P. Poulin and J. Bibette, *Phys. Rev. Lett.*, 1994, **72**, 2959.
271. O. Mondain-Monval, F. Leal-Calderon, J. Phillip and J. Bibette, *Phys. Rev. Lett.*, 1995, **75**, 3364.
272. O. Mondain-Monval, F. Leal-Calderon and J. Bibette, *J. Phys. II*, 1996, **6**, 1313.
273. R. Aveyard, B.P. Binks, W-G. Cho, L.R. Fisher, P.D.I. Fletcher and F. Klinkhammer, *Langmuir*, 1996, **12**, 6561.
274. W-G. Cho and P.D.I. Fletcher, *J. Chem. Soc., Faraday Trans.*, 1997, **93**, 1389.
275. B.P. Binks, W-G. Cho and P.D.I. Fletcher, *Langmuir*, 1997, **13**, 7180.
276. J.A. de Feijter and A. Vrij, *J. Colloid Interface Sci.*, 1978, **64**, 269.
277. H.M. Princen, M.P. Aronson and J.C. Moser, *J. Colloid Interface Sci.*, 1980, **75**, 246.
278. D. Exerowa, D. Kaschiev and D. Platikanov, *Adv. Colloid Interface Sci.*, 1992, **40**, 201.
279. O. Bélorgey and J.J. Benattar, *Phys. Rev. Lett.*, 1991, **66**, 313.
280. P. Poulin, F. Nallet, B. Cabane and J. Bibette, *Phys. Rev. Lett.*, 1996, **77**, 3248.
281. P. Poulin and J. Bibette, *Phys. Rev. Lett.*, 1997, **79**, 3290.

CHAPTER 2

Emulsion Formation

PIETER WALSTRA AND PAULINE E.A. SMULDERS

Department of Food Science, Wageningen Agricultural University, PO Box 8129,
6700 EV Wageningen, The Netherlands

2.1 Introduction

To make an emulsion, oil, water, surfactant and energy are needed. The composition of the system and the way of processing then determine emulsion type (oil-in-water or water-in-oil), droplet volume fraction (ϕ), droplet size and composition of the layer of surfactant around the droplets. These variables determine most emulsion properties, notably physical stability.[1] Consequently, knowledge of emulsion formation is of considerable importance. In this chapter, a review is given, with some emphasis on newer developments. Some aspects are left out, because they have been sufficiently discussed in earlier reviews.[2-4] For the convenience of the reader, however, important general points are recalled. Some aspects are not discussed, such as the preparation of high-internal phase emulsions, double emulsions, microemulsions and emulsions with very coarse drops. Typically, the emulsions considered have droplets of, say, a micrometre in diameter. Some specialized methods of emulsion formation will also be left out.

Making drops is easy, and it will not be discussed here; see an earlier review.[2] In many cases, however, small drops are desired, and making these is difficult or, in other words, it costs a large amount of energy and/or surfactant. This is because of the drop's *Laplace pressure* p_L, defined as the difference between the pressure inside and outside the droplet, which is given by

$$p_L = \gamma(1/R_1 + 1/R_2) \tag{2.1}$$

where γ is the interfacial tension between oil and water and R_1 and R_2 are the principal radii of curvature of the drop; for a spherical drop $R_1 = R_2 = d/2$, where d is diameter. To break up a drop into smaller ones, it must be strongly deformed and any deformation increases p_L, as illustrated in Figure 2.1. Consequently, the stress needed to deform the drop is higher for a smaller drop. Since the stress is generally transmitted by the surrounding liquid *via* agitation, higher stresses need more vigorous agitation, hence more energy.

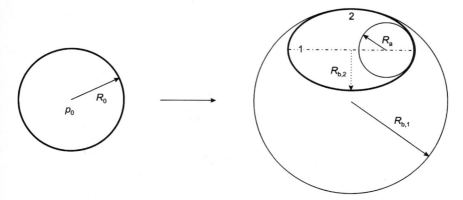

Figure 2.1 *Illustration of the increase in Laplace pressure when a drop (cross section of surface denoted by thick lines) is deformed into a prolate ellipsoid (long axis in the plane of the paper). Near 1, there is one radius of curvature (R_a), near 2 there are two ($R_{b,1}$ and $R_{b,2}$). The Laplace pressure near 1 is thus larger than that near 2 and also larger than in the undeformed drop*

The *surfactant* plays many roles, one of which is lowering interfacial tension, hence p_L, hence the stress needed to deform and break up a droplet. It is essential that the surfactant prevents coalescence of newly formed drops. The various processes occurring during emulsification, *i.e.* droplet break-up, adsorption of surfactant and droplet collisions (which may or may not lead to coalescence), are illustrated in Figure 2.2. Typically, each of these processes occurs numerous times during emulsification, and the timescale for each step is very small, for instance a microsecond.

To describe emulsion formation, we need to consider hydrodynamics and surface science. Sections 2.2 and 2.3 are especially about the hydrodynamics, the first of these describing mechanisms and processes. It follows that different régimes for emulsification can be distinguished. Section 2.3 treats the factors governing the break-up of drops in the various régimes, assuming the interfacial tension to be constant. In Section 2.4, the roles of the surfactant are discussed, integrating the interfacial aspects with the hydrodynamics, to the extent that current theory allows. Section 2.5 gives some additional considerations.

In the following, results on *droplet size distribution* will often be given. If the number frequency of droplets as a function of droplet diameter d is given by $f(d)$, the nth moment of the distribution then is

$$S_n \equiv \int_0^\infty d^n f(d) \partial d \tag{2.2}$$

Any type of average d is now given by

$$d_{nm} = (S_n/S_m)^{1/(n-m)} \tag{2.3}$$

In most cases, d_{32} (the volume/surface average or Sauter mean), or d_{43} will be used.

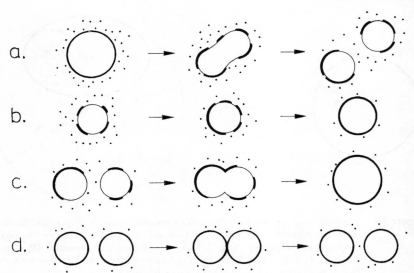

Figure 2.2   *Various processes occurring during emulsion formation. The drops are depicted by thin lines and the surfactant by heavy lines and dots. Highly schematic and not to scale*
(Redrawn with permission from ref. 4)

The width of the size distribution can be given as the variation coefficient c_m, which is the standard deviation of the distribution weighted with d^m, divided by the corresponding average d. Generally c_2 will be used, which corresponds to d_{32}.

Another variable is the specific surface area A (surface area of all droplets per unit volume of emulsion), which is given by

$$A = \pi S_2 = 6\phi/d_{32} \qquad (2.4)$$

2.2   Régimes

There are many different methods and machines to make emulsions and Table 2.1 gives an overview of some of the most important ones. The methods differ in several respects. The process can be batch-wise or continuous. In most types of apparatus, droplets can be formed as well as disrupted into smaller ones, but in some cases droplets below a certain size must already be present before starting. The most important variable may be the mechanism by which droplets are disrupted. This is achieved by external forces, that generally act *via* the continuous phase. In most cases, the forces derive from mechanical energy; disruption due to chemical energy will be briefly mentioned in Section 2.4, electrical energy will not be considered here. See ref. 2 for some other methods.

In all methods considered there is liquid flow, and it is useful to distinguish between unbounded and strongly confined flow. In *unbounded flow* any droplet is surrounded by a large amount of flowing liquid or, in other words, the confining

Table 2.1 *Some important methods and apparatus to make emulsions*

Method	Droplet formation	Régime[a]	Mode of operation[b]	Intensity[c]
Pipe flow	Yes	TI, TV	C	L
Static mixer	Yes	TI, TV, LV	C	L–M
Stirrer	Yes	TI, TV, LV	B, C	L–M
Toothed disk (*e.g.* UltraTurrax)	Yes	TI, TV, LV	B, C	M
Colloid mill	Yes[d]	(TV), LV	C	H
High-pressure homogenizer	No	TI (+ CI), TV, (LB)	C	H
Ultrasound generator	Yes	CI	B, C	M–H
Membrane emulsification	Yes[e]	Injection	B, C	–

[a]See text for notation; in parentheses means uncommon. [b]B = batch, C = continuous (flow-through) operation. [c]L = low, M = medium, H = high. [d]Not always easy. [e]No droplet break-up.

walls of the apparatus are far away from (most of) the droplets. The forces then can be frictional (mostly called viscous) or inertial. *Viscous forces* cause shear stresses to act on the interface between droplet and continuous phase; they act primarily in the direction of the interface. The shear stresses can be generated by laminar flow, and then we designate the break-up régime as LV. For Reynolds numbers (*Re*) above a critical value, the flow is turbulent. *Re* is given by

$$Re \equiv vl\rho/\eta \qquad (2.5)$$

where v is linear liquid velocity, ρ mass density of the liquid and η its viscosity; l is a characteristic length and it is given by the diameter for flow through a cylindrical tube and by twice the slit width in a narrow slit. It is thus a matter of scale of the apparatus and of the liquid viscosity whether flow is laminar or turbulent. If the turbulent eddies are much larger than the droplets, they exert shear stresses on the droplets. The régime then is designated TV.

If the turbulent eddies are very small, comparable in size to the droplets, *inertial forces* will cause disruption. The condition is that the droplet Reynolds number Re_d is larger than unity. Re_d is derived from eqn. (2.5) by putting $l = d$ and by taking v as the velocity of the liquid very near the drop. These forces are due to local pressure fluctuations in the liquid, and they act primarily in a direction perpendicular to the droplet surface. The régime is then designated TI. Pressure fluctuations can also be due to cavitation, which is often generated by intense ultrasound. This then is the régime CI. It will not be discussed; see *e.g.* Walstra[2] and Li and Fogler.[5]

In *bounded flow* other relations hold. If the smallest dimension of the part of the apparatus in which the droplets are disrupted (say a slit) is comparable to droplet size, other relations hold; the flow is always laminar. This régime, designated LB, has been studied by Kiefer.[6] A quite different régime prevails if the droplets are directly injected through a narrow capillary into the continuous phase; it may be called the injection régime. It is the basis of so-called membrane emulsification, a fairly new technique.

Within each régime, an essential variable is the *intensity* of the forces acting. Viscous stresses in laminar flow are roughly given by ηG, where G is the velocity gradient. In turbulent flow, the intensity is best expressed by the power density ε, *i.e.* the amount of energy dissipated in the flow per unit volume and unit time. (In laminar flow, $\varepsilon = \eta G^2$.) By and large, a higher intensity means smaller droplets.

The most important régimes appear to be LV, TV and TI. For water as the continuous phase, the régime is nearly always TI. For a higher viscosity of the continuous phase, say $\eta_C = 0.1$ Pa s, the régime tends to be TV. For still higher viscosity or a smaller apparatus (small l), régime LV is generally obtained. If the apparatus is very small, as in many laboratory homogenizers, the régime is nearly always LV (or even LB), also for aqueous systems. For these régimes, (semi)quantitative theory is available, giving the timescale and magnitude of local stresses, hence the resulting droplet size. Moreover, timescales of droplet deformation, adsorption of surfactant and mutual collision of droplets can be given. A summary of these is given in Table 2.2. Further explanation is given in the ensuing sections. It should be realized that these régimes represent to some extent idealized situations and that intermediate régimes may occur.

Table 2.2   *Various régimes for emulsification; see text*

Régime[c,d]	Laminar, viscous forces LV	Turbulent, viscous forces TV	Turbulent, inertial forces TI
Re, flow	$< \sim 1000$	$> \sim 2000$	$> \sim 2000$
Re, drop	<1	<1	$>1^a$
$\sigma_{ext} \approx$	$\eta_C G$	$\varepsilon^{1/2}\eta_C^{1/2}$	$\varepsilon^{2/3}d^{2/3}\rho^{1/3}$
$d \approx$ [b]	$\dfrac{2\gamma We_{cr}}{\eta_C G}$	$\dfrac{\gamma}{\varepsilon^{1/2}\eta_C^{1/2}}$	$\dfrac{\gamma^{3/5}}{\varepsilon^{2/5}\rho^{1/5}}$
$\tau_{def} \approx$	$\dfrac{\eta_D}{\eta_C G}$	$\dfrac{\eta_D}{\varepsilon^{1/2}\eta_C^{1/2}}$	$\dfrac{\eta_D}{\varepsilon^{2/3}d^{2/3}\rho^{1/3}}$
$\tau_{ads} \approx$	$\dfrac{6\pi\Gamma}{dm_C G}$	$\dfrac{6\pi\Gamma\eta_C^{1/2}}{dm_C\varepsilon^{1/2}}$	$\dfrac{\Gamma\rho^{1/3}}{d^{1/3}m_C\varepsilon^{1/3}}$

[a]For $d > \eta_C^2/\gamma\rho$. [b]Only if $\eta_D \gg \eta_C$.
[c]Symbols: Re = Reynolds number = length × density × velocity/viscosity; We = Weber number = σ_{ext}/Laplace pressure; d = drop diameter; ε = power density; γ = interfacial tension; η = viscosity; G = velocity gradient; Γ = surface excess; ρ = mass density; m = [surfactant]; τ = characteristic time; σ = stress.
[d]Subscripts: D = dispersed phase; C = continuous phase; def = deformation; ads = adsorption; ext = external; cr = critical value for break-up.

2.3   Droplet Break-up

In this section it is assumed that the interfacial tension between drop and continuous phase is constant. This would mean that either no surfactant is present or a large excess exists.

2.3.1 Laminar Flow

2.3.1.1 Flow Types

Laminar flow can be of a variety of types and Figure 2.3 gives some important examples of plane or two-dimensional flows. They vary from purely rotational to purely elongational (extensional), *i.e.* without any rotation. The well-known flow type of simple shear consists of equal parts of rotation and elongation. For simple shear, the velocity gradient G (in units of reciprocal time) equals the shear rate; for hyperbolic flow, G equals the elongation rate. Axisymmetric elongational flow can also exist, for instance if the z axis in Figure 2.3c would be the axis of symmetry. Such a type of flow occurs where a liquid is forced through a small constriction, for instance when it enters the needle in a syringe. Furthermore, intermediate types of flow may exist, and those between simple shear and elongation are especially interesting. They are described by a parameter α in the velocity vector, such that $\alpha = 0$ corresponds to pure simple shear and $\alpha = 1$ to pure hyperbolic flow.

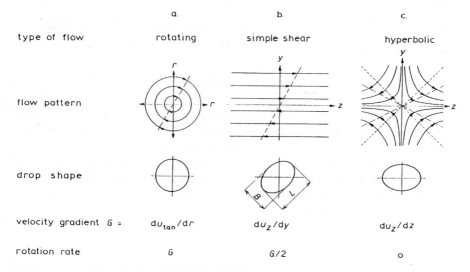

Figure 2.3   *Various types of two-dimensional flow and the effects on drop deformation and rotation*
(Redrawn with permission from ref. 2)

The strength of a flow is generally expressed by the stress it exerts on any plane in the direction of flow. It is simply given by ηG, but it should be realized that the viscosity depends on flow type. Generally, η is taken to be the shear viscosity, which is the one determined in nearly all viscometers. For elongational flow we then have

$$\eta_{el} = Tr\eta \qquad (2.6)$$

where Tr is the dimensionless Trouton number. For Newtonian liquids, $Tr = 2$ for two-dimensional uniaxial elongational flow and $Tr = 3$ for axisymmetric uniaxial flow; for biaxial flows the values are 4 and 6, respectively. This implies that elongational flows exert higher stresses for the same value of G than simple shear flow. For non-Newtonian liquids, the relations are more complicated: η then depends on G (generally decreasing with increasing G) and, moreover, the values of Tr tend to be higher. For strongly viscoelastic liquids, Tr is even much higher, greatly affecting flow pattern and droplet break-up.[2] This is not discussed here.

Deformation and bursting of drops in various types of laminar flow have been studied in much detail, beginning with the pioneering studies of G.I. Taylor and extensive theoretical and experimental work by S.G. Mason and co-workers. Often quoted extensive experimental studies are those by Grace,[7] Bentley and Leal[8] and Stone *et al.*[9] A fairly recent review is by Stone;[10] see also ref. 2. In most of these studies a single drop of fairly large size (about 1 mm or larger) in a relatively large apparatus was taken, under conditions of constant flow pattern. The increase in velocity gradient generally was very slow, to allow near equilibrium of the drop shape during its deformation, although effects of a sudden increase in G have also been investigated.

The deformation primarily depends on the ratio of the external stress over the Laplace pressure, expressed in a dimensionless *Weber number*, in the present case given by

$$We = G\eta_C d/2\gamma \qquad (2.7)$$

where η_C is the shear viscosity of the liquid surrounding the drop. (Incidentally, several workers use the term capillary number rather than Weber number for a stress caused by laminar flow.) The deformation of the drop, which may be expressed in various ways, increases with increasing We, and above a certain critical value of We the drop bursts, *i.e.* it breaks up into smaller ones. We_{cr} is observed to depend on two parameters, *i.e.* the type of flow, for instance expressed in the parameter α, and on the viscosity ratio

$$\lambda \equiv \eta_D/\eta_C \qquad (2.8)$$

where η_D is the viscosity of the material in the drop. Results are illustrated in Figure 2.4. These are for the most part experimental results, which agree rather well with theoretical predictions.

In *simple shear* ($\alpha = 0$), a drop is deformed into a prolate ellipsoid, as shown in Figure 2.3b. The longer axis of the drop attains an orientation of 45 degrees to the direction of flow. The material in the drop rotates. For small We, the deformation $D \approx We$. The deformation is defined as

$$D \equiv (L - B)/(L + B) \qquad (2.9)$$

where L is the greatest length of the drop and B its breadth. For larger We, the drop attains a slender shape with a 'waist' and eventually it breaks into two equal sized fragments at the We_{cr} given in Figure 2.4. The deformation at burst D_b then is about 2/3, for λ values of the order of unity. At small values of λ, the deformation at burst

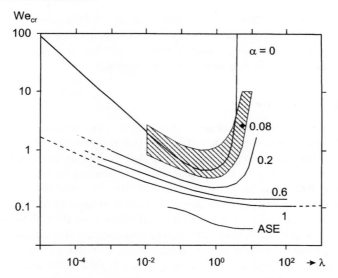

Figure 2.4   *Critical Weber number for break-up of drops in various types of flow. Single-drop experiments in two-dimensional simple shear (α = 0), hyperbolic flow (α = 1) and intermediate types,[7, 9–15] as well as a theoretical result for axisymmetrical extensional flow (ASE).[11] The hatched area refers to apparent* We_{cr} *values obtained in a colloid mill[18]*

is much larger. The deformation then is best expressed as the draw ratio, given by L/d, where d is the original drop diameter. (The relation between the two is about $(L/d)^{3/2} = (1 + D)/(1 - D)$.) For $\lambda = 10^{-3}$, the draw ratio at burst is about 12, which means that the drop is extended into a thread before breaking. At $\lambda > 4$, break-up does not occur. The drop attains a maximum D value equal to $5/4\lambda$. The reason is, broadly speaking, that the liquid in the drop, being more viscous than that around it, cannot flow as fast as the shear rate G tries to cause deformation. The result is that the drop as a whole starts to rotate (at a rate $G/2$) without being further deformed.

For *elongational flow* ($\alpha = 1$), We_{cr} is much smaller and does not greatly depend on λ. In such a flow, no rotation occurs, and drops of high viscosity ratio can also be deformed. Moreover, the Trouton number = 2, meaning a shearing stress twice as high as for simple shear at the same G. (In Figure 2.4, a calculated curve for axisymmetric flow is also given, where $Tr = 3$, which leads indeed to still smaller We_{cr}.) Moreover, in elongational flow, extension of the drop into a thread implies that the shearing force acting on the drop becomes larger as the drop becomes more deformed, which is not the case in simple shear flow. The deformation at burst is also smaller than for $\alpha = 0$; $D_b \approx 0.94$ (draw ratio about 10) for $\lambda = 10^{-3}$, and $D_b \approx 0.4$ for large λ.

It is further seen in Figure 2.4 that a fairly small elongational component in the flow pattern (small α) has a marked effect on We_{cr} and on its dependence on λ. This means that it becomes more easy to break up drops, especially at high λ. The smaller α is, the lower the maximum value of λ at which break-up is still possible.

For instance, for $\alpha = 0.2$, break-up does not occur at $\lambda > 15$, and the drop attains a fairly small deformation of about 0.1, almost independent of the value of *We*. Introducing a simple shear component in elongational flow (say, $0.5 < \alpha < 1$), has fairly little effect on the effectiveness of droplet break-up.

2.3.1.2 Complications

Not all workers in the field have obtained precisely the same results. This is because it is difficult to obtain a true steady state just before break-up. In practice, We_{cr} depends on dG/dt and on irregularities in the flow. Moreover, even trace amounts of surfactants can have a distinct effect; this is discussed in Section 2.4. It is generally seen that break-up of drops into two smaller ones also leaves some, generally three, very small 'satellite' drops. In a detailed study of this phenomenon, up to 19 satellite drops have even been observed.[12] However, these drops are of the order of 10 μm in diameter. Consequently, it is very unlikely that satellite drops are formed during practical emulsification, where droplets of about 1 μm diameter are generally desired. Any satellite droplets would then be far smaller than 0.1 μm and have an excessive Laplace pressure.

Another phenomenon often observed is called tip streaming, studied in detail by de Bruijn.[13] The deformed drop attains pointed ends, and small droplets then are shed from these tips, which may go on for a long time. Again, the droplets shed are of some 10 μm in size, and they will not be formed from small drops. Tip streaming is caused by the presence of surfactants, even if present at very low concentration.[13]

Most of the studies on single drops in laminar flow were done at small dG/dt, trying to keep the drop shape at near equilibrium. This is, however, not a situation that is likely to occur in most practical conditions. Some workers, notably Grace,[7] have also studied what happens when a drop is suddenly subjected to a stress much greater than the critical stress for break-up (*i.e.* $We \gg We_{cr}$). This has been further considered by Janssen and Meijer,[14] who distinguish between two situations, 'stepwise equilibrium' and 'transient dispersion', in practical emulsification. In the first situation, a drop that is subject to the critical stress (corresponding to We_{cr}) would be broken into two smaller ones, which then may be subject to a critical stress (which implies a higher value of G than for the parent drop) to break up, and so on, until the maximum possible *We* is too small to cause further break-up. In the other situation, a drop would suddenly be subjected to very high *We* and be extended into a long, almost cylindrical, thread, which would then be subject to Rayleigh instability (see *e.g.* Rumscheidt and Mason[15]) and break up into many small droplets; their diameter would roughly be 1.7 times that of the diameter of the cylinder. The authors concluded that the latter mechanism—which may better be called '*capillary break-up*'—is more efficient, especially in simple shear flow, and they suggested that it is the more common one.

2.3.1.3 Practical Results

These ideas were further tested by Wieringa *et al.*[16] who made emulsions in a colloid mill, an apparatus that may give fairly pure simple shear flow. They concluded that breaking up of long threads into many drops indeed occurred in a colloid mill; there will generally not be enough time for stepwise break-up in such a machine, unless flow rate is kept quite small. In a model study, they also determined the number of droplets formed from one parent drop, as a function of We_b, i.e. the We value at which the thread formed breaks into droplets, relative to We_{cr}. Examples are in Figure 2.5, and it is seen that one drop can break into many, hence very small, droplets. In a study on the efficiency of a static mixer, for emulsions of a viscosity high enough to ensure laminar flow, a similar conclusion was drawn about the break-up mechanism.[17]

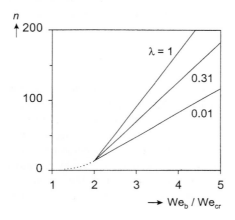

Figure 2.5 *The number of droplets* n *in which a parent drop is broken when it is suddenly extended into a long thread, corresponding to* We_b, *which is larger than the calculated value of* We_{cr}, *at various values of* λ. *Emulsions contain 80% oil in water stabilized by egg yolk*
(Redrawn with permission from ref. 16)

In practice, the resulting droplet size distribution is of greater importance than the critical drop size for break-up. One can, by taking the average size of the droplets obtained, establish apparent We_{cr} values. The capillary break-up model would then predict that We_{cr} roughly equals unity, independent of λ, at least for λ values smaller than 1 or 2. Such a study was done by Armbruster[18] with a colloid mill. The mill was designed to produce pure simple shear flow. The resulting droplet size distribution was determined for a range of conditions: viscosity of both phases, flow rate, dispersed phase volume fraction and type of surfactant. The surfactant was always present in excess and, in calculating We, the equilibrium interfacial tension between both phases was taken. The resulting d_{32} values were used to calculate apparent We_{cr} values. Results are shown in Figure 2.4; they agree reasonably well with the values for single drops in simple shear flow for λ values up to about 2. Droplet break-up was observed up to $\lambda = 10$, rather than 4 as found for single drops in model studies. This may have been due to a slight contribution of elongation to the flow field (say, $\alpha = 0.1$). It appears, however, that the apparent We_{cr} tends to increase with decreasing λ for $\lambda < 0.3$. This seems to be at variance with the conclusions of Wieringa *et al.*,[16] who have very good evidence for the capillary break-up determining the size distribution in a colloid mill at $\lambda \approx 0.03$. More detailed studies at low λ values may be needed to settle the point.

2.3.2 Turbulent Flow

2.3.2.1 Kolmogorov Theory

Turbulent flow is characterized by the presence of eddies, which means that the local flow velocity u generally differs from the time-average value $\bar{u}$. The velocity fluctuates in a chaotic way and the average difference between u and $\bar{u}$ equals zero. However, the root-mean-square average

$$u' = \langle (u - \bar{u})^2 \rangle^{1/2} \tag{2.10}$$

is finite and is an important characteristic of turbulent flow. The value of u' generally depends on direction, but for very high Re (say, $Re > 50\,000$) and at small length scales the turbulence can be isotropic, implying that u' does not significantly depend on direction. If so, relations characterizing the flow become simple. They have been worked out by Kolmogorov (see *e.g.* Levich[19] and Davies[20]) and will briefly be recalled here. It should be said from the outset that Kolmogorov theory gives scaling laws, and the relations should in fact be completed by unknown proportionality factors, although these are generally of order unity.

Turbulent flow shows a spectrum of eddy sizes (l). The largest eddies have the highest u'. They transfer their kinetic energy to smaller eddies, which have a smaller u', but a larger velocity gradient, given by u'/l. Small eddies thus have a high energy content, and they are called *energy-bearing eddies*, length scale l_e. In turn, these transfer their energy to still smaller eddies, which finally dissipate their energy into heat, of size l_0. This smallest eddy size, also called the Kolmogorov scale, is given by

$$l_0 = \eta^{3/4} \rho^{-1/2} \epsilon^{-1/4} \tag{2.11}$$

ϵ is called the power density, *i.e.* the amount of energy dissipated per unit volume of liquid and per unit time. (Note: ϵ has also been called energy density, although it is given in $W\,m^{-3}$, not $J\,m^{-3}$; some authors express it per unit mass, *i.e.* $W\,kg^{-1}$.) ϵ and η are the main parameters determining isotropic turbulence.

Local flow velocities depend on the distance scale x considered and, for a scale comparable to the size of an energy-bearing eddy, the velocity near that eddy is

$$u'(x) = \epsilon^{1/3} x^{1/3} \rho^{-1/3} \qquad x \approx l_e \tag{2.12}$$

The velocity gradient in an eddy is given by $u'(x)/(x)$ and it is seen to increase strongly with decreasing size. The eddies have a rather short lifetime, given by

$$\tau(l_e) \equiv l_e/u'(l_e) = l_e^{2/3} \epsilon^{-1/3} \rho^{1/3} \tag{2.13}$$

The local flow velocity for scales much smaller than the size of an energy bearing eddy is given by

$$u'(x) = \epsilon^{1/2} x \eta^{-1/2} \qquad x \ll l_e \qquad (2.14)$$

from which it follows that the local velocity gradient, $u'(x)/x$, now is independent of distance. It may further be noticed that in eqn. (2.12) the viscosity is not a variable (although it is a factor that determines Re and l_0, hence the validity range of the equation), whereas it is a variable in eqn. (2.14).

2.3.2.2 Inertial Forces

Kolmogorov considered droplet break-up in turbulent flow due to inertial forces; see *e.g.* refs. 2, 19, 20. By combining eqn. (2.12) with the Bernoulli relation ($p + \frac{1}{2}\rho u^2 = $ constant), local pressure fluctuations near energy-bearing eddies are given by

$$\Delta p(x) = \rho[u'(x)]^2 = \epsilon^{2/3} x^{2/3} \rho^{1/3} \qquad (2.15)$$

If Δp is larger than the Laplace pressure of a drop (eqn. (2.1): $p_L = 2\gamma/d$) near the eddy, the drop would be broken up. Break-up would be most effective if $d = l_e$. Putting eqn. (2.15) equal to eqn. (2.1) and putting $x = d_{max}$, there results the following expression for the largest droplets that are not broken up in the turbulent field

$$d_{max} = \epsilon^{-2/5} \gamma^{3/5} \rho^{-1/5} \qquad (2.16)$$

Some results are given in Table 2.3. It should be remembered that eqn. (2.16), like all equations in Kolmogorov theory, has scaling relations. Often, eqn. (2.16) is also applied for the average droplet size resulting from emulsification, say d_{32}. In that case, the proportionality constant would be smaller than for d_{max}.

The validity of eqn. (2.16) is subject to some conditions. First, the droplet size

Table 2.3 *Sample calculations for droplet disruption in isotropic turbulent flow (orders of magnitude only)*

Variable	Units	Eqn.	Values[a]					
ϵ	W m^{-3}	–	10^4	10^4	10^8	10^8	10^{12}	10^{12}
η_C	mPa s	–	1	100	1	10	1	10
$\eta_C^2/\gamma\rho$	μm	2.18	0.1	1000	0.1	10	0.1	10
l_0	μm	2.11	18	550	1.8	10	0.2	1
$d_{max}|$TI	μm	2.16	400	(400)	10	10	0.25	(0.25)
$d_{max}|$TV	μm	2.23	(3×10^4)	300	(30)	10	(0.3)	0.1

[a]Other values taken: $\rho_C = 10^3$ kg m^{-3}, $\gamma = 0.01$ N m^{-1}, Re large, $We_{cr} = 1$ in eqn. (2.23). Results in parentheses are thought to be invalid.

obtained cannot be much smaller than l_0, as given by eqn. (2.11). It follows that this relation is generally fulfilled for small η_C; see Table 2.3. Second, the flow near the droplet should be turbulent. This depends on the droplet Reynolds number, given by

$$Re_{dr} = du'(d)\rho_C/\eta_C \tag{2.17}$$

The condition is $Re_{dr} > 1$, and combination with eqn. (2.12) leads to:

$$d > \eta_c^2/\gamma\rho \tag{2.18}$$

Examples are in Table 2.3, and it is seen that a small η_C is needed, the smaller for a higher value of ϵ. Third, the droplet deformation time should be smaller than the lifetime of the eddy near it; this is discussed further on.

Provided that (i) ϕ is small, (ii) η_C is not much larger than 1 mPa s, (iii) η_D is fairly small, (iv) γ is about constant, and (v) the machine is not very small, eqn. (2.16) is observed to hold well. This was discussed before,[2] and we will only summarise the results. In pipe flow, the relation becomes

$$d \propto X_P{}^{2/5}\bar{u}^{6/5}\rho^{-3/5}\gamma^{3/5} \tag{2.19}$$

where X_P is the pipe diameter. It may be mentioned that very small droplets cannot be obtained and that they generally show a wide size distribution. For stirrers, where ϵ is proportional to $\rho X_s{}^2\nu^3$, the Hinze–Clay relation applies, given by

$$d \propto X_S{}^{-4/5}\nu^{-6/5}\rho^{-3/5}\gamma^{3/5} \tag{2.20}$$

where X_S is the stirrer diameter and ν the revolution rate (s^{-1}). In a high-pressure homogenizer, the power density is proportional to the homogenizing pressure p_H to the power 3/2, and the relation becomes

$$d \propto p_H{}^{-3/5}z^{2/5}\gamma^{3/5} \tag{2.21}$$

where z is the effective distance in the homogenizer valve over which the turbulent energy is dissipated.

Although Kolmogorov theory was derived for isotropic turbulence only, it is striking that the resulting equations also seem to hold for Re values much smaller than the mentioned 50 000.[2] The smallest droplets obtained, however, will have formed in regions of the highest power density. Since power density can greatly vary from place to place, especially if Re is not very high, the droplet size distribution can be very wide. Consider, for instance, a stirrer in a large tank. The average ϵ may be quite small, but near the stirrer tips ϵ will be much higher. If stirring goes on long enough for every droplet to pass a couple of times close to stirrer tips, fairly small droplets with little spread in size (say, $c_2 = 0.5$) may still be obtained.

2.3.2.3 Viscous Forces

One condition for droplet break-up in the régime TI is, as stated above, that the flow near the drop is turbulent. If it is laminar, break-up by viscous forces may be possible. Suppose that flow rate (u) near a drop varies greatly over a distance equal to d, local velocity gradient G. We then have a pressure difference over the drop of $\frac{1}{2}\rho\Delta(u^2) = \bar{u}\rho Gd$. At the same time a shear stress is acting on the drop, magnitude $\eta_C G$. Viscous forces will thus be predominant for $\eta_C G > \bar{u}\rho Gd$, leading to the condition

$$\bar{u}d\rho/\eta_C = Re_{dr} < 1 \tag{2.22}$$

fully equivalent to eqn. (2.17). The local velocity gradient can be found from eqn. (2.14), and we derive that $\eta_C G = \epsilon^{1/2}\eta_c^{1/2}$. Insertion into eqn. (2.17) then leads to eqn. (2.18) with the inequality sign reversed. There is thus a consistent criterion marking the transition between régimes TI and TV.

For the latter régime, combination of eqns. (2.7) and (2.14) yields

$$d_{max} = We_{cr}\gamma\epsilon^{-1/2}\eta_C^{-1/2} \tag{2.23}$$

The question is what value We_{cr} will have. Inspection of Figure 2.4, and taking the capillary break-up mechanism into account, leads to the conclusion that We_{cr} will rarely be over unity, since it is very likely that the flow will have an elongational component. This is because the flow in eddies is always, for a considerable part, circular. Effective We_{cr} may even be as low as 0.1, but since the constant implied in eqn. (2.14) is unknown anyway, it makes little sense to use eqn. (2.23) as more than a scaling law.

Some calculated results are in Table 2.3 for $We_{cr} = 1$. It is seen that for not very small η_C the resulting d_{max} is smaller in the régime TV than in TI. This is the case if

$$\eta_C > \rho^{2/5}\gamma^{4/5}\epsilon^{-1/5} \tag{2.24}$$

which is equivalent to criterion (2.18). Régime TV has not been well studied. In work by Tjaberinga *et al.*,[21] turbulent pipe flow was generated in a dilute emulsion of water in paraffin oil, measuring the resulting average droplet size. The authors concluded that the results fitted the theory reasonably well. In a high-pressure homogenizer the power density typically is $10^{11}-10^{12}$ W m^{-3} and, although the relation between average d and p_H generally agrees well with eqn. (2.21), it has been argued[2] that the régime is at the border between TI and TV. When increasing η_C from 1 to 4.5 mPa s, the resulting d_{43} at the same p_H decreased by a factor of 1.8.[22] This is at least in qualitative agreement with a shift in régime from TI to TV.

2.3.3 Effect of Droplet Viscosity

The higher the viscosity of the dispersed phase, the longer it will take to deform a drop of that liquid. By and large, the deformation time is of the order of

$$\tau_{\text{def}} = \eta_D / \sigma_{\text{ext}} \tag{2.25}$$

where σ_{ext} is the external stress acting on the drop. Two conditions should be met for eqn. (2.25) to hold. First, σ_{ext} should be clearly larger than the Laplace pressure, since the latter tends to counteract deformation, the more so as deformation increases. Second, $\eta_D \gg \eta_C$ (or $\lambda \gg 1$), since otherwise the viscous resistance of the liquid surrounding the drop will also slow down deformation. In principle, η_D in eqn. (2.25) should be replaced by $(\eta_D + c\eta_C)$, where c is an unknown constant (of order 3?).

The consequences of the fact that time is needed for droplet deformation, hence break-up, are especially important at $\lambda > 1$. They will be discussed separately for the various régimes.

2.3.3.1   Régime TI

The constraints imposed by finite τ_{def} have been discussed by Walstra.[3] It appears likely that eqn. (2.25) is indeed valid for stresses due to inertial forces. By using eqn. (2.15) for the external stress, this results in

$$\tau_{\text{def}} = \eta_D \epsilon^{-2/3} d^{-2/3} \rho^{-1/3} \tag{2.26a}$$

$$\tau_{\text{def}} = \eta_D \epsilon^{-2/5} \rho^{-1/5} \gamma^{-2/5} \qquad d = d_{\text{max}} \tag{2.26b}$$

where eqn. (2.26b) is obtained by inserting for d the value of d_{max} according to eqn. (2.16). The deformation time can, however, never be smaller than the time corresponding to the natural oscillation of the drop. This can be derived from the treatment by Lamb[23] to be of the order of

$$\tau_{\text{nat}} = \tfrac{1}{2} d^{3/2} \rho_D^{1/2} \gamma^{-1/2} \tag{2.27a}$$

$$\tau_{\text{nat}} = \tfrac{1}{2} \epsilon^{-3/5} \rho^{1/5} \gamma^{2/5} \qquad d = d_{\text{max}} \tag{2.27b}$$

although it also depends on the ratio ρ_D / ρ_C. Eqn. (2.27b) is derived by putting $d = d_{\text{max}}$ and $\rho_D = \rho_C = \rho$.

The droplet can in principle be broken up if the external stress acts for a period $> \tau_{\text{def}}$. In the régime TI this period would be given by the lifetime of the energy bearing eddies, according to eqn. (2.13). By putting $l_e = d$, we arrive at

$$\tau_{\text{eddy}} = d^{2/3} \epsilon^{-1/3} \rho^{1/3} \tag{2.28a}$$

$$\tau_{\text{eddy}} = \epsilon^{-3/5} \rho^{1/5} \gamma^{2/5} \qquad d = d_{\text{max}} \tag{2.28b}$$

It may be noted that τ_{nat} and τ_{eddy} are about equal, provided that d results from unhindered break-up in the régime TI.

Table 2.4 gives some calculated values of the various time scales, for conditions (small η_C) where $Re_{\text{dr}} > 1$, according to eqns. (2.17) and (2.18). It is seen that even for η_D as small as 0.01 Pa s, $\tau_{def} > \tau_{eddy}$, at least for large ϵ. This would mean that the droplets cannot become as small as predicted by eqn. (2.16), because there is not enough time to deform them by an eddy of a size comparable to d. They can, however, be broken up by larger eddies. By putting eddy time (eqn. 2.28a) equal to deformation time (eqn. 2.26a), and assuming further that $d_{max} = l_e$, we arrive at

$$d(\tau_{def}) = \eta_D^{3/4}\epsilon^{-1/4}\rho^{-1/2} \qquad (2.29)$$

Some calculated results are in Table 2.4. They are generally larger than the corresponding d_{max}, except for one example. Since d can never be smaller than d_{max}, this result should be discarded, but the other values of $d(\tau_{def})$ would be about correct.

Table 2.4 *Calculated examples of time scales and droplet diameters at a range of conditions (see text); approximate results*

Régime	Parameter[a]	Unit	Eqn.	η_D/Pa s	ϵ/W m^{-3}		
					10^6	10^9	10^{12}
TI	τ_{eddy}	μs	2.28b	–	160	2.5	0.04
	τ_{def}	μs	2.26b	0.01	63	4	0.25
	τ_{def}	μs	2.26b	1	6300	400	25
	d_{max}	μm	2.16	–	63	4	0.25
	$d(\tau_{def})$	μm	2.29	0.01	(32)	5.6	1.0
	$d(\tau_{def})$	μm	2.29	1	1000	180	32
TV	l_0	μm	2.11	–	180	30	
	$\tau(l_0)$	μs	2.30	–	300	10	
	τ_{def}	μs	2.31	0.01	30	1	
	τ_{def}	μs	2.31	1	3000	100	
	d_{max}	μm	2.23	–	30	1	
	τ_{br}	μs	2.32	0.01	470	16	
	τ_{br}	μs	2.32	1	2200	75	

[a]Further data: $\eta_C = 10^{-3}$ and 0.1 Pa s in the régimes TI and TV, respectively; $\rho_C = \rho_D = 10^3$ kg m^{-3}; $\gamma = 0.01$ N m^{-1}

There is, however, a discrepancy with experimental results. It was observed, for instance, that for a high-pressure homogenizer, with ϵ up to 10^{12} W m^{-3} and η_D at 30 mPa s, eqn. (2.21) was exactly obeyed.[2.24] This concerns the order of magnitude of the droplet size as well its dependence on homogenizing pressure, p_H; the slope of the log–log plot was exactly -0.6, as predicted by eqn. (2.21), rather than -0.375, as would be predicted from eqn. (2.29). The discrepancy can be partly due to the fact that all the relations derived from Kolmogorov theory are scaling laws, but it seems unlikely that the timescales would be more than a factor 10 off, and larger differences are observed. Possibly, the discrepancy can be resolved by the following argument. In the derivation of eqn. (2.26) it was implicitly assumed that the deformation and break-up of a droplet is a transient phenomenon, occurring by close approach of a droplet with one eddy. It has been argued by several authors, however, that break-up may occur due to passage of a number of energy-bearing eddies close to a droplet, whose eddies then induce the drop to oscillate, whereby the drop eventually bursts.[2] The natural oscillation time is about equal to the lifetime of the energy-bearing eddies, as we have seen above (*cf.* eqns. 2.27 and 2.28), so this would not impose a restriction.

Nevertheless, for fairly large η_D and not too small ϵ, droplet break-up must be greatly hindered by the relatively long τ_{def}. This is borne out by experimental results. In experiments with an UltraTurrax and a high-pressure homogenizer, $\epsilon = 5 \times 10^8 - 10^{12}$ W m^{-3} and $\eta_D = 3 - 150$ mPa s, Walstra[22] observed $d_{43} \propto \eta_D^{1/3}$. Results by Karbstein[25] showed, for a homogenizer, $\eta_D = 2 - 1500$ mPa s, an exponent of about +0.4. It may be noticed that these exponents are much smaller than the value $3/4$ in eqn. (2.29). This may point to a transition between droplet sizes determined by eqn. (2.16) to eqn. (2.29). Results by Pandolfe[26] in a homogenizer yielded an exponent that was on average about +0.7. He also observed that the slope of the lines of log p_H *versus* log d_{43} changed from -0.54 to -0.13 for η_D changing from 50 to 300 mPa s. The trends are as expected, but the absolute values are not. This may have to do with the fact that the homogenizer used by Pandolfe was quite small; see further on.

2.3.3.2   Régime TV

In this régime, droplets are typically broken up by eddies of size l_0. Inserting l_0 according to eqn. (2.11) in eqn. (2.13) leads for the lifetime of these eddies to

$$\tau(l_0) = \eta^{1/2}\epsilon^{-1/2} \tag{2.30}$$

i.e. independent of droplet size (provided $d \ll l_0$). It is more difficult to find the deformation time. If the mentioned mechanism of 'capillary break-up' would occur, the external stress ($\eta_C G$) is clearly larger than p_L and eqn. (2.25) may be valid. Using eqn. (2.14) to derive G, the deformation time would become

$$\tau_{def} = \eta_D\epsilon^{-1/2}\eta_C^{-1/2} \tag{2.31}$$

which probably underestimates τ_{def} unless $\lambda \gg 1$. The condition $\tau(l_0) > \tau_{def}$ (eqns. 2.30 and 2.31) is simply equivalent to $\lambda < 1$. Assuming, on the other hand, that the drop is in equilibrium during deformation, the time needed for break-up has been experimentally determined for a range of conditions, albeit not for small droplets. From results by Grace[7] it can be derived[16,21] that

$$\tau_{br} \approx 3.4\eta_D^{1/3}\eta_C^{2/3}d\gamma^{-1} \cdot \tag{2.32}$$

Calculated results are in Table 2.4. It is seen that for a large value of η_D (here 1 Pa s, leading to $\lambda = 10$, which is not very large), $\tau_{br} < \tau_{def}$; nevertheless, τ_{br} is longer than $\tau(l_0)$ by a significant margin. This means that also in this régime the break-up of very viscous drops (large λ) must be greatly hindered, implying that the resulting drops would be much larger than predicted by eqn. (2.23). Results by Karbstein[25] with an UltraTurrax in the régime TV ($\eta_C = 0.04$ Pa s) showed for λ values of 0.05, 1.5 and 31, d_{32} values of 3, 4 and 12 μm, respectively, at constant energy input; the effect of λ can thus be considerable. The authors are not aware, however, of a suitable theory or of a systematic experimental study.

2.3.3.3 Régime LV

The effect of $\lambda = \eta_D/\eta_C$ on break-up in laminar flows has already been discussed; see especially Figure 2.4. In steady and unbounded simple shear flow, break-up does not occur for large λ, but it may occur in extensional flow. Also here, there may be restrictions of time scale. It has been argued by Walstra[2] that for axisymmetric extensional flow through a sharp constriction, the duration of elongation is of the order of $2/G$. Taking eqn. (2.25) for the deformation time of a droplet, inserting an external stress of $\eta_C TrG$, where $Tr = 3$ in this type of flow, we arrive at

$$\tau_{\text{def}} = \eta_D/3\eta_C G \qquad (2.33)$$

Putting this equal to $2/G$, the condition for break-up would become $\lambda < 6$. It has indeed been observed (S.G. Mason, personal communication) that a drop of high λ passing through the centre of the constriction, and thus undergoing purely elongational flow, is not broken up; however, if the drop passes off-centre, where the flow has a shear component ($0 < \alpha < 1$), it can break up, in accordance with Figure 2.4.

In most practical situations this will mean that break-up is a matter of chance, and that a drop may have to pass through several potential break-up zones in a machine, before it is actually disrupted. This is what may happen in a static mixer, if it is long enough and η_C is high (small Re). Grace[7] has obtained droplet break-up in static mixers for λ as high as 110; d_{32} values down to 0.1 mm were then obtained. It should be realized that the break-up time for such a droplet would be of the order of a second, according to eqn. (2.32).

In conclusion, it is very difficult to make fine emulsions if λ is larger than, say, 10, and the higher λ is, the larger the smallest drops obtained. A suitable method would be to emulsify in steps, where power density increases and the effective time available for break-up decreases for each subsequent step. Another strategy is phase inversion. A high λ generally means that the emulsion is of the O/W type. It then is far more easy to make a fine W/O emulsion (small λ) and subsequently induce a phase inversion. The latter has recently been studied in detail by Vaessen[27] and reviewed by Stein.[28] Another way to make emulsions with highly viscous small drops is mentioned in Section 2.4.

2.3.4 Effects of Scale and Continuous Phase Viscosity

The effects of the magnitude of η_C have already been discussed. In régime TI it would have no effect on droplet size, in régime TV a larger value would lead to smaller drops, and this effect would be even stronger in régime LV. The equations given for d_{max} even predict that extremely small droplets can be formed for high η_C (and small λ). For instance, application of eqn. (2.23) for régime TV, with $\gamma = 0.01\,\mathrm{N\,m^{-1}}$, $\epsilon = 10^{11}\,\mathrm{W\,m^{-3}}$ and $\eta_C = 10\,\mathrm{Pa\,s}$ would yield $d_{max} = 10\,\mathrm{nm}$. However, the combination of large values for ϵ and η_C cannot be realized in practice, except at excessive cost; $\eta_C\epsilon = 10^8\,\mathrm{N^2\,m^{-3}}$ is about the highest possible value. Something similar holds for régime LV in a colloid mill. In practice $G \leqslant 10^5\,\mathrm{s^{-1}}$ and $\eta_C G \leqslant 10^4\,\mathrm{Pa}$.

The value of η_C and the size of the apparatus determine what régime prevails, *via* the effect on Re. This is especially important in high-pressure homogenizers. In a large machine and low η_C, Re is always very large and the resulting average d is proportional to $p_H{}^{-0.6}$ (eqn. 2.21). If viscosity is higher and $Re_{dr} < 1$, the régime is TV and $d \propto p_H{}^{-0.75}$ is expected. For a smaller machine, as used in a laboratory, where the slit width of the valve may be of the order of micrometres, Re will be small and the régime will be LV; the relation would become $d \propto p_H{}^{-1.0}$. If the slit width is so small as to be of the order of the droplets, the régime will be LB, where droplet break-up would not depend on the viscosity of either phase. This case has been treated by Kiefer,[6] and we have calculated[4] that $d \propto p_H{}^{-0.9}$ would result in a homogenizer. In Figure 2.6a, results on the relation between d_{43} and p_H are given for a very small and a large homogenizer applied to the same system. It appears that the régime in the small machine is in between LV and LB. It should also be realized that in a small machine for both these régimes the drops would only be broken up once, when entering the valve slit, whereas in a large machine break-up events may occur about 100 times during passage of the valve. This would imply that in a small machine the droplet size shows a wide size distribution after one passage of the valve. Repeated homogenization would then lead to a much smaller distribution width. This is illustrated in Figure 2.6b.

If η_C is increased by adding high polymers, this may readily lead to turbulence depression. This is of special importance in the régime TI. It has been studied by Walstra,[2,29] who showed that addition of polymers caused d_{32} to increase and c_2 (the spread in droplet size) to decrease. This would agree with the smallest eddies being removed from the eddy spectrum.

2.4 Roles of the Surfactant

A surfactant is a material that lowers interfacial tension, and it thereby facilitates droplet break-up. This is illustrated in Figure 2.7. It may be noted that the droplet size obtained at high m is about proportional to $\gamma^{3/5}$, as expected in régime TI (eqn. 2.16). It is also seen that the surfactant level at which a plateau value of A is reached considerably varies among surfactants.

The lowering of γ is important but is not the essential role of surfactants in emulsification. Consider, for example, making an O/W emulsion with either paraffin oil or triglyceride oil. With pure water, the former gives $\gamma \approx 50$, the latter $\gamma \approx 30\,\mathrm{mN\,m^{-1}}$. If surfactant is added to the water, to arrive at an equilibrium value of $\gamma \approx 30\,\mathrm{mN\,m^{-1}}$ with paraffin oil, an emulsion can be made; with pure triglyceride oil and water, yielding the same γ, an emulsion cannot be made. Hence, the surfactant must have another effect.

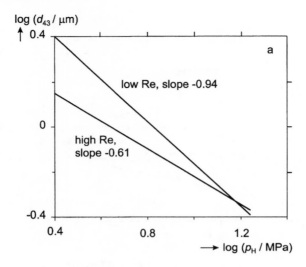

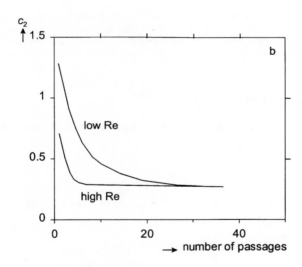

Figure 2.6 *Comparison of droplet size distributions obtained with two high-pressure homogenizers; a very small one (low* Re*) and a large one (high* Re*). (a) Average droplet size as a function of homogenizing pressure* p_H*. (b) Relative distribution width* c_2 *as a function of the number of passages through the homogenizers. The o/w emulsions contain 20% soybean oil, 30 mg mL^{-1} sodium caseinate in 75 mM imidazole buffer (pH 6.7)*

In this section, the various roles of the surfactant will be considered. A review by Lucassen-Reynders[30] would apply to régime LV. We considered earlier especially régime TI.[4] It may be useful to begin by recalling some properties of surfactants.

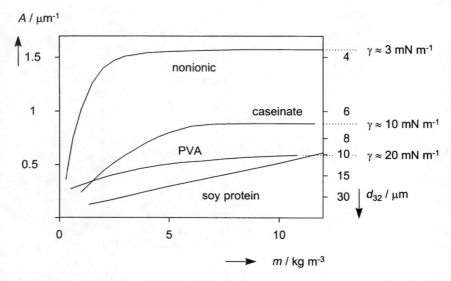

Figure 2.7   *Specific droplet surface area* A *and average droplet size* d_{32} *as a function of total surfactant concentration* m, *obtained at approximately constant conditions; o/w emulsions of* $\phi = 0.2$. *Some plateau values of* γ *are also given* (Redrawn with permission from ref. 4, after various sources)

2.4.1   Surfactant Properties

A surfactant can lower the interfacial tension γ_0 of a clean oil–water interface to a value γ and

$$\Pi \equiv \gamma_0 - \gamma \tag{2.34}$$

where Π is called the surface pressure. The dependence of Π on other properties is given by the Gibbs equation

$$d\Pi = -d\gamma = RT\Gamma d\ln a \tag{2.35}$$

where Γ is the surface excess (also called surface load or amount adsorbed, here to be expressed in moles of surfactant per unit surface area), and a is the activity of the surfactant in solution, for instance expressed as mole fraction. At high a, the surface excess reaches a plateau value; for many surfactants it is of the order of 3 mg m^{-2}. Also γ reaches a plateau value. For small concentrations, a generally equals the concentration of surfactant in solution, *i.e.* total concentration minus the amount adsorbed (or dissolved in the other phase). Eqn. (2.35) only holds for one (nonionic) surfactant and at equilibrium. An important relation with respect to emulsification is that between γ and m_C, the (molar) concentration of surfactant in the continuous phase. This relation follows from the adsorption isotherm, *i.e.* Γ as a function of a, and the equation of state, *i.e.* Π as a function of Γ.

The relations mentioned vary greatly among surfactants, which can be classified in various ways. An important distinction is between small-molecule surfactants, for convenience called 'soaps', and polymers (both synthetic polymers and proteins are used). These differ greatly in properties, as illustrated below. Another distinction is between soluble and insoluble surfactants. In emulsification, surfactants have to be soluble, unless they are rendered insoluble, by chemical modification, after adsorption onto the O–W interface.

Another characteristic is the surface activity; how much surfactant is needed, or more precisely, how large should the surfactant activity be, to reach a certain value of Γ (or of γ)? The differences between a soap and a polymer are illustrated in Figure 2.8. It is seen that the polymer is far more surface active: the value of m_C needed to obtain the same Γ is smaller by about 2 orders of magnitude for the polymer; the difference expressed in mass concentration is even more, by about 4 orders of magnitude. Soaps and polymers also differ in the applicability of eqn. (2.35). For a soap, it nearly always appears to hold, also in an emulsion. For a polymer, equilibrium is often not obtained, at least in the timescales considered; see *e.g.* ref. 1 for explanations. This is borne out by the two curves for β-casein in Figure 2.8; the one obtained by emulsification does not agree at all with the adsorption isotherm obtained by unhindered adsorption on a macroscopic interface. For a soap, the value of Γ can be calculated from the adsorption isotherm when an emulsion is made from known quantities of materials and the specific surface area is known. This is not so for a polymeric surfactant, where Γ has to be determined separately.

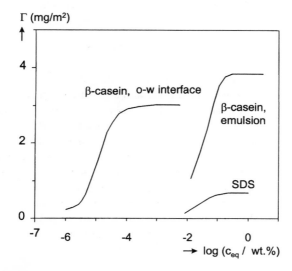

Figure 2.8   *The surface excess Γ at the oil–water interface as a function of the concentration of surfactant in solution (c_{eq}) for β-casein and sodium dodecyl sulfate (SDS). See text. The oils are: β-casein (O–W interface) toluene, β-casein (emulsions) soybean, SDS benzene*
(After various sources, including ref. 31)

The next question is how small γ can become. This depends on the equation of state. For very small values of Γ it is given by

$$\Pi = RT\Gamma \qquad (2.36)$$

which is the two-dimensional analogue of the universal gas law. Eqn. (2.36) implies that Π will be much larger for a soap (small MW) than for a polymer (large MW) at the same surface load expressed in units of mass. At higher surface excess, results according to eqn. (2.36) are very far removed from the truth; see also Figure 2.10, further on. The non-ideality is very different for large and small surfactant molecules; see *e.g.* Lucassen-Reynders[32] for a discussion. Small-molecule surfactants can, in principle, attain a very close packing in the interface, and thereby cause a low interfacial tension, say 1 mN m^{-1} at an O–W interface. For a polymeric surfactant, non-ideality sets in at a far smaller value of Γ, if expressed in mol m^{-2}. Moreover, Π does not only depend on Γ, but also on the conformation of the molecules in the interface. The lowest value of γ attainable for polymers is of the order of 10 mN m^{-1}, because a close packing of the interface with polymer segments cannot be obtained.

The value of γ will also depend on the liquids constituting both phases. As mentioned above, γ_0 at the O–W interface is 30 and 50 mN m^{-1} for triglyceride and paraffin oil, respectively. In a first approximation, the relation between Π and Γ and that between Γ and a are the same for both interfaces, implying that for triglyceride oil lower values of γ are obtained than for paraffin oil at the same value of a.

Another surface property that will be used in this chapter is the surface dilational modulus

$$E_{SD} \equiv \mathrm{d}\gamma/\mathrm{d}\ln A = -\mathrm{d}\Pi/\mathrm{d}\ln A \qquad (2.37)$$

It relates the change in interfacial tension resulting from a relative change in interfacial area, and an increase in the surface area of a drop will always occur during emulsification. E_{SD} generally is the absolute value of a complex quantity, composed of an elastic and a viscous term. In other words, E_{SD} depends on timescale, at least for soluble surfactants. This is because expansion of a surface will cause a lowering of Γ, which means that equilibrium between Γ and a no longer exists; the equilibrium will be restored by adsorption of surfactant from solution, and this takes time, the less so when surfactant concentration is very high. Consequently, E_{SD} will be small for high a; naturally, it is also small for very low a. Because of the lack or slowness of attaining equilibrium for polymeric surfactants, their E_{SD} values will generally not be the same for expansion and compression of the interface.

2.4.1.1 *Surfactant Mixtures*

In practice, the 'emulsifiers' used are generally mixtures of surfactants, often containing several different components. Moreover, specific mixtures of surfactants are often used deliberately. The properties of such mixtures are widely variable, and a vast literature exists. Here, only the main points will be briefly summarized.

- The Gibbs equation (2.35) does not hold for mixtures; relations for two surfactants exist. Also other equations, *e.g.* those for the surface dilational modulus, need modification. Generally, the presence of more than one surfactant, even if only in trace amounts, tends to increase the value of E_{SD} at high surfactant concentrations.
- The various components vary in surface activity and in the plateau value of γ attained. In principle, those giving the lowest γ value at the prevailing conditions tend to predominate in the interface. Since these may be present in a very low concentration, it may take a long time before they have reached the interface. Consequently, it often takes a long time before the interfacial tension stops decreasing.
- Some specific mixtures of surfactants give much smaller γ values than each one separately. So-called cosurfactants can further decrease γ, and so can salt for some surfactants. Some mixtures can at high concentrations yield γ values as low as $10^{-6}\,\mathrm{N\,m^{-1}}$.
- Mixtures of small-molecule surfactants and polymers, especially proteins, have received much attention; see Goddard and Ananthapadmanabhan[33] for a review. Interactions between these components readily occur, both in the interface and in solution, and these can markedly affect surface properties. As mentioned, soaps tend to give a lower γ value, if present at high concentration, than polymers. Consequently, they can displace polymers from the interface.

2.4.2 Effective Interfacial Tension

The value of interfacial tension of a droplet during break-up is the one that governs the break-up, but it may not be easy to establish its magnitude.

2.4.2.1 Depletion of Surfactant

During emulsification, surfactant is transferred from solution to the interface, leaving an ever lower surfactant activity. Consider, for instance, an O/W emulsion of $\phi = 0.4$ and $d_{32} = 1\,\mu\mathrm{m}$. According to eqn. (2.4) the specific surface area then is $2.4\,\mathrm{m^2\,ml^{-1}}$ and, for a surface excess of $3\,\mathrm{mg\,m^{-2}}$, the amount of surfactant in the interface would amount to $7.2\,\mathrm{mg\,ml^{-1}}$ emulsion, corresponding to $12\,\mathrm{mg}$ per ml aqueous phase (or 1.2%). Assuming further that the concentration of the surfactant leading to a plateau value of γ (m_{cr}) equals $0.3\,\mathrm{mg\,ml^{-1}}$, and that this is indeed the concentration left after emulsification, it would follow that the surfactant concentration decreases from 12.3 to $0.3\,\mathrm{mg\,ml^{-1}}$ during emulsification. This generally implies that the effective γ value increases during the process. If insufficient surfactant is present to leave a concentration m_{cr} after emulsification, even the equilibrium γ value would increase.

Another aspect is that the composition of the surfactant mixture in solution may alter during emulsification. If some minor components are present that give a relatively small γ value, these will predominate at a macroscopic interface, but during emulsification, as the interfacial area increases, the solution will soon become depleted of these components. Consequently, the equilibrium value of γ will increase during the process and the final value may be markedly larger than what is expected on the basis of macroscopic measurement.

2.4.2.2 Transport in the Interface

During droplet deformation, its interfacial area is increased. The drop will commonly have acquired some surfactant, and it may even have a Γ value close to equilibrium at the prevailing (local) surfactant activity. Will this surfactant be able to distribute itself evenly over the enlarged interface in the very short times available? Evening out can occur by surface diffusion or by spreading.

The rate of surface diffusion is governed by the surface diffusion coefficient D_s. Its magnitude will depend on several factors; it will anyway be inversely proportional to the effective viscosity felt. Assuming $D_s = 10^{-9}$ m^2 s^{-1} for a small molecule at an air/water interface, $D_s > 10^{-12}/\eta$ (in SI units) in almost all other cases, taking for η the value of the most viscous of the two phases. D_s would decrease with increasing molar mass of surfactant and with increasing Γ. It has been observed by Clark *et al.*[34] that for such conditions (proteins) D_s is indeed very small.

The time needed for surface diffusion would be

$$\tau_{\text{dif},s} = z^2/D_s > 10^{12}d^2\eta \tag{2.38}$$

where z is the distance in the interface; it is further assumed that on deformation of a drop $z \approx d$. Comparing eqn. (2.38) with those for the deformation times (eqns. 2.26, 2.31, 2.32), it turns out that under nearly all conditions $\tau_{\text{dif},s} > \tau_{\text{def}}$. For very large drops deformed in laminar flow, the surface diffusion rate can be an important variable; see *e.g.* Stone and Leal.[35]

Sudden extension of an interface or sudden local application of a surfactant to an interface, however, can produce a fairly large γ-gradient, and in such a case spreading of surfactant can occur. This proceeds as a longitudinal wave, according to Lucassen.[36] He derived for the linear velocity of such a wave on an air/water surface

$$v = 1.2(\eta\rho z)^{-1/3}|\Delta\gamma|^{2/3} \tag{2.39}$$

where z is the distance over which it has to travel. Applying this to a deformed droplet, the time needed for spreading would become

$$\tau_{\text{spr}} \approx d^{4/3}\eta^{1/3}\rho^{1/3}|\Delta\gamma|^{-2/3} \tag{2.40}$$

where η may best be taken as the larger one of η_D and η_C. Comparison of results with τ_{def} (eqns. 2.26, 2.31, 2.32) leads to the conclusion that nearly always

$\tau_{spr} < \tau_{def}$, unless $|\Delta\gamma|$ is quite small, for which case eqn. (2.40) is anyway invalid. It may be concluded that for fast deformation of a drop that has already acquired a relatively high surface excess, surfactant will become evenly distributed over the interface very fast.

2.4.2.3 Transport towards the Interface

It will be assumed that the surfactant is present in the continuous phase. The discussion is primarily based on the treatment by Levich.[19] Unless large drops are considered that are slowly deformed in laminar flow (very small $\eta_C G$), transport of surfactant to a droplet is primarily governed by convection, not by diffusion.

In the régime LV, the typical time for adsorption would be

$$\tau_{ads} = 6\pi(\Gamma/m_C)d^{-1}G^{-1} \tag{2.41}$$

In the régime TV, where $G = (\epsilon/\eta_C)^{1/2}$ for eddies of size l_0, this leads to

$$\tau_{ads} = 6\pi(\Gamma/m_C)\eta^{1/2}d^{-1}\epsilon^{-1/2} \tag{2.42}$$

In régime TI, transport of surfactant to the drop will be largely determined by the chaotic motion of the drop. The resulting adsorption time then is roughly

$$\tau_{ads} = (\Gamma/m_C)\rho^{1/3}d^{-1/3}\epsilon^{-1/3} = (\Gamma/m_C)\rho^{2/5}\gamma^{-1/5}\epsilon^{-1/5} \tag{2.43}$$

where the second part is obtained by inserting eqn. (2.16) for d. (Note: in earlier publications[2-4] we have used eqn. (2.42) for the régime TI, but this appears to be incorrect, since it is the drop that 'sweeps up' surfactant coming in its way, rather than surfactant being transported towards the drop; consequently, length scales $> l_0$ should be considered.)

It should further be noticed that in the derivation of these equations, it was assumed that all molecules closely approaching the droplet will indeed be adsorbed. Naturally, this cannot be true as soon as the interface is to some extent covered. Full adsorption would thus be reached after, say, 10 times τ_{ads}. It should, however, be realized that the underlying relations are scaling laws, and that the results cannot be seen as giving more than orders of magnitude.

The factor Γ/m_C is present in all equations for τ_{ads}. A reasonable value is 10^{-6} m; as mentioned, plateau values for Γ are generally about 3 mg m^{-2} and the surfactant concentration in the solution will often be about 3 kg m^{-3}. Some calculated results are given in Table 2.5. In régime TI, it appears that τ_{ads} and τ_{def} are of the same order of magnitude, except at small ϵ, where τ_{ads} is relatively shorter. This may imply that at small ϵ the effective γ during emulsification is smaller than at high ϵ. Results for a higher η_D are not given, because of the great uncertainty in the values of d obtained (see Table 2.3). It is, however, fairly certain that the ratio τ_{ads}/τ_{def} will be larger for a higher η_D. In régime TV, τ_{ads} appears to be quite a bit longer than τ_{def}, at least at fairly small η_D. At higher droplet viscosity the results are less certain, since eqn. (2.23) cannot be correct;

Table 2.5 *Calculated examples of time scales (in μs) for deformation and surfactant adsorption under various conditions; approximate results (see text)*

Régime	Parameter[a]	Eqn.	η_D/Pa s	ε/W m^{-3}		
				10^6	10^9	10^{12}
TI	d_{max}/μm	2.16	–	63	4	0.25
	τ_{ads}	2.43	–	2.5	0.6	0.2
	τ_{def}	2.26b	0.01	63	4	0.25
TV	d_{max}/μm	2.23	–	30	1	
	τ_{ads}	2.42	–	200	190	
	τ_{def}	2.31	0.01	30	1	
	τ_{def}	2.31	1	3000	100	

[a]Further data: $\eta_C = 10^{-3}$ and 0.1 Pa s in the régimes TI and TV, respectively; $\rho_C = \rho_D = 10^3$ kg m^{-3}; $\gamma = 0.01$ N m^{-1}; $\Gamma/m_C = 10^{-6}$ m.

nevertheless, deformation will be slow, and adsorption may well be able to keep up with it.

It may be concluded that in many cases adsorption is not fast enough to keep up with deformation, and that the relative difference increases during emulsification. The latter is due to (i) depletion causing m_C to decrease and (ii) τ_{ads} decreasing slower than τ_{def} with decreasing d. Consequently, Γ would typically be halfway between zero and its plateau value. This implies that Π would be less than half of its plateau value. Especially for polymeric surfactants, γ may remain fairly large. This is because (i) the plateau value of γ is relatively large (*e.g.* 10 mN m^{-1}) and (ii) it may take a considerable time after adsorption of polymer molecules, maybe even of the order of a second, to adapt their conformation and to attain the final, *i.e.* lowest, interfacial tension.

Of course, adsorption times can be decreased, and effective γ decreased, by increasing m_C.

2.4.2.4 *Very Small Interfacial Tensions*

One may try to make the effective γ very small by adding a high concentration of a suitable surfactant mixture. When succeeding in decreasing γ_{eff} by a factor of 100, the mechanical energy needed to produce an emulsion with a given droplet size will be decreased by a factor between 10 and 100, according to régime and depending on further conditions. Would it also be possible to obtain very small droplets, say < 100 nm? This turns out to be very difficult.

In the régime TI, a problem is that d cannot become significantly smaller than l_0. Combining this condition with eqns. (2.16) and (2.11) leads to

$$\gamma > \eta_C^{5/4}\varepsilon^{1/4}\rho^{-1/2} \qquad (2.44)$$

Taking *e.g.* $\eta_C = 10^{-3}$, $\epsilon = 10^{12}$ and $\rho = 10^3$, all in SI units, results in $\gamma > 6$ mN m^{-1}. Another condition is that $\tau_{eddy} > \tau_{def}$, which leads *via* eqns. (2.28) and (2.26) to

$$\gamma > \eta_D^{5/4}\epsilon^{1/4}\rho^{-1/2} \qquad (2.45)$$

which is even stricter.

In régime TV a condition like eqn. (2.44) is not imposed, but a condition about deformation time also holds, as was discussed earlier. Nevertheless, at rather small ϵ, quite small d values can be obtained, of the order of a μm.

This is applied in the so-called PIT method of Shinoda *et al.*[37,38] PIT stands for phase-inversion temperature. It is observed that many nonionic surfactants decrease in HLB number with increasing temperature. Below the PIT (which also depends on the composition of both phases), an O/W emulsion tends to be formed, but above the PIT, a W/O emulsion; see further on for an explanation. At the PIT the interfacial tension is very small, and quite small droplets result. These are unstable to coalescence, but by rapidly cooling the emulsion after emulsification a stable O/W emulsion having fine droplets can be obtained.[38] The droplet break-up is presumably in régime TV and fairly small ϵ values suffice. The method is widely applied in industrial practice.

Very small γ values, down to 10^{-5} N m^{-1}, can also be obtained with some nonionic surfactants at high salt concentration.[39]

2.4.3 Effect on Droplet Deformation

The mere presence of a surfactant, apart from its effect on the magnitude of γ, may affect the deformation and break-up of a drop. This will now be briefly discussed.

2.4.3.1 Interfacial Tension Gradients

Surfactants allow the existence of interfacial tension gradients, which may be considered to be their most important function. As depicted in Figure 2.9a, any liquid motion will be continuous across a liquid interface, if it contains no surfactant. A clean interface cannot withstand a tangential stress. If a liquid flows along an interface with surfactant, the latter will be swept downstream, as in Figure 2.9b. This causes an interfacial tension gradient and a balance of forces will be established, according to

$$\eta(dv_x/dy)_{y=0} = -d\gamma/dx \qquad (2.46)$$

If the γ-gradient can become large enough, it will virtually arrest the interface. The largest value attainable for $d\gamma$ equals about Π_{eq}, *i.e.* $\gamma_0 - \gamma_{eq}$. If it acts over a small distance, a considerable stress can develop, say 10 kPa.

On the other hand, if surfactant is applied at one site on an interface, a γ-gradient is formed that will cause the interface to move, roughly at the velocity given by

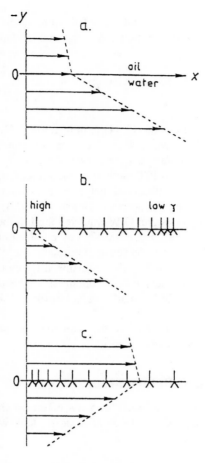

Figure 2.9 *Interfacial tension gradients and flow near an oil–water interface. (a) No surfactant, velocity gradient continuous across the interface. (b) Velocity gradient causes an interfacial tension gradient. (c) Interfacial tension gradient causes flow (Marangoni effect)*
(Reproduced by permission from ref. 4)

eqn. (2.39). The interface will then drag some of the bordering liquid with it, as depicted in Figure 2.9c. This is called the *Marangoni effect*. Eqn. (2.46) would also hold in this case, assuming inertial forces to be negligible.

The importance of the existence and the magnitude of γ-gradients will be discussed later on. One aspect may be mentioned now: the stability of thin films. This would be especially important during the beginning of emulsification, when films of continuous phase may be drawn through the disperse phase, or when collision of the still very large, hence deformable, drops causes films to form between them. Such a film will immediately break if no γ-gradient can be formed. Film stability is discussed at length by Lucassen;[40] see also ref. 2.

The magnitude of the γ-gradients and of the Marangoni effect will depend on the surface dilational modulus E_{SD}; see eqn. (2.37). For a plane interface in contact with one surfactant-containing phase, in the absence of streaming, and for a small expansion or compression of the interface, the following relations have been derived by Lucassen[40]

$$E_{SD} = \frac{-d\gamma/d\ln\Gamma}{(1 + 2\zeta + 2\zeta^2)^{1/2}} \tag{2.47}$$

$$\zeta = \frac{dm_C}{d\Gamma}(D/2\omega)^{1/2} \tag{2.47a}$$

$$\omega = d\ln A/dt \tag{2.47b}$$

where D is the diffusion coefficient of the surfactant and ω represents a timescale, in the present case the time needed for doubling the surface area; it would roughly equal τ_{def}. During emulsification, the magnitude of E_{SD} is dominated by the numerator in eqn. (2.47), because ζ remains small. The value of $dm_C/d\Gamma$ tends to go to very high values when Γ reaches its plateau value; this implies that the modulus will go through a maximum when m_C is increased. However, during emulsification, or more specifically during droplet deformation, Γ will always remain smaller, as discussed earlier in this section. Taking reasonable values for the variables, viz. $dm_C/d\Gamma = 10^2-10^4$ m^{-1}, $D = 10^{-9}-10^{-11}$ m^2 s^{-1} and $\tau_{def} = 10^{-2}-10^{-6}$ s, it is calculated that $\zeta < 0.1$, at all conditions. The same conclusion can be drawn for the value of E_{SD} in thin films, as between closely approaching drops, where not the timescale but the limited amount of surfactant in the film is determinant.[40] It may thus be concluded that for conditions that prevail during emulsification, E_{SD} increases with m_C and follows the relation

$$E_{SD} \approx d\Pi/d\ln\Gamma \tag{2.48}$$

except for very high surfactant concentrations combined with low agitation intensity.

Figure 2.10 shows for some surfactants the relation between Π and $\ln\Gamma$; E_{SD} would thus equal the slope of the curve. The difference between the soap and the polymers is striking. The soap will produce a much higher modulus during emulsification, the more so because the prevalent Γ value tends to be higher, as discussed before. Moreover, the curves as shown in Figure 2.10 are determined at equilibrium. For a small-molecule surfactant this may make little difference, because Π is determined by Γ only, but a polymer may change its conformation, and thereby the relation between Π and Γ. The two proteins shown differ greatly in conformational stability. Lysozyme is supposed to change little in conformation upon adsorption, but β-casein can expand considerably. Timescales of such changes can be long, presumably of the order of a second. This then means that for a polymeric surfactant the values of Π and E_{SD} may

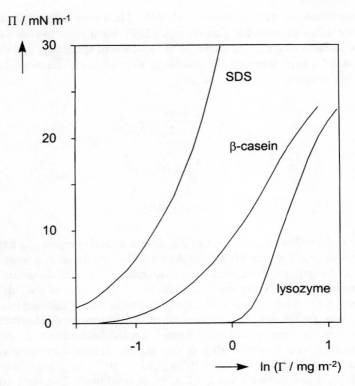

Figure 2.10 *Surface pressure Π at the air–water interface as a function of $\ln \Gamma$ for three surfactants*
(Reproduced with permission from ref. 4)

be significantly smaller than expected on the basis of curves like those in Figure 2.10.

2.4.3.2 Deformation Mode

The presence of a surfactant means that during deformation of a drop the interfacial tension need not be the same everywhere on the drop, because of the flow along the drop; see Figure 2.9. This has two consequences: (i) the equilibrium shape of the drop will be affected and (ii) any γ-gradient formed will slow down the motion of liquid inside the drop, which would in principle diminish the amount of energy needed to deform and break the drop. Almost all studies done on the effect of surfactants on deformation mode were on single drops in the régime LV.

Tip streaming was mentioned previously. De Bruijn[13] showed that it is due to accumulation of surfactant at the tips of the drop, giving locally a strong γ-gradient. As mentioned, tip streaming would only occur for fairly large drops. In a study by Groeneweg *et al.*[41] in the régime TV, it was tried to establish whether the results on single drops could be reproduced in a stirred vessel. They observed that the resulting drop size distribution was bimodal for average drop sizes of the order of 10 μm or larger; this was presumably due to small drops formed by tip streaming.

Some theoretical studies[35,42] reveal differences in the mode of deformation and the relative deformation at breaking of the drop, but the effects seem to be moderate. See also a discussion of the results by Janssen *et al.*[43] Currently, the importance for practical emulsification is not too clear.

Another effect was observed by Williams *et al.*[44] They studied break-up of drops in simple shear at a range of λ values; *cf.* Figure 2.4. For the surfactant β-lactoglobulin (a globular protein), if present at fairly high concentration, they observed that We_{cr} remained somewhat <1, for $\lambda = 10^{-4}-0.1$; this was because the drops always burst at small deformation. On the other hand, β-casein, a random coil protein, did not show such anomalous behaviour; at small λ the drops had to be deformed very strongly before bursting; hence, We_{cr} was much larger. The behaviour of β-lactoglobulin correlated with its surface dilational modulus,[45] which was high and almost purely elastic. Williams *et al.* assumed that the protein forms a gel-like network in the interface, which would also agree with its very high surface shear viscosity.[46] This may imply that eqn. (2.46) does not always hold; besides the stress due to the γ-gradient at the right-hand side of the equation, a stress induced by the elastic deformation of the surface layer may have to be added. However interesting these observations are, their importance for emulsification in practice has yet to be established. As was mentioned in Section 2.3, the effective We_{cr} may be small anyway, due to the capillary break-up mode encountered in several practical situations.

When a surfactant is present, γ is also time dependent: during deformation γ will locally change, generally increase. This means that the free energy needed for enlargement of the interface does not only consist of the term $\gamma\Delta A$, but also includes $A\Delta\gamma$.[2] This aspect has been studied in some detail by Janssen *et al.*[43,47,48] Some of their results are given in Figure 2.11 and it is clear that the stress needed for break-up is higher than predicted by Figure 2.4. In their experiments the effective γ value was shown to be very close to equilibrium. The effect of the term $A\Delta\gamma$ can be interpreted as the time-dependent surface dilational modulus itself resisting surface enlargement. Janssen *et al.* found a good correlation between experiments and results by replacing γ in eqn. (2.7) by $(\gamma + bE_{SD})$, where E_{SD} was calculated according to eqn. (2.47); the constant b was about 0.23. It has also been reasoned[49] that the additional resistance to surface enlargement should be accounted for in the surface dilational viscosity η_{SD}, by adding a term η_{SD}/d to the droplet viscosity. We agree with Janssen *et al.* that their method is fundamentally better.

This still leaves the question of the practical importance of the effect. It will be very difficult to predict the effective value of E_{SD} for most practical situations; the small value of the constant mentioned above, which was determined at fairly ideal conditions, also points to this difficulty. Another point is that the studies by Janssen *et al.* were done under near equilibrium conditions. It may be that the effect is larger for a more sudden deformation. The presence of a surface dilational modulus may also cause the deformation time to increase.

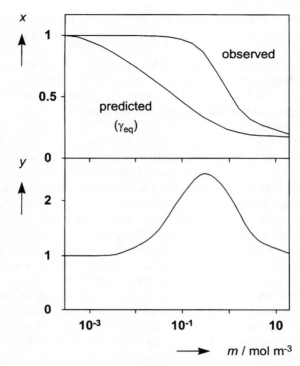

Figure 2.11 *Effect of surfactant concentration* m *on break-up of a single drop in simple shear flow.* x *is the critical droplet size for break-up relative to that in the absence of surfactant, as observed and as calculated for* γ_{eq}. y *is the ratio of* We_{cr} *obtained over its value predicted for* γ_{eq}.
(Reproduced with permission from ref. 4)

2.4.4 Preventing Recoalescence

In most cases, droplets without surfactant that closely encounter each other will coalesce. Since droplet encounters are very frequent during emulsification, coalescence would frequently occur. Moreover, the droplets will often be pressed together with great force. An essential role of the surfactant then is to prevent, or at least greatly diminish, recoalescence of newly formed droplets.

In practice, surfactant is always present during emulsification, and it is interesting to know whether coalescence then is a frequent phenomenon. To study this, a method for determining coalescence during emulsification has recently been developed in the authors' laboratory.[4,50] In brief, equal amounts of two O/W emulsions (1 and 2) are mixed, which are the same in composition and droplet size distribution, except for the oil phases, which have different refractive indices, n_1 and n_2. The mixture is then emulsified again. It is subsequently diluted with a solvent to yield for the continuous phase a refractive index of $(n_1 + n_2)/2$. This dilute emulsion will have a certain turbidity if the oils 1 and 2 are unmixed, whereas turbidity will be virtually zero when the two oils have become completely mixed by droplet coalescence. The decrease of turbidity is thus a measure of the extent of recoalescence.

The method was applied in homogenizers, régimes TI and LV or LB, for SDS[50] and proteins[4] as surfactant; ϕ was about 0.2 and the drop sizes were of the order of 1 µm. It may be concluded that coalescence did occur, but only to a very small extent, unless relatively little surfactant was present, say at 0.1 times the CMC. Even in the latter case, recoalescence was small if the homogenization of the mixture was at a clearly lower pressure than that used for making the two original emulsions.

The question now is: what mechanism is responsible for preventing recoalescence? Most authors assume that it is colloidal repulsion, caused by the adsorbed layer of surfactant. This is, however, unlikely. The pressure at which two droplets can be pressed together in the emulsifying apparatus would be ηG in régime LV and is given by eqn. (2.15) for régime TI. For 1 µm droplets the result is about 20 kPa. The repulsive stress can be derived from the steepest slope of the curve of colloidal interaction free energy *versus* separation distance, yielding the maximum repulsive force, divided by d^2. From results according to the DLVO theory (electrostatic repulsion), taking reasonable values for the parameters involved, it turns out that the repulsive stress would be rarely over 0.2 kPa, *i.e.* by two orders of magnitude smaller than the external stress. Consequently, colloidal repulsion would be far insufficient during emulsification. It may be that for some surfactants giving a very high zeta potential, or giving very strong steric repulsion (*i.e.* some polymers), colloidal repulsion may become significant. It may also be noted that both the external stress in the régime LV and the DLVO repulsion will be inversely proportional to droplet diameter; the stress in régime TI will be proportional to $d^{-1.8}$. In other words, for larger drops the colloidal repulsion would also be insufficient.

As we reasoned before,[4] the counteracting stress must be due to the formation of γ-gradients. When two drops are pushed together, liquid will flow out from the thin layer between them, and the flow will induce a γ-gradient; see Figure 2.9b. This produces a counteracting stress, according to eqn. (2.46). The resulting stress would be

$$\tau_{\Delta\gamma} \approx 2|\Delta\gamma|/\tfrac{1}{2}d \tag{2.49}$$

where the factor 2 follows from the fact that two interfaces are involved and where it is assumed that the γ-gradient extends over a distance of half the droplet diameter. Taking $\Delta\gamma = 0.01$ N m^{-1}, the stress would amount to 40 kPa, *i.e.* of the order of the calculated value of the external stress. The stress due to the γ-gradient cannot as such prevent coalescence, since it only acts for a short while, but it will greatly slow down the mutual approach of the droplets. The external stress will also act for a short time, and it may well be that the drops move apart before coalescence can occur. The effective γ-gradient will, of course, depend on the value of E_{SD}; see eqn. (2.48) and Figure 2.10.

The reasoning given above is to some extent borne out by the results of Taisne *et al.*,[50] who made emulsions with SDS at a low concentration, with low and high salt concentrations. At high salt, the thickness of the electrical double layer would be greatly reduced and repulsion would be insufficient to prevent coalescence; the emulsion showed indeed marked coalescence, contrary to the one made with low salt. However, both emulsions had the same size distribution directly after making. Consequently, there cannot have been a clear difference in recoalescence during emulsification. Electrostatic repulsion does depend on ionic strength, but the magnitude of E_{SD} and the development of a γ-gradient will not, or at least not markedly.

Closely related to the mechanism preventing recoalescence discussed above is the Gibbs–Marangoni effect, originating from van den Tempel;[51] it is depicted in Figure 2.12. Here, the depletion of surfactant in the thin film between approaching droplets is also taken into account. This may even cause a γ-gradient without liquid flow being involved, thereby causing an inward flow of liquid that tends to drive the droplets apart. Such a mechanism would only act if the drops are insufficiently covered with surfactant (Γ clearly below its plateau value), as occurs during emulsification. Another pre-requisite is that the interface can move fast enough, but eqn. (2.40) shows that this would be the case.

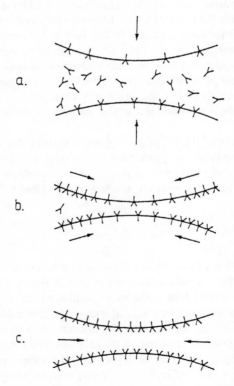

Figure 2.12   *Diagram explaining the Gibbs–Marangoni effect for two approaching droplets during emulsification. Surfactant molecules are depicted by* Y. *See text* (Reproduced with permission from ref. 4)

The Gibbs–Marangoni effect would also explain *Bancroft's rule*: when making an emulsion of two phases, the one in which the surfactant is most soluble will become the continuous phase. If the surfactant is in the droplets, a γ-gradient as depicted in Figure 2.12b would never develop, and the drops would be prone to coalescence. Hence, surfactants with an HLB value >7 tend to produce O/W emulsions; those with HLB < 7, W/O emulsions.

In our opinion, the two mechanisms cannot be clearly separated. They act jointly and at the same conditions. They greatly depend on the Gibbs elasticity of the film, which is, in principle, twice the value of E_{SD}. This points to the importance of this variable for emulsification.

It may also explain differences between soaps and polymers as surfactant. The latter give larger drops than the former, at otherwise identical conditions.[3,4] As shown in Figure 2.7, this does not only apply to near equilibrium conditions (excess surfactant), but also for lower surfactant concentration. The results would qualitatively agree with the polymers giving a far smaller value of E_{SD} at small concentration (Figure 2.10). A systematic quantitative study of the effect of surfactant properties on the resulting droplet size, taking the above considerations into account and for various régimes, has yet to be done.

In Section 2.5, a few words will be said on kinetic aspects of recoalescence.

2.4.5 'Spontaneous Emulsification'

True spontaneous emulsification would lead to microemulsions, and those are not the subject of this chapter. There is, however, a closely related topic, that of interfacial turbulence. It has been studied by Sternling and Scriven[52] and by Gouda and Joos.[53]

If two liquid phases (1 and 2) are separated by an interface of very small interfacial tension, and if surfactant is transferred from phase 1 to phase 2, because the surfactant is initially present in phase 1 and is more soluble in phase 2, an unstable situation exists. Motion of both liquids near the interface is generated in the form of so-called roll cells. The driving force arises from the difference in chemical potential of the surfactant between the two phases. The effect depends on several factors, and it appears to be more likely if $\eta_1 > \eta_2$. The instability is due to γ-gradients, causing Marangoni effects. This is illustrated in Figure 2.13. It appears that fairly long fingers of phase 1 can protrude into phase 2, and these are ultimately subject to Rayleigh instability, breaking up into small drops.

It may well be that a combination of fairly gentle stirring (*i.e.* fairly long deformation times) and the generation of interfacial turbulence may cause formation of small drops. A pre-requisite is that γ is very small, which requires a high concentration of surfactant. Moreover, the surfactant must be well soluble in the disperse phase (be it as such or in the form of micelles) and nevertheless be more soluble in the continuous phase. Methods based on these principles may provide nearly the only manner in which reasonably small drops of a very viscous oil in an aqueous phase can be made. The authors are unaware, however, of a systematic experimental study on the role of interfacial instability in practical emulsification.

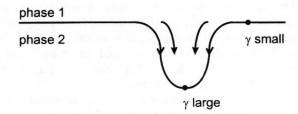

Figure 2.13 *Diagram illustrating interfacial instability. Phase 1 would originally contain the surfactant, which is, however, more soluble in phase 2*

2.5 Further Considerations

2.5.1 Effects of Volume Fraction

Up until here, it was assumed, often implicitly, that the volume fraction of disperse phase ϕ is small. In practice, variation in ϕ can have large effects. Obviously, it affects the rate of collision between droplets during emulsification, and thereby the *rate of coalescence*, if conditions are such as to allow coalescence. In a first approximation it would depend on the relation between τ_{ads} and τ_{col}, where the latter stands for the average time it takes until a droplet collides with another one. It can be derived from Smoluchowski kinetics, and we will, again, follow Levich.[19] In the various régimes, the hydrodynamic constraints are the same as for τ_{ads}; see eqns. (2.41)–(2.43). In régime LV, for instance, $\tau_{col} = \pi/8\phi G$. It turns out that for all régimes the ratio of both characteristic times would be given by

$$\kappa \equiv \tau_{ads}/\tau_{col} \propto \phi\Gamma/m_C d \qquad (2.50)$$

where the proportionality factor would be at least of order 10. For example, for $\phi = 0.1$, $\Gamma/m_C = 10^{-6}$ m and $d = 10^{-6}$ m (total surfactant concentration of the emulsion should then be about 0.5%), κ would be of order 1. For $\kappa \gg 1$, considerable coalescence is likely to occur. This is thus more likely at high ϕ. The coalescence rate would then markedly increase during emulsification, since both m_C and d become smaller during the process. If the emulsification proceeds long enough, the droplet size distribution may then be the result of a steady state of simultaneous break-up and coalescence.

The reasoning given above is an oversimplification. A far more elaborate treatment of the collision and adsorption phenomena is possible. Fundamentals of these processes have been treated, *e.g.* by van de Ven.[54] Under some conditions, *i.e.* for fairly large drops and well defined hydrodynamic conditions, such theory can be usefully applied to emulsification. It has been reviewed, *e.g.* by Stein.[28] Furthermore, the probability that two colliding drops do coalesce will not only depend on the ratio of τ_{ads} to τ_{col}; this is because the former only governs Γ, and we have seen in Section 2.4 that for the same Γ value E_{SD} may vary greatly among surfactants, and E_{SD} must be an important variable determining coalescence.

The *various effects* of an increase in ϕ can be summarized as follows.

1. τ_{col} is shorter, hence coalescence will be faster and d will be larger, unless κ remains small.

2. Emulsion viscosity η_{em} increases, hence Re decreases. This may imply a change of flow type from turbulent to laminar, hence to régime LV. For turbulent flow, Re_{dr} decreases, possibly resulting in a change from régime TI to TV. (Note: although Re_{dr} would be governed by η_C rather than by η_{em}, both may be considered about equal for the present problem, since in régime TV $d < l_0$, and the eddies thus sense the viscosity of the whole system.) The effect of these changes on the resulting droplet size will depend on various conditions.

3. In laminar flow, the effective η_C becomes higher. This may seem strange since the particles causing η_{em} to increase are also those that are disrupted, and are thus of the same size. On the other hand, the presence of many particles means that the local velocity gradients near a droplet will generally be higher than the overall value of G. Consequently, the local shear stress ηG does increase with increasing ϕ, which is as if η_C increases. It has often been observed (see *e.g.* ref. 16) that taking the viscosity of the emulsion rather than that of the continuous phase can be a reasonable approximation.

4. In turbulent flow, increase in ϕ will induce turbulence depression; in principle, this will lead to a larger d. It is often assumed that for small ϕ the root-mean-square eddy velocity $u'(x)$ (see eqn. 2.14) will be smaller by a factor $(1 + b\phi)$, where b values reported vary widely, from 0.8 to 8; see *e.g.* refs. 2, 28 and 55. In the authors' view this may be an oversimplification. As mentioned in Section 2.3, turbulence depression by added high polymers tends to remove the smallest eddies from the spectrum,[29] thereby causing the average d to be larger and the size distribution to be more narrow.

5. If the mass ratio of surfactant to continuous phase is constant, an increase in ϕ gives a decrease in surfactant concentration, hence an increase in γ_{eq}, hence an increase in d; and also an increase in κ, hence an increase in coalescence rate, hence an increase in d. These conclusions would not apply if at all stages $\kappa \ll 1$.

6. If the mass ratio of surfactant to disperse phase is constant, the changes as mentioned under 5 are reversed; again, unless $\kappa \ll 1$.

It may be clear that general conclusions cannot easily be drawn, since several of the mechanisms mentioned may come into play. Some *experimental results* will briefly be mentioned. Goulden and Phipps[56] and Walstra[57] used a high-pressure homogenizer and compared various ϕ (up to 0.4) at constant initial m_C, régime TI, probably changing to TV for higher ϕ. With increasing ϕ (>0.1), the resulting d increased and its dependence on p_H became weaker; see also Figure 2.14. This points to increased coalescence (effects 1 and 5), which would also agree with the surfactant being protein, or even protein aggregates, leading to a small value of E_{SD}. Turbulence depression (effect 4) must also have played a role.

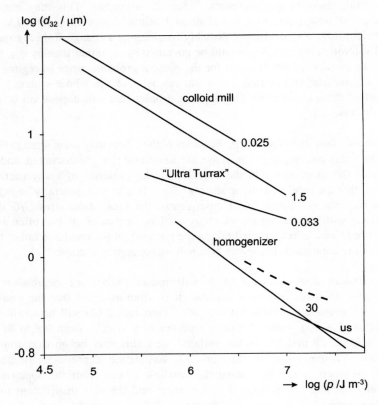

Figure 2.14   *Average droplet diameters obtained in various emulsifying machines as a function of the energy consumption* p; us *means ultrasonic generator. The numbers near the curves denote the viscosity ratio* λ. *The results with the homogenizer are for* $\phi = 0.04$ (solid line) and $\phi = 0.3$ (broken line)
(Redrawn with permission from refs. 2, 24, 25, 57 and 59)

Kumar *et al.*[55] studied droplet break-up in a stirred vessel (turbine stirrer) at $\phi = 0.1-0.8$ and with an excess of surfactant (SDS). They reasoned that in various regions in the vessel different régimes prevailed, and the régime leading to the smallest drop size would be determinant. Up to $\phi \approx 0.5$, droplet size appeared to be determined by break-up in régime TI. The resulting d_{max} was about proportional to $(1 + 4\phi)^{1.2}$, which was ascribed to turbulence depression (effect 4). For still higher ϕ, d_{max} decreased. Presumably, droplet size was then determined by break-up in the régime LV, which prevailed in a region near the impeller; now d_{max} was about inversely proportional to ϕ_{em} (effect 3).

Wieringa *et al.*[16] did studies with a colloid mill, régime LV, at high ϕ. They concluded that the drop size obtained agreed with theory, assuming that $\eta_C = \eta_{em}$ (effect 3). They had to take the shear rate thinning of the system into account, when making emulsions at various G.

Karbstein[25] compared various machines. The surfactant was always in excess and she took $\phi = 0.1-0.9$. In a colloid mill, she observed no significant effect of ϕ on d_{32}, provided that the energy expenditure per unit volume p was the same. It should be realized that p would be about proportional to $\eta_{em}G$, implying that d_{32} was about proportional to η_{em}^{-1} (effect 3). She also used an UltraTurrax, where the régime must have been TV. The resulting trends were as for the colloid mill, which would imply that also in régime TV effective η_C roughly equals η_{em}.

2.5.2 Emulsification Efficiency

The manufacture of emulsions may cost much energy. When making an emulsion of $\phi = 0.1$, in a high-pressure homogenizer at 10^7 Pa, resulting $d_{32} = 0.6$ μm, $\gamma = 0.01$ N m^{-1}, the net increase in surface free energy $A\gamma = 6\phi\gamma/d_{32} = 10^4$ J m^{-3}. The making of such an emulsion would actually take 10^7 J m^{-3}, implying an energy efficiency of 0.1%. The other 99.9% is dissipated into heat.

As mentioned, the intensity of the process, or the effectiveness in making small droplets, is often governed by the net power density ε. The net energy consumption per unit volume of emulsion then is given by

$$p = \int \varepsilon(t)\, dt \qquad (2.51)$$

where t is the time during which emulsification occurs. Break-up of drops will only occur when ε is high, which means that energy dissipated at low ε levels is wasted. Batch processes will generally be less efficient than continuous ones, since it is very difficult to realize a high value of ε throughout the whole apparatus. A stirrer in a large vessel is a good example of a situation where by far most of the energy applied is dissipated at a low intensity level. The larger the vessel relative to the stirrer, the longer the process must last before all the emulsion droplets have finally been broken up into small ones. Even for a reasonable ratio of the volume in or near the stirrer and that of the vessel, the total energy consumption may be higher by an order of magnitude than in a continuous-flow apparatus. The best strategy would thus be to create a situation in which the energy is dissipated in the shortest possible time, although the time must of course be long enough for droplets to be deformed and broken up.

In a high-pressure homogenizer, p simply equals the homogenizing pressure p_H. In most other machines, ε has to be integrated over time according to eqn. (2.51), which may be far from easy. In the régime LV, $\varepsilon = \eta G^2$. Another way to arrive at the energy consumption is from the increase in temperature during processing, since virtually all energy is generally dissipated, though it is not easy to determine unequivocally the temperature rise. The method most often applied is by measuring the electrical power consumption W and dividing it by the volume flow rate Q. This gives the overall energy consumption; the difference with the net energy consumption will generally be by at least 10% and may be (much) larger for a smaller machine.

Some results are shown in Figure 2.14. Those for the colloid mill and the Ultra Turrax-like machine are from Karbstein and Schubert,[25,58] who did an extensive practical study on the suitability of various machines for emulsification under a range of conditions. The other results are by Walstra,[2,24,57] who mostly made fairly dilute emulsions with proteins as surfactants. It is seen that the homogenizer is generally superior, although it should be mentioned that it concerns the net energy consumption in these cases, and overall energy consumption for Karbstein and Schubert's results. This does somewhat exaggerate the difference, although it will still be considerable. This stands to reason, since the homogenizer precisely does what is desirable: dissipate the available energy in the shortest possible time.

There are, of course, various restrictions. The ultrasonic emulsification result looks very efficient, but the method is expensive and only suitable for small quantities. Furthermore, although d_{32} is small, the relative width of the size distribution is large. This may also be the case in some other machines. It is often the size of the largest droplets rather than the average that is important, for instance when sedimentation or creaming of the emulsion is undesirable. A homogenizer does not readily handle high volume fractions or disperse phases of high viscosity, and some stirrer types may then be more suitable. A colloid mill is only suitable for fairly high η_C and η_D/η_C should be below about 5. The maximum possible intensity varies greatly among machines. In a high-pressure homogenizer, ε may be as high as 10^{12} W m^{-3}, in an Ultra Turrax type machine at most 10^{10} W m^{-3}, and in most stirrers far smaller. In a colloid mill, the highest shear stress attainable is about 10^4 Pa.

What measures can be taken to enhance efficiency if an emulsion of given properties has to be made? These may include the following.

- Optimize efficiency of agitation by increasing ϵ and decreasing dissipation time. This often means increasing the ratio of the 'active volume' (where ϵ is quite high) to the inactive volume.
- Make the emulsion at a high ϕ and dilute afterwards, since by far most of the energy is dissipated without being useful, whose amount is about proportional to the total volume. At very high ϕ, excessive recoalescence may occur, posing a limit. Moreover, high ϕ may cause a change of régime or turbulence depression.
- Add more surfactant, thereby creating a smaller γ_{eff} and possibly diminishing recoalescence. This measure can be readily combined with the previous one. Of course, adding more surfactant does increase costs.
- If possible, dissolve the surfactant in the disperse rather than in the continuous phase. It often leads to smaller droplets; see Section 2.4, Spontaneous Emulsification.
- It may be useful to emulsify in steps of increasing intensity, especially for emulsions with a highly viscous disperse phase.

2.5.3 Membrane Emulsification

The high energy expenditure mentioned in classical emulsification has induced several workers to try other methods, where only the energy needed to produce the interfacial area has to be provided. A fairly recent technique is that of membrane emulsification.[59,60] In this method, the disperse phase is pressed through a membrane and droplets leaving the pores are immediately taken up by the continuous phase. The membrane is commonly made of porous glass or of ceramic materials. The general configuration is a membrane in the shape of a hollow cylinder; the disperse phase is pressed through it from the outside, and continuous phase is pumped through the cylinder (cross-flow). The flow also causes detachment of the protruding droplets from the membrane.

Some pre-requisites are that (i) the membrane is hydrophilic for a hydrophobic disperse phase and *vice versa*, since otherwise the drops cannot be detached; (ii) the pores must be sufficiently far apart from each other to prevent the drops that come out touching each other and coalescing; and (iii) the pressure over the membrane should be sufficient to achieve drop formation. The latter pressure should be at least of the order of the Laplace pressure of a drop of diameter equal to the pore diameter. For pores of 0.4 μm and $\gamma = 5$ mN m^{-1}, this would amount to 50 kPa, but larger pressures are needed in practice, for instance 3×10^5 Pa, also to obtain a significant flow rate of the disperse phase through the membrane.

It is certainly possible to make emulsions in this way, with fairly simple apparatus and without applying intense agitation, but the often claimed monodispersity has not been substantiated. The smallest drop size obtained is about three times the pore diameter. A main disadvantage is that the process is extremely slow. The amount of disperse phase produced is, for instance, 10^{-5} m^3 per m^2 of membrane surface per second. This implies that very long circulation times are needed to produce even small volume fractions. It must also lead to considerable energy expenditure. Further study of the method would be desirable.

2.6 Concluding Remarks

Most of the important basic aspects of emulsion formation appear to have been identified and may be reasonably well understood, although the quantitative importance of recoalescence under various conditions has yet to be established. Even if that would be achieved, it does not mean that further research is not needed. There are two reasons for this. First, there is a wide range of variables: initial drop size, régime and flow type, agitation intensity, viscosity of either phase, type and concentration of surfactant. All of these need consideration and they often affect each other's effect in a different way under different conditions. Second, the situation in a machine producing emulsions is often very chaotic: flow conditions vary with location and the various processes during emulsification (Figure 2.2) occur simultaneously. Predicting the result then calls for the application of sophisticated mathematical process simulation, even if the process conditions are known as a function of location.

Hopefully, this review will provide the reader with some understanding of the aspects involved in emulsion formation and, moreover, stimulate further work.

2.7   References

1. P. Walstra, in 'Encyclopedia of Emulsion Technology', ed. P. Becher, Dekker, New York, 1996, vol. 4, p. 1.
2. P. Walstra, in 'Encyclopedia of Emulsion Technology', ed. P. Becher, Dekker, New York, 1983, vol. 1, p. 57.
3. P. Walstra, *Chem. Eng. Sci.*, 1993, **48**, 333.
4. P. Walstra and I. Smulders, in 'Food Colloids: Proteins, Lipids and Polysaccharides', eds. E. Dickinson and B. Bergenståhl, The Royal Society of Chemistry, Cambridge, 1997, p. 367.
5. M.H. Li and H.S. Fogler, *J. Fluid Mech.*, 1978, **88**, 513.
6. P. Kiefer, Ph.D. Thesis, University of Karlsruhe, 1977.
7. H.P. Grace, *Chem. Eng. Commun.*, 1982, **14**, 225.
8. B.J. Bentley and L.G. Leal, *J. Fluid Mech.*, 1986, **167**, 241.
9. H.A. Stone, B.J. Bentley and L.G. Leal, *J. Fluid Mech.*, 1986, **173**, 131.
10. H.A. Stone, *Annu. Rev. Fluid Mech.*, 1994, **226**, 65.
11. D. Barthès-Biesel and A. Acrivos, *J. Fluid Mech.*, 1973, **61**, 1.
12. M. Tjahjadi, H.A. Stone and J.M. Ottino, *J. Fluid Mech.*, 1992, **243**, 297.
13. R.A. de Bruijn, *Chem. Eng. Sci.*, 1993, **48**, 277.
14. J.M.H. Janssen and H.E.H. Meijer, *J. Rheol.*, 1993, **37**, 597.
15. F.D. Rumscheidt and S.G. Mason, *J. Colloid Sci.*, 1962, **17**, 260.
16. J.A. Wieringa, F. van Dieren, J.J.M. Janssen and W.G.M. Agterof, *Trans. Inst. Chem. Eng.*, 1996, **74-A**, 554.
17. R.A. de Bruijn, *Procestechnologie*, **1997** (5), 16.
18. H. Armbruster, Ph.D. Thesis, University of Karlsruhe, 1990.
19. V.G. Levich, 'Physicochemical Hydrodynamics', Prentice-Hall, Englewood Cliffs, 1962.
20. J.T. Davies, 'Turbulence Phenomena', Academic Press, London, 1972.
21. W.J. Tjaberinga, A. Boon and A.K. Chesters, *Chem. Eng. Sci.*, 1993, **48**, 285.
22. P. Walstra, *Dechema Monographie*, 1974, **77**, 87.
23. H. Lamb, 'Hydrodynamics', 6th edn., Dover Press, New York, 1945.
24. P. Walstra, *Neth. Milk Dairy J.*, 1975, **29**, 279.
25. H. Karbstein, Ph.D. Thesis, University of Karlsruhe, 1994.
26. W.D. Pandolfe, *J. Disp. Sci. Technol.*, 1981, **2**, 459.
27. G.E.J. Vaessen, Ph.D. Thesis, Eindhoven University of Technology, 1996.
28. 29. P. Walstra, *Chem. Eng. Sci.*, 1974, **29**, 882.
30. E.H. Lucassen-Reynders, in 'Encyclopedia of Emulsion Technology', ed. P. Becher, Dekker, New York, 1996, vol. 4, p. 63.
31. D.E. Graham and M.C. Phillips, *J. Colloid Interface Sci.*, 1979, **70**, 415.
32. E.H. Lucassen-Reynders, *Colloids Surf. A*, 1994, **91**, 79.
33. E.D. Goddard and K.P. Ananthapadmanabhan, eds., 'Interactions of Surfactants with Polymers and Proteins', CRC Press, Boca Raton, 1993.
34. D.C. Clark, M. Coke, P.J. Wilde and D.R. Wilson, in 'Food Polymers, Gels and Colloids', ed. E. Dickinson, The Royal Society of Chemistry, Cambridge, 1991, p. 272.
35. H.A. Stone and L.G. Leal, *J. Fluid Mech.* 1990, **220**, 161.
36. J. Lucassen, *Trans. Faraday Soc.*, 1968, **64**, 2221.

37. K. Shinoda and H. Kunieda, in 'Encyclopedia of Emulsion Technology', ed. P. Becher, Dekker, New York, 1983, vol. 1, p. 337.
38. K. Shinoda and H. Saito, *J. Colloid Interface Sci.*, 1969, **30**, 258.
39. J.A.M.H. Hofman and H.N. Stein, *J. Colloid Interface Sci.*, 1991, **147**, 508.
40. J. Lucassen, in 'Anionic Surfactants: Physical Chemistry of Surfactant Action', ed. E.H. Lucassen-Reynders, Surface Science Series, vol. II, Dekker, New York, 1981, p. 217.
41. F. Groeneweg, F. van Dieren and W.G.M. Agterof, *Colloids Surf. A*, 1994, **91**, 207.
42. W.J. Milliken and L.G. Leal, *J. Colloid Interface Sci.*, 1994, **166**, 275.
43. J.J.M. Janssen, A. Boon and W.G.M. Agterof, *AIChE J.*, 1997, **43**, 1436.
44. A. Williams, J.J.M. Janssen and A. Prins, *Colloids Surf. A*, 1997, **125**, 189.
45. A. Williams and A. Prins, *Colloids Surf. A*, 1996, **114**, 267.
46. E. Dickinson, B.S. Murray and G. Stainsby, *J. Chem Soc., Faraday Trans. 1*, 1988, **84**, 871.
47. J.J.M. Janssen, A. Boon and W.G.M. Agterof, *Colloids Surf. A*, 1994, **91**, 141.
48. J.J.M. Janssen, A. Boon and W.G.M. Agterof, *AIChe J.*, 1994, **40**, 1929.
49. E.H. Lucassen-Reynders and K.A. Kuijpers, *Colloids Surf.*, 1992, **65**, 175.
50. A. Taisne, P. Walstra and B. Cabane, *J. Colloid Interface Sci.*, 1996, **184**, 378.
51. M. van den Tempel, *Proc. 3rd Int. Congr. Surf. Activity*, 1960, **2**, 573.
52. C.V. Sternling and L.E. Scriven, *AIChE J.*, 1959, **5**, 514.
53. J.H. Gouda and P. Joos, *Chem. Eng. Sci.*, 1975, **30**, 521.
54. T.G.M. van de Ven, 'Colloidal Hydrodynamics', Academic Press, London, 1988.
55. S. Kumar, R. Kumar and K.S. Gandhi, *Chem. Eng. Sci.*, 1991, **46**, 2483.
56. J.D.S. Goulden and L.W. Phipps, *J. Dairy Res.*, 1964, **31**, 195.
57. P. Walstra, in 'Gums and Stabilizers for the Food Industry', eds. G.O. Phillips, D.J. Wedlock and P.A. Williams, IRL Press, Oxford, 1988, vol. 4, p. 323.
58. H. Karbstein and H. Schubert, *Chem. Eng. Proc.*, 1995, **34**, 205.
59. K. Kandori, in 'Food Processing: Recent Developments', ed. A.G. Gaonkar, Elsevier, Amsterdam, 1995, p. 113.
60. R. Katoh, Y. Asano, A. Furaya, K. Sotayama and M. Tomita, *J. Membr. Sci.*, 1996, **113**, 131.

CHAPTER 3

Emulsion Formation by Nucleation and Growth Mechanisms

BRIAN VINCENT, ZOLTAN KIRALY* AND TIM M. OBEY

School of Chemistry, University of Bristol, Bristol BS8 1TS, UK

3.1 Introduction

Emulsions, in general, are thermodynamically unstable systems, as are most dispersions of one phase in a second (liquid) phase. This situation is illustrated in Figure 3.1. For most systems of two liquids, having only slight mutual solubilities, the lowest free energy state is the two co-existing liquid phases, α and β. In order to form an emulsion of, say, phase α in phase β (hereafter, referred to as an α/β emulsion), work has to be put in to increase the free energy (F) of the system, by an amount ΔF_{form}:

$$\Delta F_{form} = \gamma \Delta A - T \Delta S \tag{3.1}$$

In eqn. (3.1) the first term on the right-hand side refers to the free energy change associated with the increase in interfacial area (ΔA), where γ is the α/β interfacial tension. It is a *positive* term. The second term on the right-hand side refers to the entropy of dispersion of the droplets. This latter term is generally small and insignificant, such that in most systems it can generally be neglected, in comparison with the $\gamma \Delta A$ term. However, it is also *positive* and, hence, if $\gamma \Delta A$ is actually small enough, such that $\gamma \Delta A < T \Delta S$, then ΔF_{form} is overall *negative* and emulsification then becomes a spontaneous process. This condition requires that γ is very small (typically $\ll 10^{-2}$ mN m^{-1}), which is the situation pertaining to *microemulsions*, which are thermodynamically *stable*. However, for most binary liquid systems, γ is significantly higher, even in the presence of most surfactants or polymers. Hence, ΔF_{form} is indeed positive, and emulsions formed by comminution methods are, therefore, thermodynamically unstable. In principle, the emulsion should break,

*Current address: Department of Colloid Chemistry, Attilla Joszef University, Szeged, H-6720, Hungary

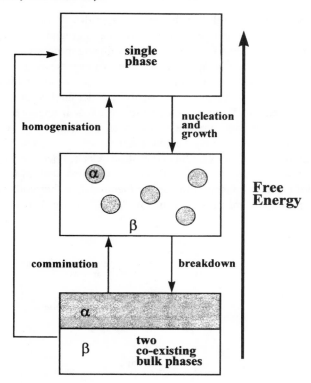

Figure 3.1 *Schematic representation of emulsion formation*

and phase separation ensue. However, *kinetic* stability may be conferred on the emulsion, as is discussed elsewhere in this book.[1]

An alternative, general procedure for preparing dispersions, including emulsions, is also illustrated in Figure 3.1. This involves phase separation, by a nucleation and growth mechanism, from an initially *single-phase* system. This may be achieved in one of the following two ways.

(1) By some change in thermodynamic conditions (*i.e.* temperature, pressure or concentration), such that the two phases α and β (whether previously emulsified, or not), become homogenised to one phase. Reversal of that change in temperature, pressure or concentration leads to the required phase separation, by nucleation and growth. This involves critical-size nuclei of the separating phase being formed, and these entities undergoing a series of aggregation and coalescence steps, *via* intermediates of colloidal dimensions, until phase separation is complete. In order to form a colloidal dispersion (of particles, droplets or bubbles) it is necessary to prevent the complete breakdown of the system, and to effectively 'trap' it at some intermediate stage, *i.e.* the dispersion is kinetically stable. This may occur naturally, due to interfacial charge being developed [see (2) below], or can be invoked by the addition to the system of some 'stabiliser', such as a surfactant or a polymer, which induces charge- or steric-stabilisation of the intermediate colloidal entities. Clearly,

the average particle size, and also the particle size distribution, that emerges will depend, *inter alia*, on the relative rates of: (i) the aggregation/coalescence of the extant particles and (ii) the adsorption of the stabilising moieties.

A common example to illustrate this sequence of events would be the formation of a foam (bubble in liquid dispersion) on opening a can of lager. Inside the can the system is one-phase, CO_2 having been dissolved under applied pressure. When the pressure is released, small gas bubbles form, which will aggregate and coalesce, until sufficient protein, also present in the beer, adsorbs at the growing bubble/aqueous solution interface to give a stable foam.

The formation of emulsions by this type of process has been less common. Such cases are reviewed in Section 3.2. In this case it is obviously more convenient to use temperature and/or concentration as the manipulative thermodynamic variable. The general principles are illustrated in Figure 3.2. Figure 3.2(a) shows a typical

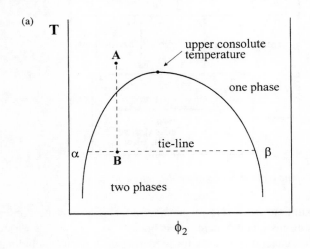

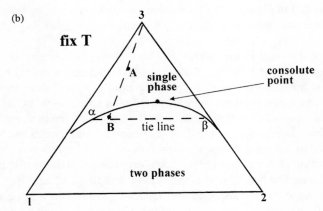

Figure 3.2 *Phase separation in liquid systems*: (a) *two liquid components, as a function of temperature*; (b) *three liquid components, at a fixed temperature*

temperature/composition phase diagram for a binary liquid mixture, in this case one showing an *upper* consolute temperature, but the same principles apply to other types of phase diagrams, *e.g.* ones with lower consolute temperatures, closed loops, *etc.* Figure 3.2(b) shows a corresponding ternary phase diagram for a three-component system, at a fixed temperature. In both cases, the dashed line (AB) shows one possible route from the one-phase to the two-phase region. In the case of the binary system (Figure 3.2a) this corresponds to changing the temperature, at fixed overall composition of the binary mixture. (Note: one could equally well move 'horizontally' across the phase diagram to point B, from the one-phase region, by a change in overall composition, keeping the temperature fixed. This could be evoked by the preferential evaporation of one of the components.) The composition of each of the two coexisting phases (*i.e.* α and β) is given by the points at the end of the tie-line shown in the figure. In order to form an emulsion, clearly the phase separation process has to be stopped at some intermediate stage, again by the presence of a suitable 'stabiliser' (surfactant or polymer which adsorbs at the emergent liquid/liquid interface). An interesting question to consider at this stage is: which type of emulsion might one expect to see, α/β or β/α? Although factors such as the relative amounts of the two primary liquid components, and their relative viscosities, may play some role (all else being equal, the continuous phase would, in general, be the one which is richer in the component present in greater quantity, and/or the one having the greater viscosity), these effects are usually completely dominated by the nature of the surfactant stabiliser which is added. Any liquid/liquid interface containing a close-packed layer of surfactant molecules will have a 'natural curvature' which is dictated by the structure of the surfactant, in particular the ratio of the effective head-group cross-sectional area to that of the tail-group. This concept is discussed elsewhere in detail in this book,[2] but is illustrated, in a simplistic way, in Figure 3.3 here. The natural curvature of the interface will determine, at the nucleation stage, whether these initial droplets, and hence the emergent emulsion, is α/β or β/α.

(2) Nucleation and growth may also be induced by a chemical reaction, or a series of reactions, usually carried out isothermally by having the reaction mixture well thermostatted. If a reaction of the type

$$A + B \rightarrow C \tag{3.2}$$

takes place, where A and B are miscible in a given solvent, but C is immiscible, then C will, of course, 'precipitate'. Mostly, this type of chemistry is used in the preparation of solid/liquid dispersions. A common example is ion-association in water, *e.g.* $Ag^+ + I^- \rightarrow AgI$. This forms the basis of the preparation of aqueous AgI sols. Because of the different solubilities of these two ions in solution (Ag^+ is more hydrated), then by carefully choosing the relative initial concentrations of Ag^+ and I^-, the particles develop a net charge, which leads to stabilisation of the AgI dispersion. No additional added 'stabiliser' is required in this case; the dispersion is kinetically stable. Latex formation, although involving much more complex polymerisation chemistry, would be another example where charge stabilisation plays a role in stabilising the growing polymer particles.[3] It is

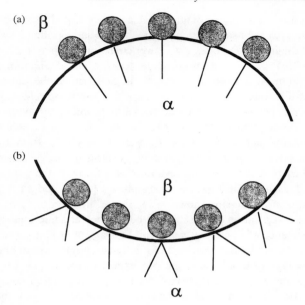

Figure 3.3 *'Natural' interfacial curvature, as determined by molecular 'geometry'*

interesting, in this case, that for much of the polymerisation process the latex 'particles' are soft (monomer swollen) and therefore somewhat liquid-like. If suitable chemistry could be devised such that the *final* 'particles' were truly liquid, then this type of approach could be used for preparing emulsions. In terms of eqn. (3.2) above, C, in the condensed phase, simply has to be a liquid, rather than, as is more commonly the case, a solid. Some examples of this type of approach will be discussed in Section 3.3.

A major advantage of using a nucleation and growth method over comminution methods is that, in general, one should have much tighter control over the average droplet size, and the droplet size distribution, in the former case. This is true whether the nature of the phase separation process is essentially *physical* [(1) above] or *chemical* [(2) above], and whatever the nature of the resultant dispersion: particles, droplets or bubbles. In order to achieve a narrow size distribution, it is necessary for the nucleation steps, and any early aggregation/coalescence, to be over before growth effectively sets in. Such growth may occur by further 'precipitation' from the continuous phase onto/into the existing particles/droplets, or by continued 'reaction' within the droplets, as occurs in latex (emulsion) polymerisation. So-called 'secondary nucleation' steps must be avoided. The particle/droplet *number* concentration is then effectively fixed once the primary nucleation/aggregation steps are over. The final average particle/droplet size is determined by the quantity of reactant material available.

Examples of the two primary routes, *physical* and *chemical*, for preparing emulsions by the· nucleation and growth route are described respectively in the following two sections.

3.2   Physical Methods

The general principle by which emulsions may be formed by a nucleation and growth mechanism, which involves crossing a phase separation boundary, was elucidated in the previous section by reference to Figure 3.2. The boundary lines shown there separate a homogeneous one-phase region from a two-phase region. Moving from point A to point B (Figures 3.2a or 3.2b) involves crossing that boundary line. It is interesting to note, in passing, that so-called 'critical opalescence' is often observed close to the boundary, but still in the one-phase region, if the A → B line crosses the boundary in the vicinity of the actual consolute (critical) point. The opalescence results from transient (unstable) nucleation of the new phase, and associated light scattering, prior to the actual boundary being reached, beyond which binodal phase separation occurs. Conversely, *supersaturation* may occur with the single phase being retained, even though the boundary has been crossed. The point is that if nucleation is *homogeneous*, then nuclei (of the separating phase) of a minimum size need to form before growth is spontaneous; this minimum size decreases the further from the boundary one locates point B. The limit to any supersaturation occurs when point B is the spinodal point: homogeneous nucleation and growth are both spontaneous at and beyond the spinodal point. On the other hand, if 'dust' is present in the system, *heterogeneous* nucleation (on the dust particles) may well occur very close to the boundary itself, *i.e.* at or very close to the corresponding binodal point. The walls of the containing vessel may also induce *heterogeneous* nucleation.

In the case of Figure 3.2a, moving from point A to point B involves a change in temperature. Many illustrations could be given of such systems, where the two coexisting phases are both liquids. For example, the *n*-hexane + nitrobenzene system has an *upper* consolute temperature at 19 °C (phase separation on *cooling*), whereas triethylamine + water has a *lower* consolute temperature at 18.5 °C (phase separation on *heating*), whilst nicotine + water has a 'closed loop' two-phase region with the upper consolute temperature at 215 °C and the lower one at 60 °C.

Figure 3.2b is concerned with three-component systems, at a fixed temperature. In this case, moving from point A to point B involves a change in concentration. More specifically, the actual AB line illustrated involves the removal of component 3, without removing any of 1 or 2. This condition could be achieved if 1 and 2 were essentially *involatile* liquids whilst 3 was a *volatile* one, and could, therefore, be selectively removed by slow evaporation of the mixture, say in a rotary evaporator. It is clearly going to be more difficult to find systems which are suitable for this form of emulsification than using temperature as the variable, as in Figure 3.2a. Surprisingly little work, however, has been carried out pursuing either of these routes. An extensive (but as yet unpublished) study of emulsification using both these routes has been made by Kiraly and Vincent. They worked with mixtures of polyethylene oxide (PEO) and polydimethylsiloxane (PDMS), both of low molecular weight. The molecular weights of the PEO samples used were 222, 500, 1000 and 2000; those of the PDMS samples were 311, 770, 1000 and 2000. All these (commercial) samples are *liquids* at room temperature, but, being polymers, they are essentially involatile.

The temperature–composition phase diagrams of these PEO + PDMS mixtures all show an upper consolute temperature and selected examples are given in Figure 3.4. Figure 3.5, on the other hand, shows selected examples of the *three-component* phase diagrams, with toluene as component 3, at a fixed temperature of 30 °C.

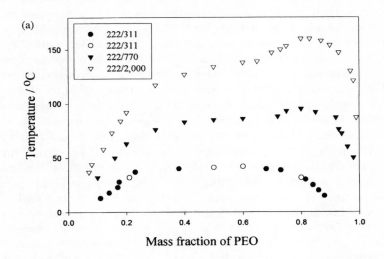

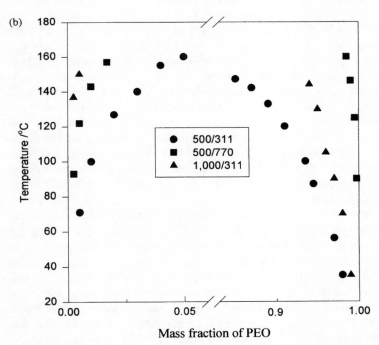

Figure 3.4 *Temperature–composition phase diagrams for various PEO + PDMS mixtures; key refers to molecular weights of PEO and PDMS respectively*

As outlined in the Introduction, in order to form (kinetically) stable emulsions on crossing the phase boundary line by cooling (Figure 3.4), or by toluene evaporation (Figure 3.5), it was necessary to add a surfactant stabiliser. In this case, PEO-*b*-PDMS block copolymers were used as the macrosurfactant. These were prepared in house, by coupling of the PDMS and PEO moieties, which themselves had been prepared by anionic polymerisation, followed by termination with certain functional groups which subsequently couple together. The details are given in Kiraly and Vincent.[4] Two particular molecular weight PEO-*b*-PDMS blocks were used: 2000-*b*-2000 and 2000-*b*-5000. These block copolymers are expected to adsorb at the emergent interface between the two separating liquid phases (PEO-rich + PDMS-rich) on crossing the phase boundary line. Steric stabilisation will then occur of the droplets which form by nucleation, followed by growth and, maybe, some coalescence and/or Ostwald ripening.

The amount of droplet growth will, in principle, be controlled by the total amount of surfactant available. Basically the amount of surfactant determines the total interfacial area that is sustainable (which in turn depends on the close-packed area occupied by each macrosurfactant molecule at the liquid/liquid interface). Control of the nucleation step is therefore essential. Nucleation should be fast if mono-dispersity is an issue. The number of initial nuclei formed then controls the number concentration of droplets. This, together with the total amount of macrosurfactant available, then determines the average size of the droplets formed. Coalescence may occur if the rate of adsorption of the macrosurfactant is insufficient to maintain *saturation* coverage of the emergent/growing interface of all the droplets. Coalescence reduces the number concentration of the droplets and widens the polydispersity of the final emulsion. In addition, Ostwald ripening may play a role in increasing the average droplet size in the emulsion. This is likely if components 1 and 2 have significant solubilities in each other, as in the case of PDMS + PEO.

Emulsions were formed using both the temperature reduction route for the binary system, and the evaporation route for the ternary system containing toluene. Typically, the total macrosurfactant concentration used was up to 5% (by wt), for emulsions which spanned the whole PDMS + PEO concentration range: 1% for those containing 10% (by wt) of one of the polymers; 5% for a 50/50 mixture of the two. At these relatively low concentrations it was found that the presence of the macrosurfactants did not significantly change the phase diagrams shown in Figures 3.4 and 3.5.

The temperature-variation route for forming emulsions with the PEO + PDMS mixtures, plus macrosurfactant PEO-*b*-PDMS, involved first heating a stirred mixture of the polymers and surfactant to a temperature above the corresponding phase boundary line shown in Figure 3.4, to form a homogeneous solution. The system was then cooled back into the two-phase coexistence region. It was found that rapid cooling to a given temperature, accompanied by rapid stirring, led to emulsions with the narrowest size distributions and average droplet sizes <1 μm. However, it was also found that such reasonably narrow size-distribution emulsions were only formed for *dilute* mixtures (up to ~10%) of PDMS in PEO (giving rise to PDMS-in-PEO emulsions), or of PEO in PDMS (giving rise to PEO-in-PDMS emulsions). There was no doubt that these lower volume fraction emulsions formed

(a)

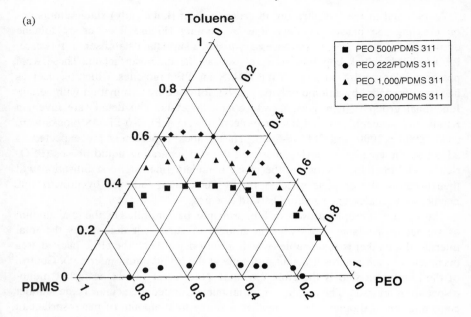

(b)

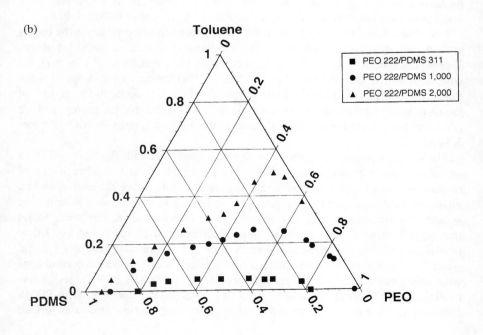

(c)

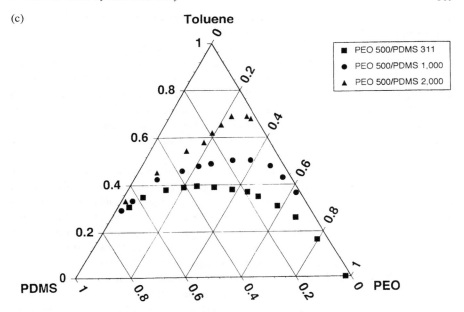

Figure 3.5 *Toluene + PDMS + PEO phase diagrams (30 °C)*

by this temperature variation, nucleation and growth route were much less polydisperse, and had smaller average droplet sizes, than those formed by simple comminution methods (*e.g.* mechanical stirring or the use of valve homogenisers) at the corresponding final temperature.

One suspects that the reason why it was difficult to form reasonable emulsions at higher volume fractions, say in the 50/50 PEO/PDMS region, by this method is that nucleation and initial growth are too fast for the macrosurfactant to adsorb quickly enough (even at the higher concentrations employed) to prevent substantial coalescence of the growing droplets.

The evaporation of a third (volatile) component method proved much more amenable for producing PDMS/PEO emulsions of higher volume fraction. In this method, the two polymers and surfactant were dissolved in toluene at 30 °C until the system was homogeneous. The container was then attached to a vacuum line and the toluene rapidly evaporated, whilst maintaining rapid stirring with a magnetic follower. Again, rapid cooling and rapid stirring were both conducive to forming reasonably monodisperse emulsions, typically in this case in the 1 to 2 μm diameter range. An interesting question is: which type of emulsion was formed PEO-in-PDMS or *vice versa*? This could be simply determined by attempting to dilute the emulsion, as formed, with either PDMS or PEO. Dilution occurred readily when the diluent corresponded to the continuous phase. If the diluent corresponded to the dispersed phase, then, if sufficient were added, *phase inversion* was observed. In the region of the phase inversion the systems became highly viscous, as is usually observed with traditional oil/water emulsions. It may well be that lamellar or other associated structures form in the region of the phase inversion concentration, giving rise to the high viscosity.

One aspect that needs further clarification in the work by Kiraly and Vincent reported above is exactly how the cooling/evaporation rate, and the stirring rate, affect the droplet size and distribution. Clearly on crossing the binodal point, somewhere along the A → B line, then as point B (Figure 3.2a and 3.2b) is approached, different tie lines are crossed continuously. This means that the compositions of the two separating phases are also seeking to change continuously. Hence, the actual nucleation events are complex! This aspect is poorly understood at present.

It is surprising that very little other work has been reported in the literature on emulsification by crossing phase boundaries. However, Shinoda and Friberg[5] have discussed the formation of reasonably monodisperse macroemulsions by temperature 'jumps' in one-phase microemulsion systems, such that the solubilisation phase boundary (SPB) is crossed. A recent paper on this topic has appeared by Morris *et al.*[6] who investigated temperature-*reduction* induced transitions from the one-phase oil/water microemulsion region (L_1 phase) into the two-phase (L_1 + oil) region, in the system decane + water + $C_{12}E_5$ nonionic surfactant. They studied mixtures in which the surfactant: decane ratio was kept constant ($\sim$1:1 wt ratio), and the (total) concentration of (decane + surfactant) was increased in water from 0 to 40% (by wt). The SPB was found to be independent of this concentration and to occur at 25.0 °C. With decreasing temperature above 25.0 °C the microemulsion droplets become smaller within the single L_1 phase. At 25.0 °C the natural curvature of the microemulsion droplets is reached (cf. Figure 3.3), so that on cooling below 25.0 °C (the SPB) one might expect oil to be rejected from the microemulsion droplets and to form a separate phase. This does not happen, however, until the temperature is 21.9 °C. Only below this temperature do oil droplets appear and the corresponding macroemulsion is formed, as evidenced by an increase in the turbidity of the system with time. It would seem, therefore, that a *metastable* region exists between 25.0 °C and 21.9 °C. In this region the microemulsion droplets just swell and supersaturate, beyond the limit dictated by their natural curvature, which increases (*i.e.* their natural radius becomes *smaller*) as the temperature is lowered. Morris *et al.*[6] also consider the question of how the macroscopic emulsion droplets actually form below 21.9 °C. In a homogeneous nucleation they suggest that some of the existing microemulsion droplets would have to grow to macroscopic sizes. However, there is an apparent dilemma in that the driving force for the phase separation is for the droplets to become smaller, as discussed above, to optimise their curvature, and not to grow! They suggest that the droplet growth must therefore occur by Ostwald ripening in the system.

Clearly this whole area of emulsification by *physically* crossing phase boundary lines is one which warrants much more research effort.

3.3   Chemical Methods

There are also surprisingly few examples where chemical reactions have been used to form emulsions by a nucleation and growth process. All that is required, in principle, is that, in terms of eqn. (3.2), the product C is a liquid in the condensed state, which is also largely immiscible in the liquid medium. The authors were not able to locate any example of such chemistry in the literature, where A, B and C are 'simple' molecules, although it ought not to be too difficult to devise some examples! Emulsion or dispersion polymerisation reactions come close, but most such polymerisation reactions are usually allowed to continue to near 100% conversion, such that very little monomer is left, and because most polymers are solid, the resultant particles are solid. Clearly, if a liquid polymer [*e.g.* low molecular mass poly(ethylene oxide) (PEO) or polydimethylsiloxane (PDMS)] were to be used then, for example, a PEO/hydrocarbon or a PDMS/water emulsion could be produced.

Obey and Vincent reported the production of PDMS/water emulsions by a step-growth (condensation), dispersion polymerisation reaction.[7] The basic chemistry is very similar to that used for preparing monodisperse silica particles, introduced by Stöber *et al.*,[8] which involves the hydrolysis of, for example, tetraethoxysilane (TEOS) in ethanolic solution, using ammonia and water. Minehan and Messing[9] have described a 'spontaneous emulsification' process for producing silica particles by mixing an ethanolic solution of partially hydrolysed TEOS into water. The size of the resultant silica particles (0.1 to $\sim$50 μm) depends on the degree of pre-hydrolysis of the TEOS.

In the case of PDMS droplets, TEOS is replaced by dimethyldiethoxysilane (DMDEOS), and the base-catalysed hydrolysis is carried out in water, since PDMS is soluble in ethanol. The basic chemistry is set out below.

$$\text{Me}_2\text{Si(OEt)}_2 \xrightarrow[\text{NH}_3]{\text{H}_2\text{O}} \text{Me}_2\text{Si(OH)}_2 \xrightarrow{-\text{H}_2\text{O}} -(\text{Me}_2\text{SiO})-_x$$

The reaction is a step-growth, self-condensation polymerisation. Beyond a certain chain length the oligomeric PDMS precipitates to form droplets. These droplets continue to grow over several weeks. After one day they are typically about 1 μm in size, growing to about 3 to 5 μm after 4 weeks, depending on the initial monomer (DMDEOS) concentration, but with only a slight dependence on the concentration of ammonia used. A typical optical micrograph of a PDMS-in-water emulsion is shown in Figure 3.6.

The narrow particle size distribution, evident from Figure 3.6, is maintained with time as the droplets grow. This suggests that this growth occurs by further polymerisation in solution, followed by deposition on existing droplets, without re-nucleation. Also, little or no coalescence or Ostwald ripening occurs in these emulsions.

Obey and Vincent[7] carried out extensive ^{1}H and ^{29}Si NMR studies on the PDMS ('silicone' oil) phase that formed. They showed that the product was predominantly the *cyclic* 'D$_4$' species (where D is a $-\text{Me}_2\text{SiO}-$ repeat unit), although some short, linear homologues of the form $^-\text{O}-(\text{Me}_2\text{SiO})_x-\text{Me}_2\text{SiO}-$ are also formed ($x < \sim 6$). Because of their anionic nature, these linear species would appear to accumulate at the silicone oil/water interface, and hence give rise to charge stabilisation of the droplets. The electrophoretic mobility of the (dialysed) droplets,

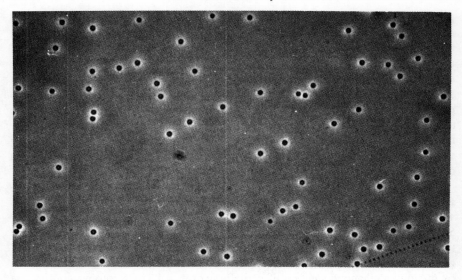

Figure 3.6 *A typical PDMS-in-water emulsion, prepared by the hydrolysis of dimethyl-diethoxysilane. Magnification:* ×1000
(Reproduced with permission of Academic Press from ref. 7)

in 10^{-3} M KCl aqueous solution, increased with increasing ammonia concentration used, up to values $\sim 2 \times 10^{-8}$ m^2 s^{-1} V^{-1}. It was found that the ratio of linear:cyclic species could be increased by adding ethanol to the reaction medium (up to the solubility boundary for PDMS); Table 3.1 shows these data.

Table 3.1 *Mol% of linear PDMS, as a function of wt% ethanol in the medium*

Wt% monomer	Wt% ethanol/water	Mol% linear PDMS
5	0	11
5	40	40
35	55	86

It would seem that, for a given initial monomer concentration, the kinetics of polymerisation must increase the higher the concentration of ethanol present, such that *inter*molecular propagation (leading to linear chains) becomes favoured over *intra*molecular propagation ('tail biting', which leads to closed rings).

The fact that the emulsions are free of added surfactant or polymeric stabilisers is of some significance in that the droplet/solution interface is in this case a truly fluid one, whereas those bearing adsorbed stabilisers are invariably viscoelastic, exhibiting Gibbs–Marangoni effects. This should be of interest to those wishing to carry out experiments to test the various hydrodynamic theories of liquid droplets (*e.g.* diffusion, sedimentation, viscosity, electrophoresis), compared to solid particles. The internal liquid flow pattern, induced in dispersed droplets when they are translating (or rotating) relative to a liquid continuum, is disrupted by the presence of viscoelastic adsorbed films at the oil/water interface.

Anderson et al.[10] subsequently studied the effect of deliberately adding a surfactant to the reaction mixture in which the PDMS droplets are formed. Addition of sodium dodecyl sulfate (up to 1% wt/vol) had some effect on reducing the droplet size produced, but the effect of incorporating a nonionic macrosurfactant, such as Synperonic F108™ (an ethylene oxide/propylene oxide/ethylene oxide triblock copolymer), or Tegopren TG 5851 or 5863™ (a PDMS backbone, with ethylene oxide side-chain, brush copolymer), was more dramatic. Indeed, above the critical micelle concentrations of these macrosurfactants, PDMS-in-water micro-emulsions were formed. With Tegopren TG 5863™ these increased in diameter over the size range 20 to 80 nm, as the percentage of ammonia in the reaction mixture was increased from 5 to 60%. It seems that the micelles solubilise the precipitating PDMS.

Goller et al.,[11] from the same group, have studied the effect of using a mixture of methylalkoxysilane monomers, rather than DMDEOS alone. Addition of the monofunctional monomer trimethylethoxysilane would act as a chain-stopper, with reduction in molecular weight, but addition of the trifunctional monomer methyltriethoxysilane (MTEOS) should lead to cross-linking within the droplets. This is what Goller et al. found. The viscosity of the silicone oil phase increased sharply on increasing the ratio MTEOS:DMDEOS until, above a ratio of about 1:1 of the two monomers, solid-like particles, resembling silica particles, were formed. At small values of this ratio the droplets were essentially viscoelastic in nature, and may be regarded as inorganic microgel particles. The authors studied the solubilisation into, and the associated swelling in diameter of, these microgel particles in the presence of various organic solvents, as a function of the MTEOS content (i.e. the degree of internal cross-linking within the droplets). For example, with the non-crosslinked droplets (0% MTEOS), when n-heptane was placed in contact with the aqueous phase the droplets just grew in size continuously with time, until eventually they became unstable and coagulation, followed by rapid coalescence, occurred (presumably because the area per interfacial charge group increased as the diameter of the droplets increased, reducing electrostatic repulsion between the droplets). However, cross-linked droplets containing MTEOS swelled in diameter to a limiting value dependent on the amount of MTEOS present. The results are shown in Figure 3.7.

Dispersions of such microgel particles may have potential use as controlled release species, e.g. for active agrochemical molecules which are only sparingly water-soluble, and which have correspondingly high partition ratios inside the oil droplets. The absence of added surfactants in such formulations has to be a 'plus' feature. Goller and Vincent[12] have also investigated the potential use of PDMS droplets, of the type described here, as pressure release species (e.g. for use as flavour or perfume carriers). In this case the PDMS droplets (or cross-linked microgel particles) are encapsulated within a shell of silica, of controlled thickness.

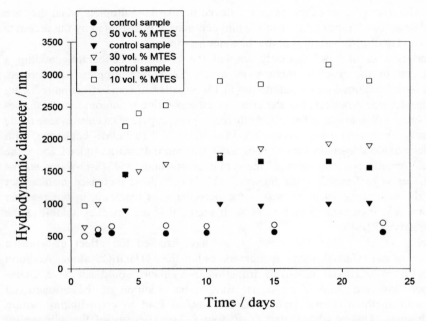

Figure 3.7 *The time-dependent swelling behaviour of cross-linked PDMS microgel droplets in the presence of* n-*heptane (open symbols). The samples contained varying vol% of methyltriethoxysilane (the 'cross-linker'), as indicated. The filled symbols represent the control samples,* i.e. *continuing to grow without n-heptane present*

3.4 References

1. B.P. Binks, in Chapter 1 of this book.
2. A.S. Kabalnov, in Chapter 7 of this book.
3. See, *e.g.* R.G. Gilbert, 'Emulsion Polymerisation', Academic Press, London, 1995.
4. Z. Kiraly and B. Vincent, *Polym. Int.*, 1992, **28**, 139.
5. K. Shinoda and S.E. Friberg, 'Emulsions and Solubilisation', Wiley-Interscience, New York, 1986.
6. J. Morris, U. Olsson and H. Wennerstrom, *Langmuir*, 1997, **13**, 606.
7. T.M. Obey and B. Vincent, *J. Colloid Interface Sci.*, 1994, **163**, 454.
8. W. Stöber, A. Fink and E. Bohn, *J. Colloid Interface Sci.*, 1968, **26**, 62.
9. W.T. Minehan and G.L. Messing, *Colloids Surf.*, 1992, **63**, 181.
10. K. Anderson, T.M. Obey and B. Vincent, *Langmuir*, 1994, **10**, 2493.
11. M.I. Goller, T.M. Obey, D.O.H. Teare, B. Vincent and M. Wegener, *Colloids Surf. A*, 1997, **123**, 183.
12. M.I. Goller and B. Vincent, *Colloids Surf. A*, in press.

CHAPTER 4

Emulsion Flocculation and Creaming

MARGARET M. ROBINS AND DAVID J. HIBBERD

Institute of Food Research, Norwich Research Park, Colney, Norwich
NR4 7UA, UK

4.1 Introduction

4.1.1 Instability Processes

An emulsion may be defined as a heterogeneous system, consisting of at least two immiscible liquids or phases, one of which is dispersed in the form of droplets in the other. Emulsions are generally unstable with respect to their component bulk phases. Rearrangement from the droplet form to the two bulk liquids will occur with a net reduction in interfacial area and this is energetically favourable. However, it is relatively simple to erect kinetic barriers to this process to achieve metastable states that are for all practical purposes completely stable.

Figure 4.1 is a schematic diagram illustrating the major instability processes that emulsions undergo. Figure 4.1(a) represents the metastable state before the onset of instability. The droplets can be polydisperse and in real systems are animated by thermal events in the continuous phase (Brownian motion).

Figure 4.1(b) depicts flocculation, an aggregation process. Flocculation occurs when there is a net attractive force between droplets which is large enough to overcome thermal agitation and cause persistent aggregation. This is distinct from being close together because of a chance encounter due to Brownian motion. Whilst the droplets are held together they are not in intimate contact and are usually separated by a layer of polymer or surfactant. Figure 4.1(b) is divided into two sections, to show that the particles may be condensed into discrete aggregates called flocs, or be connected into a single expanded structure (or network) that fills space. The morphology depends on a number of factors, including droplet concentration, droplet size and the strength of flocculation. In general, flocculation may be reversed by the input of much less energy than was required to disperse the droplets in the initial emulsion.

When the surfactant or polymer layer separating the droplets fails, intimate

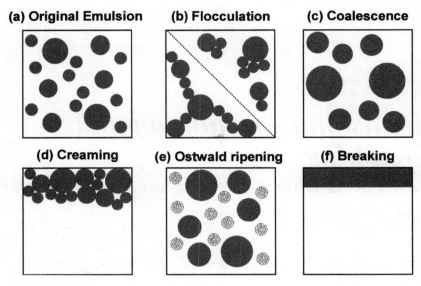

Figure 4.1 *Emulsion instability processes*

contact is achieved between the droplets and their contents flow together to form a new larger droplet. This process is called coalescence, shown in Figure 4.1(c). The rate of coalescence is dependent on droplet encounter rate and the properties of the droplet surface.

There are two instabilities that may occur without the need for droplet encounter. Figure 4.1(d) shows creaming, which occurs when there is a difference between the density of the dispersed and continuous phases. Under the influence of gravity, separation occurs with the most dense phase collecting at the bottom and the less dense phase at the top.

Figure 4.1(e) depicts a process called Ostwald ripening where the dispersed phase is transported through the continuous phase from smaller droplets to larger ones.[1] Since larger droplets have a lower surface to volume ratio than their smaller counterparts, this process occurs with a net reduction in interfacial energy. However, the dispersed phase must be significantly soluble (or solubilised by surfactant) in the continuous phase.

The eventual equilibrium state for emulsions, shown in Figure 4.1(f), is completely separated into the two phases. This process, called breaking, may take from a few minutes to several years to occur. The transition from the initial state (a) to the final state (f) must proceed *via* one of the mechanisms shown in (c) or (e). However, in practice, a complex set of inter-relationships exists between these instability mechanisms and often all will occur on the way to the final state. Frequently the onset of one of the instability mechanisms will modify the rate at which the others occur.

For example, when creaming occurs, this extra motion (in addition to Brownian motion) may increase the rate at which droplets encounter one another. This is likely to affect the rate of flocculation or coalescence as both are initiated by droplet encounter. Whilst Ostwald ripening does not require droplet encounter, the rate at which this process occurs will be affected by the distance between the droplets since dispersed material must be transported between them. Hence this process will be accelerated when droplets are either flocculated or packed into a cream layer.

4.1.2 Strategies for Stability

The complex interdependencies between colloid instability mechanisms make predicting the timescale and outcome of colloidal instability extremely difficult. When seeking to stabilise a colloidal system, it is important to eliminate as many of the instability mechanisms as possible.

To reduce *creaming* the simplest approach is to reduce the density difference between the dispersed and continuous phases. This may be achieved either by selection of density-matched bulk phases or by appropriate additions of weighting agents to either phase. The next best strategy is to reduce the droplet size, as in the homogenisation of milk. For isolated droplets in an infinite medium, Stokes' law states that the rate of creaming increases with the square of the diameter. Furthermore, as the droplet size is reduced, the rate of self-diffusion increases to a point where very small droplets may be kept from creaming by diffusional mixing, despite there being a large density difference between the dispersed and continuous phases.

When neither of the above strategies is practical, a popular approach is to increase the viscosity of the continuous phase. This may be achieved by the addition of high molecular weight polymers. These act to increase the viscosity of the continuous phase to such an extent that creaming rates become negligible. Another advantage to this approach is that the additional viscosity of the product is often perceived by customers as increased 'quality'. However, these polymers also act to reduce the stability of colloidal systems to flocculation, which can in turn lead to more rapid creaming.

The final and most difficult approach is to engineer the emulsion to be rigid enough (or solid-like) to resist the forces of creaming. In some emulsions an acceptable solution to the problem of creaming stability is achieved by the addition of gelling agents which set in the continuous phase and immobilise the dispersed phase. However, more commonly it is a requirement that the system be rigid towards creaming but display liquid-like behaviour during processing or final application. In rheological terms this equates to having either a large viscosity or yield stress at low shear rate, and a low viscosity at high shear rate. High molecular weight polymers are used as they may impart these properties. An alternative to modifying the viscosity of the continuous phase is to flocculate the dispersed phase to form a load-bearing network. These networks may be stable for extended periods but are subject to catastrophic collapse at the end of their period of stability, as shown later.[2]

An additional useful property of polymeric stabilisers is that, even when they fail to render the products stable, it is frequently found that the concentrated layers that arise from creaming of flocculated emulsions may be easier to re-disperse by gentle agitation than those that form when the emulsions cream as discrete droplets.

Coalescence and *flocculation* are strongly related to one another. Reducing the rate of drop coalescence is one of the main purposes of surfactants in emulsion formulations. The reduction in the interfacial tension on addition of surfactant itself increases the stability of the emulsion with respect to its component phases. However, the role of surfactants is much greater than this and they act to reduce droplet coalescence in several additional ways.

Ionic surfactants, as their name suggests, have an ionisable hydrophilic group. When appropriately charged surfactants are adsorbed in large numbers to the interface of emulsion droplets they increase the surface charge on them. Since the charge is the same sign on all the droplets, electrostatic forces are developed that oppose (and thus reduce the rate of) droplet encounter.

Nonionic surfactants cannot increase the surface charge and may actually reduce it. They may also screen the droplets' charge from one another. These surfactants stabilise emulsions by a *steric mechanism*. Nonionic surfactants always have large hydrophilic groups which extend away from the droplet surface a significant distance and impose a barrier to encounter.

Both ionic and nonionic surfactants may additionally increase stability to droplet coalescence by promoting Gibbs and Marangoni 'self-healing' effects which stabilise the 'film-like' layer that arises when two droplets collide.

Other than by addition of surfactants, the most obvious route to increasing stability towards coalescence is to reduce the rate and duration of droplet encounters. Since flocculation may be viewed as a prolonged droplet encounter, the strong relationship between stability to flocculation and stability to coalescence is clear. In order to reduce the rate and duration of droplet encounters the drops need to be rendered mutually repulsive, or only weakly attractive (*i.e.* experiencing forces of attraction that are small when compared to the average thermal perturbations).

4.1.3   Interaction Potential Between Droplets

The stability of an emulsion system towards flocculation and coalescence may be better understood by considering the forces between emulsion droplets. These forces arise from a range of phenomena and vary from system to system. The most ubiquitous of these forces is the van der Waals force of attraction, which arises from momentary fluctuations in the charge distribution across molecules, giving them a 'flickering dipolar' nature. The induction of complementary dipoles in adjacent molecules leads to a weak attractive force between them. A similar attraction occurs between colloidal particles, and the resulting potential decays with the inverse square of the separation between the droplets, as shown schematically in Figure 4.2.

In aqueous systems the most important repulsive potential arises from electrostatic repulsions. Droplets are usually charged to a greater or lesser extent, and their charged surfaces attract counter-ions from the solvent, forming a diffuse electrical double layer around each drop. The double layers around adjacent drops interact to

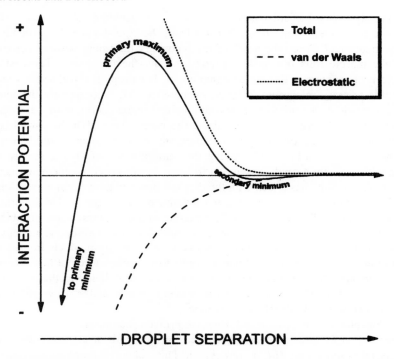

Figure 4.2 *Schematic interaction potential between two drops showing van der Waals and electrostatic potentials*

give a repulsive potential between them, varying with drop separation as shown in Figure 4.2.

A total interaction potential may be obtained by summing the van der Waals and electrostatic potentials. At low separations this total potential is dominated by the van der Waals attraction forming a deep attractive potential well called the *primary minimum*. Moving to larger separations, the curve then rises to the *primary maximum*, the highest point on the total potential curve, at separations where the electrostatic potential dominates. Continuing to still larger separations the curve once again descends, crossing the axis to form a second attractive potential well called the *secondary minimum*. This occurs because whilst van der Waals attractions are weak at high separations, they propagate further out than the electrostatic potential. Finally the curve once again rises to meet the *x*-axis as both potentials diminish towards zero.

Armed with the total interaction potential, the stability of simple emulsions may be understood. The droplets exhibit constant Brownian motion due to random thermal fluctuations. This leads to random encounters between droplets with the average magnitude of the force driving the encounter being of the order of the thermal energy of the system, kT, where k is the Boltzmann constant and T the absolute temperature. The outcome of the encounter then depends upon the form of the total interaction potential.

To visualise this, the encounter event may be considered as a marble rolling along the total interaction curve, starting from the right-hand side, with kinetic energy equal to kT. When the depth of the secondary minimum is small compared to kT the marble rolls into this minimum and out the other side. The first real obstacle it encounters is the primary maximum. For droplets stabilised to coalescence this primary maximum is large compared to kT, consequently the marble rolls up the side of the primary maximum, losing kinetic energy until it has stored it all in potential energy in climbing above the x-axis. Thereafter the particle rolls back the way it came, converting potential energy to kinetic energy, and leaves the encounter at the right-hand side having the same amount of kinetic energy as it started with, save for a small loss due to friction. At this point the encounter is over and a *stable* outcome has arisen with the droplets being in their original positions.

With this analogy in mind, other scenarios may be investigated. Had the primary maximum been small compared to kT the marble would have traversed this obstacle and then accelerated down into the primary minimum. To continue the analogy the primary minimum must be considered as a very deep well from which droplets may not roll out. At this point the encounter event is over once more but this time with an *unstable* outcome. The particles are in intimate contact, held together by van der Waals forces that are so large that the process is essentially irreversible. This outcome corresponds to droplet coalescence.

The flocculated state is achieved when the primary maximum is once more large with respect to kT, preventing coalescence, but the depth of the secondary minimum is now significant (*i.e.* a few kT's). This time the marble rolls into the secondary minimum but cannot surmount the primary maximum. Instead it rolls back into the secondary minimum but having lost a little kinetic energy, to friction, does not have sufficient energy to escape the right-hand side. Hereafter it rolls backwards and forwards in the secondary minimum, continually losing kinetic energy to friction and though it may receive extra pushes, kT packets, from further thermal events these are insufficient to give it enough kinetic energy to escape the secondary minimum. This outcome is clearly distinct from coalescence since the droplets are not in intimate contact. They are held at a close separation by forces sufficient to withstand average thermal perturbations, but should a really large thermal event occur it is possible that they may become separated (*i.e.* the process of flocculation is reversible). To add to this point, since the secondary minimum is a small feature created by the superposition of two much larger potentials, it is clear that a small change in conditions (*e.g.* a decrease in electrolyte concentration) could diminish the magnitude of the secondary minimum, releasing any droplets trapped therein.

4.1.4   Flocculation Due to Polymers

Emulsions may be flocculated by the addition of polymers. Excluding those cases where the addition of a polymer affects the van der Waals or electrostatic forces directly (*e.g.* the addition of polyelectrolytes), the process of polymer-induced flocculation may proceed by two mechanisms, bridging or depletion. These are depicted schematically in Figure 4.3.

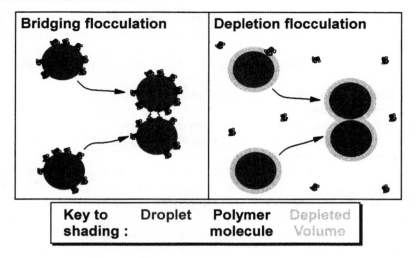

Figure 4.3 *Flocculation mechanisms for drops in the presence of polymer*

Bridging flocculation is the easier to visualise. This occurs when the surfaces of the droplets are attractive to the added polymer and polymer adsorption occurs at the droplet surface. When insufficient polymer is added to coat completely the droplets' surfaces, encounters may occur between droplets orientated such that an area with adsorbed polymer on one drop meets an area free from polymer on a second drop. Under these circumstances the polymer molecule(s) already adsorbed to one drop may further adsorb to the second drop, creating a polymer *bridge*. The resilience of this bridge then maintains the droplets at a close droplet–droplet separation (*i.e.* flocculated).

For bridging flocculation to occur, several conditions must be met. The most obvious is that the droplet surfaces must be subject to polymer adsorption. In emulsions, the presence of surfactants at the interface, which are necessary for coalescence stability, means that only polymers with sufficient surface affinity to displace the surfactant from the surface, or a specific affinity for the surfactant itself, may adsorb. A second condition places a lower limit on the size of the polymers in a given system. Clearly the partially coated droplets must encounter one another, coming close enough for a polymer molecule to span the gap between the droplets. This means that the polymer molecules must be able to extend past the primary maximum in the total interaction potential. The absolute limit on polymer size will be defined by the position of the primary maximum and the radius of gyration of the polymer molecules. However, in practice, when polymer molecules become adsorbed to drop surfaces the conformation of the polymer molecule frequently alters to allow multiple attachments to the drop surface, which may reduce the distance the polymer extends into solution.

A final condition constrains the amount of polymer that may be added to the system and leads to an estimate of the optimum amount of polymer to maximise flocculation. If sufficient polymer is added to the system to coat the surfaces of the drops entirely, then further adsorption cannot occur and the system is re-stabilised towards bridging flocculation. As a first approximation, the amount of flocculation will be greatest when the surfaces of the droplets are only 50% covered with polymer prior to flocculation and will reduce to zero as the surface coverage tends to either 100% or 0%.

The process of adsorption of polymers from solution may take several hours to reach equilibrium conditions, during which time the tendency to flocculate may change radically. Consequently many systems undergoing bridging flocculation are often in a non-equilibrium state with respect to polymer adsorption, which makes the process difficult to model.

In contrast to bridging flocculation, depletion flocculation occurs when the added polymer cannot adsorb to the surfaces of the droplets. Under these conditions a layer around the droplets exists where the polymer concentration is *depleted* as compared to the bulk solution (see the *depleted volume* in Figure 4.3). The origin of the depleted volume lies with a geometrical restraint imposed by the finite volume of the polymer molecule.[3] This may be understood by modelling the polymer molecule as a sphere of radius r_g, whose position is defined by its centre. Since the polymer may not adsorb to the droplet surface, the closest it may approach the droplet is to touch the surface. In this arrangement the centre of the polymer will be a distance r_g from the droplet surface. It follows that polymer centres cannot get closer than this to the droplet surface and consequently do not contribute to the segment density there.

If we now consider the whole emulsion (*i.e.* many droplets), the overall volume of solvent within the system that is depleted in polymer may be reduced if the droplets get close enough to one another (*i.e.* flocculate) to overlap their depleted regions. By overlapping depleted regions, solvent is released to the bulk where it dilutes the polymer residing there. As this is an energetically favourable process, drops are driven in the direction of increasing flocculation.[4]

4.2   Emulsion Creaming

4.2.1   Introduction and Theoretical Principles

The prevention of creaming in emulsions is an important aspect of commercial product formulation. In this section we discuss the theoretical aspects of droplets moving under gravity, and introduce a modification to the simple theory to allow for the effect of back-flow in concentrated emulsions creaming in a closed container.

The creaming velocity, U_s, of a spherical droplet in an infinite fluid medium is given by Stokes' law:

$$U_s = \frac{\Delta \rho d^2 g}{18 \eta} \tag{4.1}$$

where $\Delta \rho$ is the density difference between the continuous and dispersed phases, d is the droplet diameter, g the acceleration due to gravity and η the viscosity of the fluid. It follows that when individual droplets within a polydisperse emulsion are subjected to gravitational forces, the droplets fractionate according to their size, and hence from an analysis of the creaming profiles for the emulsion, information about the sizes may be determined.

This process is illustrated in Figure 4.4(a), which shows volume-fraction profiles for a monodisperse emulsion undergoing creaming. Since all the droplets are the same size, fractionation does not occur. A sharp boundary develops at the base of the sample and travels to the top with the progress of time, *t*. Clearly the creaming velocity of the drops is given by the rate at which the sharp boundary rises; applying eqn. (4.1), the hydrodynamic size of the droplets may be inferred from this rate.

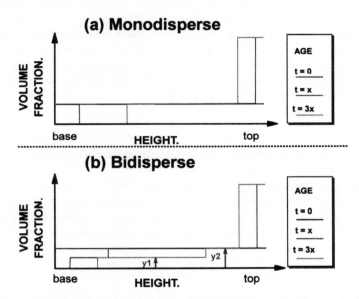

Figure 4.4 *Variation of volume fraction with height during creaming of* (a) *ideal mono-disperse emulsion and* (b) *ideal bidisperse emulsion*

In the bidisperse case, Figure 4.4(b), fractionation does occur. The large droplets cream faster than the small ones and two sharp boundaries form at the base and rise to the top at two discrete rates. The two creaming rates allow two hydrodynamic sizes to be inferred from eqn. (4.1). The rates at which the boundary rises at two volume fractions (ordinates y_1 and y_2) are sufficient to define completely the cumulative size distribution of a bidisperse dispersion. Polydisperse dispersions are treated as an extension of the bidisperse case, the number of ordinates examined being increased as required until the size distribution is sufficiently well defined. However, this simplistic analysis is only applicable to very dilute emulsions, where Stokes' law is valid (*i.e.* at infinite dilution in an infinite medium). In closed concentrated emulsions, droplets will interfere with one another and the effect of back-flow by the continuous phase becomes significant.

The effects of droplet concentration on the creaming rate are well documented, but not yet well understood. In concentrated systems the streamlines around the droplets interfere with each other and droplet–droplet encounters occur. Most real concentrated systems are also bounded. The presence of boundaries introduces edge effects and leads to back-flow of the continuous phase to fill space vacated by forward-flowing droplets.

For small droplets the effect of droplet diffusion becomes significant. Gradient diffusion, which occurs across droplet concentration gradients induced by creaming, causes a flux of particles which is opposed to the direction of creaming. The effects of droplet self-diffusion are less simple to predict but upset the simple creaming behaviour. In concentrated systems, droplets may undergo co-operative motion, and *slip-stream* in each others' wake. In addition to all these effects, if the temperature of the system is not constant then thermal convection currents will act to complicate the creaming process further. Batchelor[5] has obtained a rigorous analysis for volume fractions up to 5%, but the treatment of higher volume fractions is the subject of considerable debate. Empirically, Richardson and Zaki[6] obtained a power-law relationship between the sedimentation rate, U, of monodisperse drops and their volume fraction, ϕ

$$U = U_s (1 - \phi)^{4.65} \tag{4.2}$$

but clearly this relationship will fail as the volume fraction, ϕ, approaches a close-packing limit (well below $\phi = 1$).

Carter *et al.*[7] have attempted to allow for the effects of concentrated dispersions by using an effective viscosity in place of the medium viscosity, η, in eqn. (4.1). They chose the theoretical mean-field viscosity, η_{BR}, of a concentrated dispersion of non-interacting hard spheres given by Ball and Richmond[8]

$$\eta_{BR} = \eta_0 (1 - \phi/\phi_{max})^{-\frac{5}{2}\phi_{max}} \tag{4.3}$$

where η_0 is the low shear-rate continuous phase viscosity, ϕ is the volume fraction of the drops and ϕ_{max} is the close-packed volume fraction. Additionally, an effective density can be used along with an allowance for the effect of back-flow in a closed container.[9] The validity of these additional modifications is demonstrated by comparing the extended theory with the empirical relationship of Richardson and Zaki, as in Figure 4.5. The agreement indicates that for this emulsion, deviation from Stokes' creaming behaviour can be explained using an effective viscosity, effective density and simple back-flow calculation alone. In polydisperse systems the corrections may all be calculated using the concentration of the droplet fraction being contoured.

4.2.2 Ultrasonic Monitoring of Creaming

The practical measurement of creaming is hindered by the opacity of most emulsions. If there is any variation of the speed of the droplets during creaming, due to polydispersity or density variation, the slower-moving fraction obscures the movement of the faster droplets. Analysis of creaming rates thus needs a knowledge

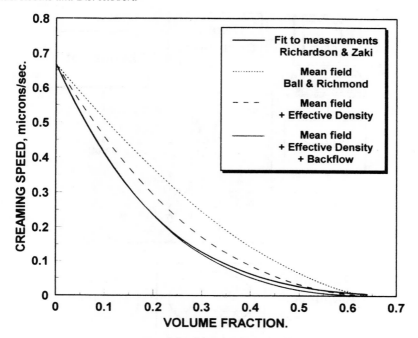

Figure 4.5 *Predicted and experimental creaming velocity as a function of volume fraction for an emulsion containing droplets with diameter 1 μm, density difference of 693 kg m⁻³, and continuous phase viscosity 1 mPa s*

of the droplet concentration with height and time. Direct sampling is possible, but time-consuming and may disturb the creaming process. Of the non-intrusive techniques proposed,[10,11] we favour the use of an ultrasonic method that we have developed at the Institute of Food Research (IFR). The advantages of ultrasonic techniques are that, in general, high frequency sound propagates through highly concentrated emulsions, the instruments are comparatively cheap and robust, and have none of the safety hazards associated with techniques using radiation such as gamma rays.

The velocity of ultrasound through a dispersion is sensitive to changes in composition. This is the principle of the IFR creaming monitor,[12] shown in Figure 4.6, which measures the ultrasonic velocity through an emulsion (or suspension) as a function of height. The time-of-flight of a pulse of ultrasound is measured across a rectangular sample cell. The time-of-flight data are converted to ultrasonic velocity by reference to measurements made in two calibration fluids, to give the sample path length and constant delay factors at each height. In complex multi-component systems the form of the velocity profile thus gained gives an indication of sample uniformity. When measuring simple emulsions the velocity data may be used to calculate the volume fraction of dispersed phase using simple mixing theory.

The emulsion, held in a rectangular perspex cell, is installed in a thermostated water bath. A carousel holds up to five such cells and rotates, under the control of a

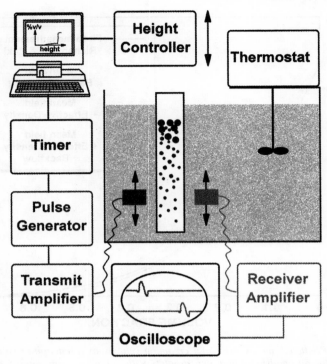

Figure 4.6 *Schematic representation of the IFR ultrasonic creaming monitor*

microcomputer, to offer each cell up to the measurement system as required. The water bath serves two functions; firstly it enables accurate temperature control ($\pm 0.02\ ^\circ\text{C}$) which is necessary since the speed of ultrasound is dependent upon temperature, and secondly the water it contains acts as an ultrasonic couplant between the ultrasonic transducers and the sample cell. Two ultrasonic transducers, a transmitter and receiver pair, are diametrically mounted in a rigid carriage so that they may be passed up and down the sample cell (see Figure 4.6). The carriage is moved by a stepper motor and may be located at any position up the height of a cell to a spatial resolution of less than 0.01 mm (but typically 1 mm). The transducers are 5 mm in radius, critically damped, and operate at a nominal frequency of 5.4 MHz. This corresponds to a typical wavelength of 235 µm in the samples studied. The width of the ultrasonic beam is estimated at 2 mm.

The measurement of time-of-flight is made using a pulse-echo overlay technique. The microcomputer controls the movement of the carousel and transducer carriage, keeping track of its height by counting stepper-motor steps and monitoring limit switches. The computer is interfaced to a counter-timer and logs the time-of-flight, sample height and absolute time under software control. By taking the mean of 500 000 time-of-flight measurements, a precision of <1 ns in 25 µs is achieved.

Measurements of time-of-flight are made at many heights in a sample to define a profile of ultrasonic velocity with respect to height in the emulsion. A scan defines the profile of a sample at any one time. To characterise creaming behaviour, a series of profiles are required at appropriate time intervals.

The speed at which sound propagates through a dispersion is a complex function of the particle volume-fraction, size, shape, the excitation frequency and the physical properties of the particulate and continuous phases. However, when the particles are much smaller than the wavelength of ultrasound and there is a significant difference between the speed of sound in the bulk dispersed and continuous phases, the effect of volume fraction greatly outweighs all the other effects so that it may be calculated by assuming the system behaves like a simple mixture using the equation:[13]

$$V = \sqrt{\frac{V_c^2}{\left(1 - \phi\left(1 - \frac{\rho_d}{\rho_c}\right)\right)\left(1 - \phi\left(1 - \frac{\rho_c V_c^2}{\rho_d V_d^2}\right)\right)}} \qquad (4.4)$$

where V is the speed of ultrasound through the dispersion, ρ_d, ρ_c, V_d and V_c are the densities and speeds of ultrasound through the dispersed and continuous phases, respectively, and ϕ is the volume fraction of the dispersed phase. A calibration curve for an oil-in-water emulsion is shown in Figure 4.7.

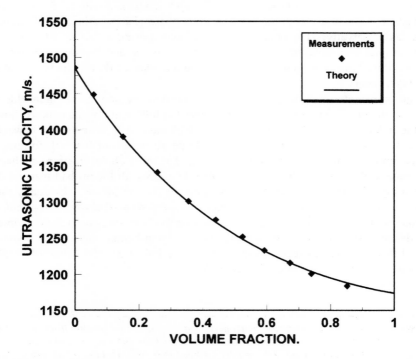

Figure 4.7 *Ultrasonic velocity versus oil volume fraction. The measured data are compared with the predictions of Urick[13] which are based on the density and ultrasonic velocity of each phase alone*

To calculate volume-fraction from velocity data, eqn. (4.4) must be solved for ϕ. Equation (4.4) ignores the effect of scattering of sound by the dispersed phase on the speed of sound. It has been shown[14,15] that in many emulsions and suspensions the effects of scattering are significant and that using the Urick equation to compute the volume fraction introduces significant errors. In such systems the predictions of simple mixing theory may be brought back in line with measurements by using an effective speed of sound and compressibility for the dispersed phase in place of the true values. This procedure has been described by Pinfield *et al.*[16] where they argue that, in the absence of size fractionation, use of *effective* parameters may allow for the effects of scattering. The use of ultrasonic scattering to obtain further information on emulsions is described in Section 4.3.

Below we describe the use of the ultrasonic monitor to detect creaming in a polydisperse concentrated emulsion, and to characterise flocculation from the creaming behaviour. The effects of added polymers on the flocculation and creaming processes are also described.

4.2.3 Creaming of Polydisperse Emulsions without Added Polymer

Experiments have been conducted on emulsions containing 20% v/v of a mixture of heptane (90% v/v) and hexadecane (10% v/v), emulsified with 0.35% nonionic surfactant in the aqueous phase. The aqueous phase also contained 0.2% sodium azide as preservative. The emulsions were prepared using a Waring Blender, and their droplet size distribution revealed a polydisperse distribution, with weight mean radius 2 μm. The emulsion was stable to coalescence during the timescale of the experiments.

The oil volume-fraction profiles of the emulsion are shown in Figure 4.8(a). Initially the oil was uniformly distributed over the full height of the cell, as shown by the initial horizontal line measured after 0.09 days. After a few days, however, the data show that the oil began to accumulate at the top of the cell, forming a concentrated cream layer of over 70% oil. The oil volume-fraction gradually decreased at the base of the cell, until, after 37.1 days, all the oil had reached the cream layer. In the case of polydisperse droplets, if they move individually then there is a range of relevant Stokes' velocities, as discussed above. This leads to diffuse boundaries in the creaming emulsion, as the faster-moving (larger) droplets are depleted from the base at an early stage, leaving behind the smaller fraction. The diffuse boundaries seen in Figure 4.8(a) are indicative of a range of droplet velocities, due to a range of droplet diameters.

Figure 4.8 *Creaming of emulsion without added polymer.* (a) *Volume fraction profiles measured using the ultrasonic creaming monitor showing the progression from uniform (horizontal line) to fully creamed state over 37 days. (Inset) Droplet size distribution inferred from the creaming data.* (b) *Height of eight ordinates y1—y8 with time. The gradient of each ordinate is related to the Stokes' velocity of the size fraction represented by each volume fraction*

(a)

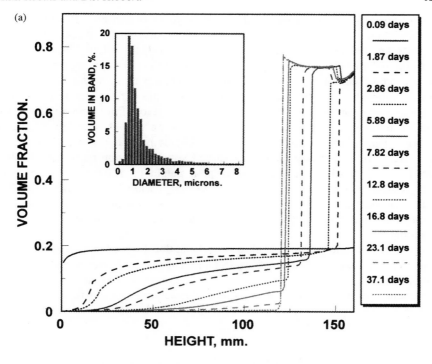

(b)

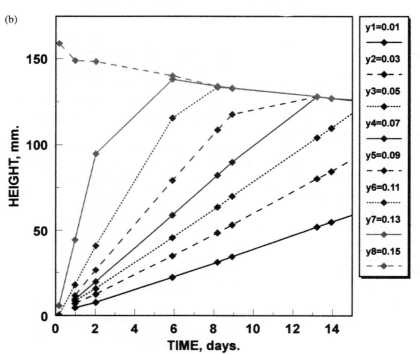

The observed creaming rates can be analysed to obtain the effective droplet size distribution in the creaming emulsion. This enables us to check whether the droplets are indeed moving individually, or whether flocculation has occurred. With the polydisperse droplets exhibiting a range of creaming rates, our approach is to examine contours of constant oil concentration as these move with time. This is equivalent to examining a number of ordinates y_1, y_2, *etc.*, in Figure 4.4. Each ordinate corresponds to a volume fraction of oil from zero to the initial volume fraction. By selecting various ordinates we convert the data in Figure 4.8(a) to height/time graphs as in Figure 4.8(b). It is clear that each fraction of oil represented by the contours of a different volume fraction rises with constant speed until it reaches the cream layer at a height of 120 mm. The speed of each contour is a measure of the Stokes' velocity (with the hindrance factor) of each size fraction. The linearity of the contours indicates that the creaming droplets are not changing size by flocculation or coalescence on the timescale of creaming.

The speed of each fraction of oil (contour) can thus be used to infer the effective hydrodynamic size distribution of the droplets. The inset to Figure 4.8(a) shows the droplet diameters obtained from the creaming contours. The size distribution is consistent with that measured using a laser diffraction particle sizer on the initial emulsion, showing that the assumption of individual droplet movement according to Stokes' velocity is justified.

4.2.4 Creaming of Emulsions with Low Concentrations of Non-adsorbing Polymer

Figure 4.9 shows the creaming behaviour of an emulsion whose continuous phase contained 0.02% of a polysaccharide, hydroxyethylcellulose (HEC). This emulsion had the same primary droplet size distribution as the previous emulsion and was also stable to coalescence. The creaming behaviour, however, showed two distinct populations of droplets, one fraction moving rapidly to the top of the container and leaving behind about half the oil which creamed in a manner more similar to the polymer-free emulsion described above.[17]

As before, the data can be analysed in terms of the rate of rise of contours of constant oil concentration. These were linear, as in the unflocculated case, showing the constant creaming rates of each size fraction. Analysis of the rates yielded the effective droplet size distribution shown inset in Figure 4.9. In addition to the primary size distribution, there was a second fraction of droplets whose inferred diameters were much larger. These are droplets that flocculated and then creamed as small aggregates. The interesting point is that only part of the emulsion became flocculated, and that a flocculated fraction was in coexistence with the primary drops. This phenomenon, however, is predicted to occur when non-adsorbing polymer is added to an otherwise stable suspension or emulsion.[18,19]

By varying the polymer concentration, in this case in the range 0.01–0.04% HEC, the proportion of the droplets that are flocculated can be varied. The creaming profiles enable the amount of flocculation to be determined accurately.

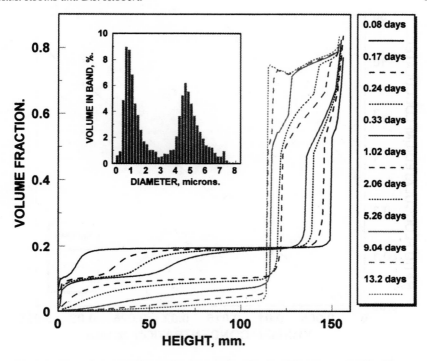

Figure 4.9 *Creaming of emulsion with 0.02% hydroxyethylcellulose (HEC). Volume fraction profiles measured using the ultrasonic creaming monitor showing the progression from initial to fully creamed state. (Inset) The droplet size distribution inferred from the creaming ordinate velocities. The larger fraction represents the equivalent hydrodynamic diameter of small flocs*

Figure 4.10 shows that the ratio of flocculated to unflocculated droplets varies exponentially with the concentration of HEC. Data for similar emulsions of 20% alkane with the same droplet size distribution but containing xanthan are also shown. The same dependence on polymer concentration is observed, although the larger xanthan molecules act as an effective flocculant at lower concentration than HEC. Both are behaving as non-adsorbing (depletion) flocculants.

4.2.5 Creaming of Emulsions with High Concentrations of Non-adsorbing Polymer

When the concentration of HEC was increased to 0.04% the emulsion creamed apparently as a single entity, with a sharp boundary that was clearly visible. Figure 4.11 shows the data from the ultrasonic monitor. The rapid creaming resulted in a cream layer that formed initially at a volume fraction of only ~68%, but with time the cream layer compacted to an even higher concentration than in the polymer-free emulsion.

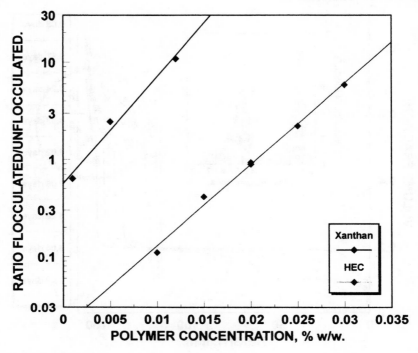

Figure 4.10 *Degree of flocculation* versus *polymer concentration in the coexistence regime for emulsions containing hydroxyethylcellulose (HEC) or xanthan. The data are expressed as the ratio of the volume fraction of flocculated droplets to the volume fraction of individual droplets*

This behaviour may be interpreted as indicating the presence of a flocculated network. Each droplet is presumed to be in 'contact' with two or more neighbouring droplets, and the resultant network creams by a compaction mechanism. The individual droplets do not coalesce; if a sample is removed from the emulsion and diluted for sizing, the original droplet size distribution is obtained.

As the polymer concentration is further increased, the creaming becomes very slow, and may be inhibited completely on the timescale of the experiments. Figure 4.12(a) shows the concentration profiles of an emulsion containing 0.95% HEC. The creaming is slow, but there is also a significant delay before creaming starts. This is shown most clearly by plotting the boundary height as a function of time, as in Figure 4.12(b). The delay has also been observed for emulsions containing xanthan[2] and Figure 4.13 shows that as the concentration of polymer increases, the delay period increases dramatically (up to 3 months in the case of 0.5% xanthan).

The reasons for the delay are not yet well established, but a possible explanation invokes a yield stress in the flocculated network. The inter-droplet attraction is presumed to be strong enough to hold the droplets in the structure. However, we speculate that the droplets in the flocculated network are continually rearranging to form more compact structures, and that after a certain time, the delay period, the

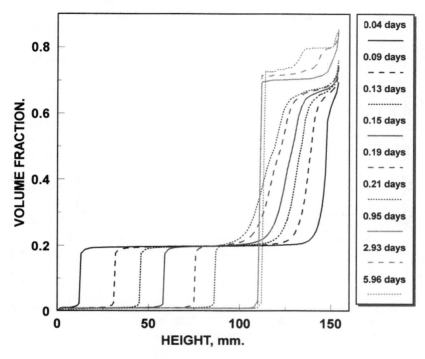

Figure 4.11 *Volume fraction profiles during creaming of an emulsion with 0.04% hydroxy-ethylcellulose (HEC), showing a flocculated network*

yield stress of the structure is reduced below the level needed to overcome the buoyancy of the droplets. The network then collapses under its own 'weight' and the continuous phase appears as a layer at the base of the sample. However, the 'yield stress' is predicted to be very small and so far has not been detectable in these emulsions.[20] More experiments and structural characterisation are needed to establish the cause of the delay.

Although it is predicted[19] that at very high polymer concentrations the emulsion will become restabilised, there is little practical evidence of this phenomenon, possibly due to the difficulty in dispersing high concentrations of viscous polymers.

4.2.6 Creaming of Emulsions with Adsorbing Polymer

So far only non-adsorbing polymers have been considered, which flocculate by a depletion mechanism. Other molecules, which are slightly hydrophobic, may be able to be adsorbed at the oil–water interface.[21] To prepare an emulsion a surfactant is also needed, but the surfactant is slowly displaced at the interface by the polysaccharide molecules. Adsorbing polymers are well known in colloid science, and their effect on the dispersion stability depends on the surface coverage of the polymer. At high concentrations, the drop surfaces are completely covered and

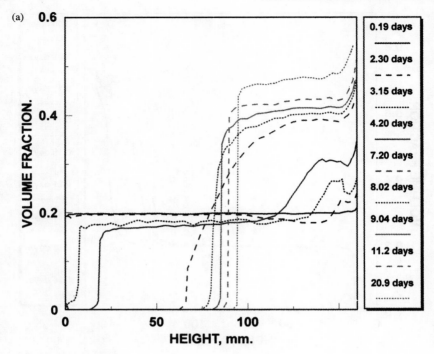

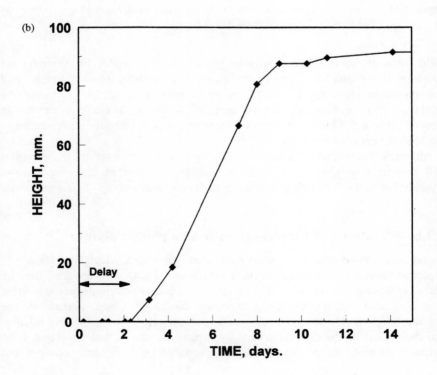

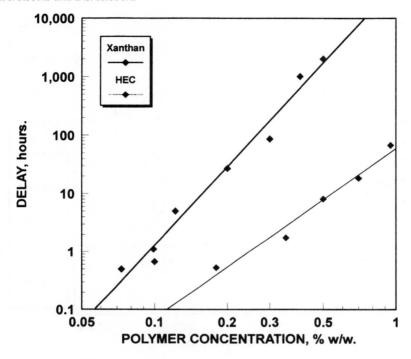

Figure 4.13 *Delay time* versus *polymer concentration for emulsions containing hydroxyethylcellulose (HEC) or xanthan*

flocculation is usually prevented. However, when the polymer is only able to cover part of the surface, then the polymer molecules may bridge between two droplets and cause persistent flocculation.[22]

The adsorption process is likely to be slow compared with the timescale for depletion flocculation, where the depletion potential is present as soon as the polymer is added. Figure 4.14(a) shows the concentration profiles of an alkane-in-water emulsion containing 0.9% of a slightly hydrophobic polymer, hydroxypropylcellulose. The volume-fraction profiles resembled those of the unflocculated system, with no sharp boundaries to indicate a flocculated network. However, when the contours of constant oil volume-fraction were derived from the creaming data, as in Figure 4.14(b), they were distinct from those previously observed [Figure 4.8(b)]. The contours did not rise linearly with a constant rate, showing the steady terminal velocity of droplets or flocs of unchanging size. In contrast, the rate of rise of the contours increased with time, which suggests that the creaming entities increased in size during the creaming process. Using laser diffraction sizing it was

Figure 4.12 *Creaming of emulsion with 0.95% hydroxyethylcellulose. (a) Volume fraction profiles showing a flocculated network. (b) Height of the base of the network with time, showing delay period*

(a)

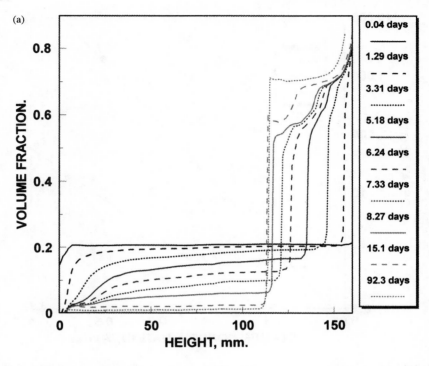

(b)

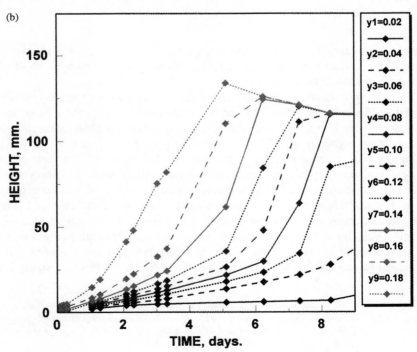

shown that the primary droplet distribution did not change, so there was no coalescence occurring, and thus the creaming was attributed to flocs that increased in size during the experiment. The creaming rates could not be used to derive a unique droplet size distribution, because the sizes were not constant. This behaviour is consistent with an emulsion that is simultaneously undergoing flocculation and creaming, due to slow adsorption of the polymer at the oil—water interface.

At higher polymer concentrations a network was formed, but usually it creamed a small amount first. It is not yet established whether there is a delay period in the case of bridging flocculation. The explanation offered above for the delay in depletion flocculation may also be relevant to bridging, but in the latter case it is likely that the droplets are less able to rearrange by rotating around each other.

Flocculation by a bridging mechanism is well known for particle suspensions, where it is straightforward to adsorb polymers onto surfactant-free interfaces, but it is less commonly observed in emulsions. An exception is a protein-stabilised emulsion[22] where the protein acts as a surfactant, but can cause bridging when there is incomplete surface coverage.

4.3 Non-intrusive Determination of Flocculation in Emulsions

4.3.1 Introduction and the Principle of Ultrasonic Scattering

In the previous section we demonstrated the use of ultrasonic velocity measurements to characterise creaming, and indirectly to characterise flocculation. However, there is more information to be obtained from an emulsion using ultrasonic spectroscopy. This involves measurement of phase velocity and attenuation of ultrasound as a function of frequency after propagation through the emulsion. There are a number of mechanisms by which ultrasound is attenuated by the emulsion, resulting in characteristic ultrasonic properties. Figure 4.15 shows the principal mechanisms of absorption.

The two intrinsic losses represent the attenuation that occurs in each of the bulk phases. The attenuation through the emulsion will be related to some volume-averaged combination of the two attenuation coefficients. The other losses are due to scattering. Reflectional losses occur when there are differences between the acoustic impedance of the two phases. The reflected sound is not absorbed by the system, but is scattered out of the path of the forward travelling wave and thus lost to the receiver.

Thermal losses (above those contributing to the intrinsic attenuation of each phase) occur when there is a difference between the specific heat capacities of the two phases. Heat is generated and dissipated by the cyclic pressure fluctuations within the sound wave. When the specific heat capacities of the two phases are

---

Figure 4.14 *Creaming of emulsion with 0.9% hydroxypropylcellulose. (a) Volume fraction profiles measured using the ultrasonic creaming monitor. (b) Height of eight ordinates y1—y8 with time showing increasing creaming velocities with time*

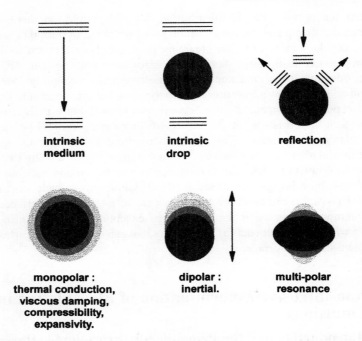

intrinsic
medium

intrinsic
drop

reflection

monopolar :
thermal conduction,
viscous damping,
compressibility,
expansivity.

dipolar :
inertial.

multi-polar
resonance

Figure 4.15 *Schematic diagram of the principal ultrasonic absorption mechanisms in emulsions*

different, sharp temperature gradients occur at the boundaries between the phases. Thermal conduction between the two phases then results in loss of energy from the wave. A second thermal loss mechanism may cause losses even in the absence of differences between the specific heat capacities of the two phases. In this case the temperature difference about the interface may be zero, as the drop and super-vening fluid fluctuate in temperature in unison. However, if the thermal expansion coefficients of the two phases are different then the drops may expand and contract relative to the fluid as the temperature cycles. This expansion and contraction radiates energy in all directions (in the form of ultrasound) away from the drop, and is lost to the straight-through wave.

A loss mechanism similar in form to thermal expansion losses arises when there is a difference between the compressibility of the drop and the supervening fluid. Once more the drop expands and contracts relative to the fluid, re-radiating energy, but this time as a direct consequence of the fluctuating pressure field.

When there is a density difference between the two phases, visco-inertial scattering occurs. The density difference leads to relative motion between the drop and the supervening fluid. The viscosity of the fluid damps this relative motion, dissipating the kinetic energy as heat. Additionally, the relative motion between the drop and the fluid re-radiates energy in the forward and backward directions, which constitutes a further loss mechanism. For high viscosity fluids, viscous damping effects are large, but accordingly the magnitude of the relative motion between the two phases is reduced which in turn reduces the re-radiative losses. The opposite effect is seen for low viscosity fluids.

The final loss mechanism shown in Figure 4.15 is called resonance. This behaviour is caused by Rayleigh waves (*i.e.* surface waves) interacting with the spherical resonance modes (*i.e.* standing wave patterns) of drops. Resonance occurs at frequencies for which the drop circumference is equal to a whole number of standing wave periods. At resonance the exaggerated motions of the drop surface radiate energy in various directions according to the spherical resonance mode number that is excited (*i.e.* $n = 1$, dipole; $n = 2$, quadrupole; $n = 3$, octupole; *etc.*).

The existence of all these loss mechanisms means that the propagation of ultrasound through emulsions may be very complex. It depends on the concentration and size of the droplets, and the thermophysical properties of both phases. However, Allegra and Hawley[23] have developed a theory for the scattering of ultrasound in simple two-component dispersions. The theory considers the particles to be independent scattering centres, randomly distributed in the liquid continuous phase. It has been shown to fit experimental data for monodisperse emulsions at low concentration.[24,25] The original scattering theory breaks down for emulsions at high volume fraction, as multiple scattering effects are not included. This is allowed for in current theories using either the equations of Waterman and Truell[26] or those due to Lloyd and Berry.[27]

Additionally, this theory assumes that the scattering droplets are randomly positioned within the continuous phase and makes no allowance for droplet–droplet interactions. It is thus reasonable to expect deviations between theory and experiment for systems where the particles are flocculated. Recent work by McClements[28] demonstrated that the attenuation and velocity of ultrasonic waves changed significantly when a sufficient concentration of surfactant micelles was added to an emulsion. Visual observations of the emulsion suggested that the ultrasonic changes were caused by the micelles acting as depletion flocculants.

A disadvantage of many ultrasonic techniques previously applied to unstable colloidal systems has been their inability to collect data over a range of frequencies fast enough to be confident the sample had not changed significantly during the measurement. However, these experiments become feasible with the development by Challis *et al.* of rapid broad-band ultrasonic spectroscopy.[29] In this technique the ultrasonic properties are determined over a wide frequency range by propagation of a single pulse through the sample. The frequency dependence of phase velocity and attenuation in the sample is obtained using fast Fourier transformation of the received signal. We have recently used this technique to detect flocculation in oil-in-water emulsions.[30]

4.3.2 Ultrasonic Scattering from Unflocculated Emulsions

These emulsions contained 1-bromohexadecane at 5% v/v as the dispersed phase, stabilised against coalescence using a nonionic surfactant. Figure 4.16(a) shows the microscopic appearance of the emulsion. The ultrasonic properties of the emulsion were measured using a broad-band spectrometer at 25 °C. Figure 4.16(b) shows the

(a)

(b)

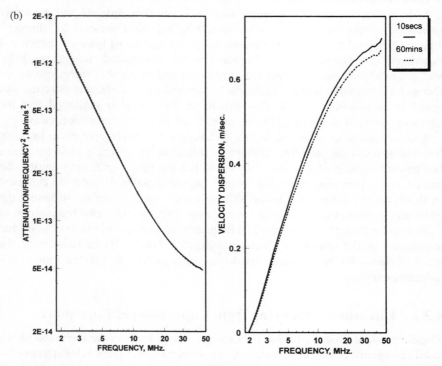

Figure 4.16 *Emulsion of 5% v/v 1-bromohexadecane without polymer.* (a) *Microscopic appearance of the emulsion.* (b) *Measured ultrasonic spectra as a function of time*

attenuation of ultrasound (transformed to attenuation divided by the square of frequency) and ultrasonic velocity dispersion (velocity at each frequency after subtraction of the velocity at 2 MHz) over the same frequency range. Two measurements are shown captured immediately after sample preparation and after 60 minutes. There was no significant change in properties over that timescale, and the data were consistent with the predictions of Allegra and Hawley's scattering theory.[29]

4.3.3   Ultrasonic Scattering from Emulsions during Depletion Flocculation

The emulsion was flocculated by the addition of 0.5% HEC to the continuous phase. The final appearance of the flocculated emulsion, shown in Figure 4.17(a), revealed the presence of a connected network of droplets, which was observed to form over a period of several hours.

The initial ultrasonic properties are very similar to the unflocculated sample, but after 60 minutes dramatic changes in the ultrasonic attenuation and velocity are apparent (Figure 4.17b). At high frequency the attenuation is much stronger and at low frequencies the attenuation is significantly weaker. At a frequency of $\sim$ 6 MHz the attenuation is unchanged. The velocity dispersion of the flocculated emulsion also shows significant changes over 60 minutes. The velocity dispersion decreases with time with the magnitude of the decrease being greatest at high frequency.

Both the attenuation and velocity ultrasonic data show differences between the samples over a timescale that can only be attributed to flocculation. The theory of

(a)

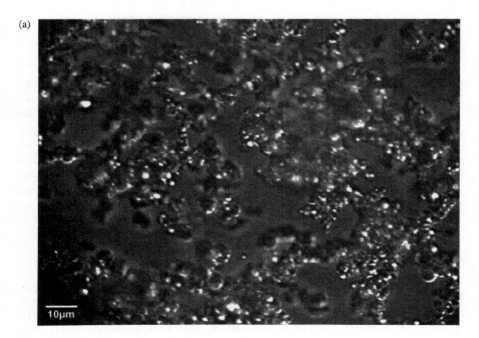

10µm

Figure 4.17(a) *Continued overleaf*

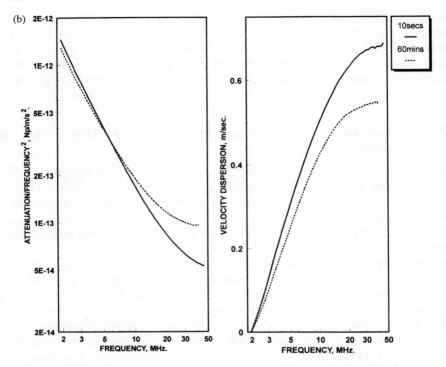

Figure 4.17 *Emulsion of 5% v/v 1-bromohexadecane with 0.5% HEC. (a) Final microscopic appearance of the emulsion, showing droplets flocculated in a network. (b) Measured ultrasonic spectra as a function of time*

ultrasonic propagation predicts a strong dependence of ultrasonic velocity and attenuation on droplet size, but it proved impossible to model the ultrasonic properties using the scattering theory for spherical drops, even allowing for larger droplets. As discussed above, the theories take no account of drop distribution, and also tend to fail at high volume fraction. It is not surprising, therefore, that when flocculation causes an uneven distribution of droplets and localised regions of high drop concentrations, the ultrasonic properties are not readily modelled using existing theories. However, it is clear that the flocculation process can be detected readily using ultrasonic scattering, before the creaming behaviour betrays its presence.

4.4 Summary

The physical stability of emulsions is an important factor in their shelf-life. Frequently they separate under gravity to form an unsightly layer of oil at the top, and/or clear aqueous phase at the base. This separation, caused by the density difference between the oil and aqueous phases, may be detected at an early stage using the IFR ultrasonic creaming monitor.

The ultrasonic creaming instrument provides detailed profiles of oil concentration as a function of height and time. As well as early detection of separation, the concentration profiles enable us to estimate the size of the creaming droplets. This is an advantage when structures are sensitive to the intrusive sampling and dilution procedures needed for conventional drop sizing.

In the presence of polymers the droplets are frequently flocculated, either by a depletion or bridging mechanism. The ultrasonic creaming data enable the degree of flocculation to be determined, and to distinguish between the behaviour of emulsions flocculated by the two mechanisms.

Flocculation can also be detected directly using ultrasonic methods. During the flocculation process the ultrasonic properties change significantly and are not consistent with theory for spherical drop scattering. This is because a network has been formed, and at present there is no suitable theory to describe the propagation of ultrasound through such a structure.

The ultrasonic techniques provide an insight into the properties of complex multiphase systems, and are applicable to a wide range of commercial dispersions.

Acknowledgments

The work described here was carried out at the Institute of Food Research by the authors with their colleagues Paul Gunning, Martin Garrood and Annette Fillery-Travis, and in collaboration with Andrew Holmes and Richard Challis at Keele University. We are grateful to the UK Ministry of Agriculture, Fisheries and Food, the Biotechnology and Biological Sciences Research Council, Shell UK and Système Bioindustries, France for past funding.

4.5   References

1. A.S. Kabal'nov, A.V. Pertzov and E.D. Shchukin, *Colloids Surf.*, 1987, **24**, 19.
2. A. Parker, P.A. Gunning, K. Ng and M.M. Robins, *Food Hydrocolloids*, 1995, **9**, 333.
3. S. Asakura and F. Oosawa, *J. Polym. Sci.*, 1958, **33**, 183.
4. P.R. Sperry, *J. Colloid Interface Sci.*, 1982, **87**, 375.
5. G.K. Batchelor, *J. Fluid Mech.*, 1972, **52**, 245.
6. J.F. Richardson and W.N. Zaki, *Trans. Inst. Chem. Eng.*, 1954, **32**, 35.
7. C. Carter, D.J. Hibberd, A.M. Howe, A.R. Mackie and M.M. Robins, *Prog. Colloid Polym. Sci.*, 1988, **76**, 37.
8. R.C. Ball and P. Richmond, *Phys. Chem. Liq.*, 1980, **9**, 99.
9. D.J. Hibberd, PhD Thesis, University of East Anglia, 1997.
10. H.A. Kearsey and L.E. Gill, *Trans. Inst. Chem. Eng.*, 1986, **41**, 296.
11. R.A. Williams, C.G. Xie, R. Bragg and W.P.K. Amarasinghe, *Colloids Surf.*, 1990, **43**, 1.
12. A.M. Howe, A.R. Mackie and M.M. Robins, *J. Disp. Sci. Technol.*, 1986, **7**, 231.
13. R.J. Urick, *J. Appl. Phys.*, 1947, **18**, 983.
14. D.J. McClements and M.J.W. Povey, *J. Phys. D*, 1989, **22**, 38.
15. V.J. Pinfield, E. Dickinson and M.J.W. Povey, *J. Colloid Interface Sci.*, 1994, **166**, 363.
16. V.J. Pinfield, M.J.W. Povey and E. Dickinson, *Ultrasonics*, 1995, **33**, 243.

17. A.J. Fillery-Travis, P.A. Gunning, D.J. Hibberd and M.M. Robins, *J. Colloid Interface Sci.*, 1993, **159**, 189.
18. D.H. Napper, 'Polymeric Stabilization of Colloidal Dispersions', Academic Press, London, 1983.
19. W.B. Russel, D.A. Saville and W.R. Schowalter, 'Colloidal Dispersions', Cambridge University Press, Cambridge, 1989.
20. P. Manoj, Institute of Food Research, unpublished data.
21. R.J. Hunter, 'Foundation of Colloid Science', Oxford University Press, Oxford, 1992, vol. 2.
22. E. Dickinson, M. Golding and M.J.W. Povey, *J. Colloid Interface Sci.*, 1997, **185**, 515.
23. J.R. Allegra and S.A. Hawley, *J. Acoust. Soc. Am.*, 1971, **51**, 363.
24. C. Javanaud, N.R. Gladwell, S.J. Gouldby, D.J. Hibberd, A. Thomas and M.M. Robins, *Ultrasonics*, 1991, **29**, 331.
25. R.C. Chivers, *Acoustica*, 1991, **74**, 8.
26. P.C. Waterman, and R. Truell, *J. Math. Phys.*, 1961, **2**, 512.
27. P. Lloyd and M.V. Berry, *Proc. Phys. Soc.*, 1967, **91**, 678.
28. D.J. McClements, *Colloids Surf. A*, 1994, **90**, 25.
29. R.E. Challis, J.A. Harrison, A.K. Holmes, and R.P. Cocker, *J. Acoust. Soc. Am.*, 1991, **90**, 730.
30. D.J. Hibberd, A.K. Holmes, M.J. Garrood, A.J. Fillery-Travis, M.M. Robins and R.E. Challis, *J. Colloid Interface Sci.*, 1997, **193**, 77.

CHAPTER 5

Rheology of Emulsions—The Relationship to Structure and Stability

ERIC DICKINSON

Procter Department of Food Science, University of Leeds, Leeds LS2 9JT, UK

5.1 Introduction

Rheological properties of emulsions are important from both the basic science and the applied viewpoints. At a fundamental level, the determination of small-deformation rheological properties can give detailed information about emulsion structure, states of droplet aggregation, and strengths of interdroplet interactions. At the industrial or commercial level, the measurement of large-deformation rheological properties is useful for manufacturers of consumer-based products for describing textural attributes like 'creaminess', 'spreadability', 'body' and 'skin feel', and also for controlling flow properties during industrial operations such as mixing, pumping and pouring. Types of rheological behaviour range widely—from the simple Newtonian flow of a dilute emulsion like homogenized milk, right through to the highly complex viscoelasticity exhibited by materials like mayonnaise, skin cream, or bitumen. Even though the latter examples obviously differ enormously in terms of chemical composition, there are in reality only a few key physical factors controlling rheology of all emulsion systems. The key factors are: average droplet size, droplet-size distribution, oil volume fraction, the nature of the interfacial layer, and the nature of the forces between the droplets.

Emulsion rheology is, of course, related to emulsion stability. In general, there are five main destabilization mechanisms to consider—coalescence, creaming (sedimentation), flocculation, phase inversion and Ostwald ripening.[1] In this article the emphasis is on the relationship between rheology and flocculation. The rheology of systems exhibiting severe instability (coalescence, ripening or phase inversion) is undoubtedly important industrially (*e.g.* in bitumen emulsions,[2] multiple emulsions[3,4]). However, it is more difficult to determine the experimental behaviour of highly unstable systems in a reproducible manner, or to interpret their rheology in fundamental colloid science terms; hence they will not be discussed in detail here. We shall focus instead on behaviour of simple emulsions (mainly of the oil-in-water type) that are stable in most respects—except that they may be flocculated. One of the most obvious manifestations of flocculation of a fine emulsion is enhanced creaming, leading eventually to phase separation. Hence, where it illuminates understanding of emulsion rheology, we shall also be interested in the creaming (sedimentation) behaviour.

Emulsion stability with respect to flocculation (and coalescence) is typically achieved by making droplets small.[1] However, even fine emulsions may be flocculated to some degree, and this flocculation may have a predominant effect on emulsion structure and rheology. In keeping with standard practice,[5-7] the word 'flocculation' is used here to describe all types of aggregation in which the local dispersed phase region remains intact, *i.e.* stopping short of droplet coalescence (or clumping). The nature of the interdroplet interactions determines the structure of the emulsion, and structure in turn determines the rheology. By 'structure' is here meant the spatial organization of the droplets on the short to medium length scale.[8]

Even simply for the purpose of classifying rheological behaviour, it is convenient to distinguish between flocculated and non-flocculated systems. This is because, firstly, the theoretical position is much more well developed as a function of oil volume fraction for dispersed (non-flocculated) systems than for flocculated ones, and, secondly, the experimental behaviour of many emulsion systems can be interpreted most effectively at the mechanistic level from a detailed consideration of the type and extent of flocculation. Much of the experimental work published recently on emulsion rheology has been concerned with the role of water-soluble polymers in controlling the structure and stability of flocculated systems. Of particular importance in such systems is the viscoelasticity of the polymer-containing aqueous continuous phase and the nature of the interaction between polymer and emulsion droplets.

This chapter first reviews general theoretical concepts and then proceeds to discuss relevant experimental results. It is assumed that readers are already reasonably familiar with basic rheological principles and techniques as set out in various texts.[9-12] Any review of this sort, of course, has some inevitable bias in coverage. In this instance it is towards the rheology of milk protein-stabilized emulsions—a topic of particular interest to the author.

5.2 Theoretical Background

Theoretical treatments of emulsion rheology are less well developed than those devised for particulate dispersions. It therefore seems appropriate to give a brief survey of classical work on particulate dispersion rheology as a prelude to discussing the rheology of dispersions of droplets. A further justification for this approach is the observation that, in practice, many a fine emulsion system can be reliably treated in rheological terms as if it were a dispersion of solid spherical particles.

5.2.1 Dilute Dispersions of Spherical Particles

5.2.1.1 Hard Spheres

Let us first consider a dilute system of rigid spherical particles dispersed at a volume fraction ϕ in a Newtonian liquid of shear viscosity η_0. According to Einstein,[13,14] the Newtonian viscosity of the whole system, η, is increased by the presence of the particles according to the equation

$$\eta_r = \eta/\eta_0 = 1 + 2.5\phi \qquad (5.1)$$

where the ratio η_r is called the relative viscosity. In deriving eqn. (5.1), Einstein assumed that there is a 'no-slip' boundary condition at the surface of the particles, and that particles are far enough apart, on average, to ensure that each acts independently in terms of its effect on the hydrodynamic flow within the dispersion medium. Therefore, the Einstein equation is strictly applicable only at high dilution ($\phi < 0.01$) for a system of non-aggregated particles interacting solely through a hard-sphere interparticle potential. The equation implies that the dispersion viscosity is independent of particle size and shear-rate.

The coefficient of 2.5 in eqn. (5.1) is the intrinsic viscosity $[\eta]$ for hard spheres:

$$[\eta] = \lim_{\phi \to 0} \frac{\eta_r - 1}{\phi} \qquad (5.2)$$

The value of $[\eta]$ will tend to differ from 2.5 if any of the assumptions of Einstein's theory are violated. This may occur if particles are electrically charged, or if they are non-spherical or deformable, or if the particle surface is coated with a layer of molecules (solvent, surfactant or polymer) of thickness significant compared to the particle size.

Particles dispersed in an aqueous medium invariably carry an electric charge. Thus they are surrounded by an electrical double-layer whose thickness κ^{-1} depends on the ionic strength of the solution. Flow causes a distortion of the local ionic atmosphere from spherical symmetry, but the Maxwell stress generated from the asymmetric electric field tends to restore the equilibrium symmetry of the double-layer. This leads to enhanced energy dissipation and hence an increased viscosity. This phenomenon was first described by Smoluchowski,[15] and is now known as the primary electroviscous effect.[16] For a dispersion of charged hard spheres of radius a at a concentration low enough for double-layers not to overlap ($\phi \ll 8a^3\kappa^3$), the intrinsic viscosity defined by eqn. (5.2) increases to 2.5β, where β is a parameter which depends on the charge on the particles and the value of κa.[16] Under conditions of constant electrolyte concentration, the value of β is proportional to $(\zeta/a)^2$, where ζ is the electrokinetic (zeta) potential. Hence, the smaller the particles are, the larger is the increase in viscosity.

The presence of an adsorbed layer of polymer (or surfactant) of thickness δ at the particle surface (or even a solvation layer of immobilized solvent molecules and ions) can lead to an enhancement in the viscosity of a dilute dispersion by increasing the effective volume fraction of dispersed phase to the value:

$$\phi_{\text{eff}} = \phi(1 + \delta/a)^3 \approx \phi(1 + 3\delta/a) \qquad (\delta \ll a) \qquad (5.3)$$

The effects of adsorbed layers or solvation layers become more important with decreasing size of the dispersed particles.

5.2.1.2   Liquid Droplets

In a dispersion of liquid droplets in flow, the transmission of tangential stress across the oil–water interface from the continuous phase to the dispersed phase causes liquid circulation in the droplets. Energy dissipation is therefore less than for hard spheres, and so the viscosity is lower than that given by eqn. (5.1). Under conditions where the interfacial tension is high enough to keep the droplets close to spherical, the intrinsic viscosity is given by[17]

$$[\eta] = 2.5 \frac{\eta_i + 0.4\eta_0}{\eta_i + \eta_0} \qquad (5.4)$$

where η_i is the (Newtonian) viscosity of the dispersed phase. The value of $[\eta]$ from eqn. (5.4) approaches the 'no-slip' Einstein result of 2.5 for $\eta_i \to \infty$ (rigid particles). With complete slip at the interface (gas bubbles, $\eta_i \to 0$), eqn. (5.4) gives $[\eta] = 1$. The Taylor equation [eqn. (5.4)] does not take account of the presence of any surface rheological effects at the oil–water interface, and so it is applicable only to highly dilute emulsions in which the adsorbed layer has a very low surface viscosity, *e.g.* an oil-in-water emulsion stabilized by ionic surfactant.

For the case of a purely viscous interfacial layer, Oldroyd obtained[18] the following expression for the intrinsic viscosity:

$$[\eta] = 2.5 \frac{\eta_i + 0.4\eta_0 + \xi}{\eta_i + \eta_0 + 0.4\xi} \qquad (5.5)$$

The quantity ξ is defined by

$$\xi = (2\eta^s + 3\mu^s)/a \qquad (5.6)$$

where η^s is the surface shear viscosity and μ^s is the zero-frequency surface dilatational viscosity. Typically, however, the adsorbed layer has elastic as well as viscous character. For the Einstein equation to be applicable, it is necessary only that the limiting moduli in shear and dilatation both do not vanish. This means that emulsion droplets may be regarded as hard spheres if the adsorbed layer has a finite two-dimensional yield stress. This is usually the case for emulsion systems stabilized by viscoelastic layers of nonionic surfactants or polymers (especially proteins).

In the presence of a shear field, a single spherical droplet is simultaneously rotated and deformed into an ellipsoidal shape. The extent of deformation is proportional to shear-rate D and interfacial tension γ.[16] Weak flows compress droplets in the principal directions of positive and negative straining, and for simple shear flow this means that the ellipsoidal major axis is oriented at 45° to the flow direction. Owing to the tensile nature of the first normal stress difference, droplets become elongated in the direction of flow at high shear-rates. Strong flows $(D \gg \gamma/a\eta_0)$ cause bursting of droplets.

5.2.1.3   Brownian Motion and Interparticle Interactions

Away from the high dilution limit, a proper theory must take account of effects of Brownian motion and interparticle interactions. The smaller the emulsion droplets, the more important are the contributions of Brownian motion and colloidal interactions. Brownian forces tend to randomize the positions of the colloidal particles, leading to formation of temporary doublets, triplets, *etc.* The hydrodynamic interactions are of longer range than the colloidal interactions, and so they come into play at rather low volume fractions $(\phi > 0.01)$, cause ordering of particles into layers and tend to destroy temporary aggregates. Hydrodynamic forces increasingly overwhelm Brownian motion at higher shear-rates and this leads to a lowering of the viscosity (shear-thinning behaviour).

The externally imposed flow field and the interparticle forces are opposed by the continuous random motion of colloid-sized droplets, as characterized by the translational thermal energy $(3k_B T/2)$ and the time t_r for a droplet to move a distance comparable to its radius,

$$t_r = \frac{6\pi\eta_0 a^3}{k_B T} \tag{5.7}$$

where k_B is Boltzmann's constant and T is the temperature. The time t_r is the characteristic relaxation time of the emulsion microstructure, and the influence of the shear flow field on the structure is determined by the quantity Dt_r, which is the ratio of the relaxation time to the characteristic time of the flow (D^{-1}). With small droplets or at low shear-rates $(Dt_r \ll 1)$, there is little disturbance of the equilibrium structure and so the emulsion appears Newtonian. However, non-Newtonian behaviour does appear when the balance between Brownian motion and interparticle forces becomes substantially disrupted $(Dt_r \approx 1)$. For emulsions with droplet sizes larger than *ca.* 1 μm, the effect of Brownian motion on the rheology is generally rather small.

The relative viscosity of a suspension of hydrodynamically interacting hard spheres can be expressed in the form

$$\eta_r = 1 + k_1\phi + k_2\phi^2 + \vartheta(\phi^3) \tag{5.8}$$

where k_1 is the Einstein coefficient (2.5), k_2 is a coefficient related to two-body hydrodynamic interactions, and other higher-order terms relate to multi-body hydrodynamic interactions. For the case where Brownian motion is predominant (small particles, low shear-rate) and the particle pair distribution function is perturbed only slightly from the isotropic equilibrium state,[19] the two-body coefficient takes the value $k_2 \approx 6$. Eqn. (5.8) with $k_2 \approx 6$ gives reasonably good agreement with experiment for hard sphere suspensions of volume fraction up to $\phi \approx 0.15$. For large particles in the absence of Brownian motion (corresponding to the high frequency limiting viscosity), the value of k_2 is slightly lower (≈ 5).

For charged emulsion droplets at low ionic strength, surface–surface electrostatic repulsive forces prevent pairs of droplets approaching closely during flow. The increased perturbation of the shear field causes the viscosity to exceed that for hard spheres, leading to an increasing value of k_2 with increasing double-layer thickness.[20] This phenomenon is generally referred to as the secondary electroviscous effect[21] to distinguish it from the primary electroviscous effect which affects the coefficient k_1.

5.2.2 Concentrated Dispersions of Spherical Particles

5.2.2.1 Hard Spheres

The problem of incorporating the combined effects of structure and multiparticle hydrodynamic interactions in the treatment of the rheology of hard sphere dispersions has been tackled successfully using Stokesian dynamics computer simulations.[22,23] However, as the simulations are extremely complex and time-consuming, there continue to be developed simpler, but still fundamental, theories based on various *ad hoc* approximations to the hydrodynamic interactions.[24,25] Although deficiencies do still remain, such theories give valuable insight into the underlying physics involved, as well as providing reasonable agreement with simulations and experiments.

Many semi-empirical expressions are available for describing the shear viscosity of concentrated dispersions of hard spheres. Probably the most widely used is the functional form suggested by Krieger and Dougherty:[26]

$$\eta_r = (1 - K\phi)^{-2.5/K} \tag{5.9}$$

Another such expression is the Mooney equation:[27]

$$\eta_r = \exp \frac{2.5\phi}{1 - K\phi} \tag{5.10}$$

Both of the above equations reduce to the Einstein limit [eqn. (5.1)] at low concentrations ($\phi \to 0$) and to $\eta_r \to \infty$ as $\phi \to K^{-1}$. The crowding factor K can be treated either as an adjustable parameter for fitting experimental data or as a theoretical parameter equal to the reciprocal of the volume fraction at which η_r diverges to infinity. For random close packing of monodisperse hard spheres, we have $\phi_{max} = 0.64$ and $K = 1.56$. An alternative theoretical route to K is to expand eqn. (5.9) as a power series in ϕ and equate the coefficient of ϕ^2 with the value of $k_2 = 5.2$ in eqn. (5.8) for non-Brownian hard spheres.[28] This gives $K = 1.66$ which implies a viscosity divergence at $\phi_{max} = 0.60$.

Measured values of the low shear viscosity for concentrated dispersions of hard spheres may be described reasonably well by the equation[25,28]

$$\eta_r = \frac{1 + 1.5\phi(1 + \phi - 0.189\phi^2)}{1 - \phi(1 + \phi - 0.189\phi^2)} \quad (5.11)$$

This equation gives the correct dilute limiting behaviour to order ϕ^2, and η_r diverges to infinity at $\phi_{max} = 0.64$. Another empirical equation fitting many experimental data sets is the formula:[29]

$$\eta_r = 1 + 2.5\phi + 10.05\phi^2 + 0.002\,73\,\exp(16.6\phi) \quad (5.12)$$

Non-Newtonian behaviour can be allowed for in the Krieger–Dougherty formalism by allowing K to vary with the shear-rate. The justification for this is the formation at higher shear stresses of a more ordered microstructure, involving strings or layers of particles, characterized by a different value of ϕ_{max}. The effect of shear stress τ on the crowding factor can be represented by an equation of the form

$$\frac{K - K_\infty}{K_0 - K_\infty} = \frac{1}{1 + (B\tau)^m} \quad (5.13)$$

where K_0 and K_∞ are the values of K at $\tau = 0$ and $\tau \to \infty$, respectively, and B and m are parameters whose values depend on the strengths of the restoring forces (due to Brownian motion and interparticle interactions) which drive the microstructure towards the isotropic situation in the absence of flow.

The rheology of a concentrated dispersion of hard spheres is sensitive to the degree of polydispersity of the system. The broader is the particle-size distribution, the larger is the value of ϕ_{max}, and therefore the lower is the viscosity. This effect is most pronounced with bimodal distributions. When two monodisperse systems of different particle size but the same volume fraction are mixed together, the resulting dispersion exhibits a minimum viscosity at a certain composition.[30]

5.2.2.2 Liquid Droplets

The effective viscosity η of a coarse dispersion of two liquids of viscosities η_i and η_0 can be described by the equation for two-phase flow:[31]

$$\frac{(\eta_i - \eta)\phi}{(\eta_i + 1.5\eta)} + \frac{(\eta_0 - \eta)(1 - \phi)}{(\eta_0 + 1.5\eta)} = 0 \quad (5.14)$$

However, eqn. (5.14) implies that droplets are highly deformable, which is generally not the case for stable emulsions.

The semi-empirical approach to concentrated systems of liquid droplets is to take the expression for the hard sphere dispersion and allow for particle deformability by adjusting the crowding factor K. That is, it is assumed that the packing of deformable droplets is more efficient because their shapes can be distorted during flow to accommodate neighbouring droplets. The viscosity of highly concentrated emulsions is controlled by the nature of the adsorbed layers and the thin liquid films of continuous phase between the droplets. The rheological properties are related to the value of the ratio γ/a, where γ is the interfacial tension and a is the droplet radius in the undeformed spherical state.[32] Particle deformability decreases with reduction in average droplet size, and so coarse emulsions tend to give more deviation from the hard-sphere Krieger–Dougherty equation than do fine emulsions. In addition, since droplets in fine emulsions tend to be less polydisperse, and the relative contribution of the adsorbed layer to the effective volume fraction ϕ_{eff} in eqn. (5.3) is higher, it is nearly always the case that emulsion viscosity decreases with increasing particle size.[33]

5.2.3 Flocculated Emulsions

5.2.3.1 Types of Flocculation

Flocculation is defined as the association of droplets due to attractive forces acting between them with no change in the primary droplet-size distribution. Depending on the strength of the interdroplet interaction, the flocculation may be weak (reversible) or strong (permanent). In weakly flocculated emulsions containing droplets of colloidal size ($\leqslant 1$ μm) there is a dynamic equilibrium between dispersed and flocculated states which is somewhat analogous to the gas–liquid phase equilibrium in simple fluids.[4] In strongly flocculated emulsions the properties are predominantly determined by the nature of the aggregation process and the kinetics of restructuring.[34]

In protein-stabilized systems, flocculation typically occurs when the interaction free energy between a pair of protein-covered droplets becomes appreciably negative at some separation. This may arise for one of a number of possible reasons.

(a) *Reduced surface charge density.* Substantial reduction of the surface charge density of the adsorbed protein layer produces loss of electrostatic stabilization.[35] This occurs, for example, on approaching the protein isoelectric point (pH $\approx$ 5) or in the presence of calcium ions.[36,37]

(b) *Increased ionic strength.* Addition of electrolyte screens out the double-layer repulsion and therefore reduces electrostatic stabilization.[35] Hence, emulsions containing commercial milk protein ingredients of high salt content may be flocculated whereas similar systems containing purified proteins of low salt content are stable.[36,38]

(c) *Reduced solvent quality.* Under poor solvent conditions (*e.g.* with addition of ethanol), protein–protein interactions tend to predominate at the expense of protein–solvent interactions. This leads to a loss of steric stabilization.[39]

(d) *Bridging flocculation.* When insufficient protein is available during emulsification to cover fully all the newly created oil–water interface, bridging of droplets by protein may occur.[40–42] Furthermore, once an emulsion is made, a net attractive interaction between the adsorbed protein layer and any other biopolymer component (additional protein or polysaccharide, possibly in the presence of surfactant) will typically induce flocculation by a bridging mechanism.[43,44] Covalent cross-linking between adsorbed protein molecules on different droplets may also cause irreversible bridging flocculation.[45,46]

(e) *Depletion flocculation.* The presence of a sufficiently high concentration of non-adsorbing polymer (typically a polysaccharide) generates a reversible depletion attraction between adjacent droplets due to exclusion of polymer from the gap between the surfaces.[47]

5.2.3.2   Effect of Floc Structure

At relatively low volume fractions, where aggregates are not joined together into a percolating network structure, the main effect of flocculation is to reduce the number of dispersed phase entities and to increase their average size. At the most elementary theoretical level, for rigid aggregates of well-defined size, the particle volume fraction ϕ in the Krieger–Dougherty equation [eqn. (5.9)] may simply be replaced by an effective floc volume fraction ϕ_f. That is, in terms of their effect on the flow properties, aggregates are represented[48] as a set of monodisperse spheres of hydrodynamic radius r_h and volume fraction

$$\phi_f = \tfrac{4}{3}\pi n_f r_h^{\,3} \tag{5.15}$$

where n_f is the floc number density.

Structures of some flocculated colloidal systems are observed to have a fractal character similar to that generated in computer simulation models.[34] The number of particles within a floc of radius r_f is given by

$$n(r) = n_0(r_f/a)^{d_f} \tag{5.16}$$

where d_f is the fractal dimensionality and n_0 is a prefactor close to unity. The floc radius r_f in eqn. (5.16) is a geometrical distance corresponding to the radius of the smallest sphere that can contain the total aggregate. Assuming that all the particles present are contained in flocs, a relationship in terms of d_f can be obtained[48] between the hydrodynamic radius r_h and the effective floc volume fraction ϕ_f:

$$r_h = a[(\kappa^{-d_f}\phi_f)/\phi]^{1/(3-d_f)} \tag{5.17}$$

The parameter κ is the ratio r_h/r_f; it can be obtained by fitting to experiment.[48,49]

The use of the Krieger–Dougherty equation in conjunction with eqns. (5.15) to (5.17) is valid only at low floc volume fractions where flocs do not interact strongly or become connected together in the flow. In this approach the system may exhibit non-Newtonian behaviour if the floc radius is dependent on the shear-rate D. A full description of the behaviour then requires a hydrodynamic model of floc disruption in flow.[50] For undeformable primary particles in the limit $D \to \infty$, the behaviour tends towards that of the completely unflocculated state ($\phi_f \to \phi$, $r_h \to a$). In phenomenological terms the rheology is described by a functional form similar to that of eqn. (5.13).

For a flocculated polydisperse emulsion exhibiting phase separation into floccu-lated regions and droplet-free regions, the secondary morphology of the emulsion may be represented[51] with an imaginary interfacial boundary separating the two regions of the system. The representation shown in Figure 5.1 is derived from a model of the rheology and dynamics of immiscible polymer blends.[52] The concept of an imaginary interface around the floc periphery was postulated previously[53] in connection with Brownian dynamics simulation of a large colloidal floc in shear flow. It is as if the flocculated region is surrounded by an imaginary flexible skin permeable to solvent but not to droplets,[53] with the tension related to the inter-droplet interactions in the same way that the 'normal' surface tension is related to intermolecular interactions.[54]

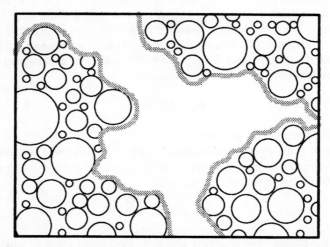

Figure 5.1 *Schematic illustration of microstructure of part of a flocculated polydisperse emulsion. The thick grey line indicates the locus of an imaginary interfacial boundary surrounding the partially connected flocculated regions* (Reproduced with permission from ref. 51)

A floc of oil droplets immobilizes a considerable amount of aqueous continuous phase, and this hydration effect has a considerable effect on the rheology[55] because the effective volume fraction ϕ_f of the emulsion is much higher than the dispersed phase fraction. The apparent viscosity of an emulsion with secondary morphology like that illustrated in Figure 5.1 is given by[51]

$$\eta = \eta_0(1 - K_H K_F \phi)^{-A} - (\gamma' q_{xy}'/D) \qquad (5.18)$$

where K_H and K_F are the so-called hydration and flocculation factors, A is a parameter related to the viscosities of the two bulk phases, and γ' and q_{xy}' are the tension and anisotropy tensor at the imaginary interface. In eqn. (5.18) the term $(K_H K_F \phi)$ is actually just another way of representing the floc volume fraction ϕ_f.[56] The term $(\gamma' q_{xy}')$ is the excess stress arising from the presence of the imaginary interface.[52]

5.2.3.3 Suspensions Flocculated by Polymer Bridging and Cross-linking

When a high-molecular-weight adsorbing polymer induces flocculation of colloidal particles or emulsion droplets, the destabilization is normally attributed to a bridging mechanism. The polymer chains become attached to more than one particle at polymer concentrations well below that required for saturation surface coverage.[57] The type of rheology exhibited by suspensions flocculated by bridging polymers is very sensitive to the strength of the polymer–particle interaction and the presence of any small-molecule surfactant which may compete with the adsorbing polymer for sites at the particle surface.[58] When the particle and polymer concentrations exceed some critical levels, suspensions flocculated by irreversible bridging form three-dimensional networks which respond elastically to small deformations at very low frequencies. When the fraction of polymer segments in trains is very substantially reduced by surfactant adsorption, the polymer adsorption is reversible, the bridging is reversible, and the flow behaviour is close to Newtonian at low shear-rates. Under intermediate conditions, where the particle surface is covered with both surfactant and polymer, the rheological behaviour can be represented as a combination of shear-thinning flow due to irreversible bridges and Newtonian flow due to reversible bridges.[58]

In concentrated emulsions stabilized by adsorbed polymers, strongly viscoelastic behaviour is associated with gel formation at two levels within the structure.[16] There is, firstly, the assembly of particle chains and aggregates into a particle gel network structure (*macro*-gelation).[59] Also there is the possibility that the adsorbed polymers around adjacent droplets may form a localized network structure (*micro*-gelation) which contributes to the total effective pair potential between the droplets.[60] The relationship between structure, rheology and interparticle interactions in monodisperse particle gel systems can be conveniently tackled using computer simulation.[34,61] Structural and small-deformation rheological properties have recently been reported[62] for a three-dimensional Brownian dynamics simulation of particle gels formed from a model of soft spherical particles incorporating a combination of flexible irreversible bonds and non-bonded interparticle interactions (attractive or repulsive). The structure of simulated particle gels of this type over intermediate length scales can be characterized in terms of a single fractal dimensionality.[63] Simulation results show[62] that there is a linear relationship between the high-frequency elastic modulus and the density of bonds (polymer cross-links) in the gel network.

5.2.3.4   Depletion Flocculation

Whenever an emulsion contains non-adsorbing entities of much smaller size than the droplets, there is the possibility of flocculation taking place as a result of an attractive depletion interaction between the droplets. Historically, the reversibility of depletion flocculation to dilution was first demonstrated[64] from observations of the creaming of natural rubber in the presence of water-soluble polymers.

The simplest theory for the depletion flocculation of spherical particles of radius *a* treats the smaller depleting entities as hard sphere solutes of diameter *d*. When the gap *h* between particle surfaces is less than *d*, there is a net interparticle attraction arising from a lower osmotic pressure in the gap than in the bulk continuous phase. The resulting depletion potential has the form[65]

$$u_d(h) = \begin{cases} -(3kTa\phi_s/d^3)(d - h)^2 & (h < d) \\ 0 & (h \geqslant d) \end{cases} \tag{5.19}$$

where ϕ_s is the solute volume fraction. More recent calculations of the hard-sphere depletion interaction to second order in solute volume fraction ϕ_s have indicated[66,67] that the depletion potential also has a positive maximum value of

$$u_d(h_{max}) = 12kTa\phi_s^2/d \qquad \text{at} \qquad h_{max} = d(1 - 3\phi_s/2) \tag{5.20}$$

and a minimum value at contact ($h = 0$) of

$$u_d(h_{min}) = -(3kTa\phi_s/d) - (3kTa\phi_s^2/5d) \tag{5.21}$$

The first term on the right of eqn. (5.21) (to order ϕ_s) is identical to the Asakura–Oosawa result [eqn. (5.19) with $h = 0$]. Although the potential energy barrier at $h \approx h_{max}$ is indicative of a kinetic stabilization effect ('depletion stabilization'), the height of the barrier is predicted to be of no practical significance unless the solute volume fraction is very high.

The interaction for a flexible polymeric solute is greater than for a hard-sphere solute because the polymer configurational entropy is reduced considerably near the particle surface.[68,69] A theoretical expression for the depletion potential in this case has the form[70]

$$u_d(h) = -2\pi a\Pi(\Delta - \tfrac{1}{2}h)[1 + (2\Delta/3a) + (h/6a)] \tag{5.22}$$

where Π is the osmotic pressure of the polymer solution of concentration ϕ_s, and Δ is the thickness of the depletion layer (roughly equal to the radius of gyration of the unadsorbed polymer in dilute solution). Hence, in dilute solution (where $\Pi \propto \phi_s$), the strength of the depletion interaction from eqn. (5.22) is directly proportional to the particle radius a ($\gg \Delta$) and to both the concentration and the molecular size of the non-adsorbing polymer. In more concentrated solutions (beyond the critical overlap concentration), the depletion layer is more compressed, the osmotic pressure is a stronger function of polymer concentration, and so the depletion interaction is stronger and of shorter range.

5.3 Experimental Investigations

Rheological measurement is a valuable tool for understanding the properties of emulsions during storage and processing. Consider, for instance, the case of a commercial salad cream. Rheology at very low applied stresses and shear-rates is relevant to shelf-life (creaming and serum separation) during extended storage in the container. Rheology at intermediate stresses and shear-rates is relevant to pouring or shaking out of the container prior to use. And rheology at high stresses and shear-rates is relevant to pumping the salad cream down a pipe in the factory immediately prior to filling the container, as well as to textural assessment in the mouth during eating. One of the most powerful ways of controlling the viscoelasticity of an emulsion like salad cream over such a wide range of flow conditions is to control the state of flocculation of the droplets.

There is no shortage of published experimental data on emulsion rheology. In terms of understanding the relationship of rheological data to structure and stability, however, some studies are more informative than others. The examples selected for consideration below are taken from the recent literature with the specific purpose of illustrating some of the key theoretical issues mentioned above.

5.3.1 Non-flocculated Emulsions

5.3.1.1 Oil-in-water Emulsions without Added Polymer

Let us first consider some experimental data from Pal's laboratory[71] for mineral oil-in-water emulsions stabilized by the nonionic emulsifier Triton X-100 (iso-octylphenoxypolyethoxyethanol). Samples were prepared with fixed volume–surface average droplet diameter $d_{32} \approx 10$ µm using a rotor/stator-type mixer with oil volume fractions in the range $0.2 \leqslant \phi \leqslant 0.7$ and 1 vol% surfactant dissolved in the aqueous phase. Apparent shear viscosities at 25 °C were measured as a function of shear-rate using a rotational viscometer with a concentric cylindrical double-gap measuring geometry (0.5 mm gap width). Figure 5.2 shows the effect of oil volume fraction ϕ on the log–log plot of shear stress τ against shear-rate D. Straight line fits to the data indicate that the rheology can be represented by a power-law model

$$\tau = \lambda D^n \qquad (5.23)$$

where λ and n are power law constants. Samples with $\phi \leqslant 0.4$ are Newtonian ($n \approx 1$) whereas those with $\phi > 0.4$ are pseudoplastic ($n < 1$). Assuming that the relative viscosity can be described by an expression of the Krieger–Dougherty type [eqn. (5.9)], the shear-thinning behaviour at high ϕ implies that K decreases with increasing τ (ϕ_{max} increases with increasing τ). At constant stress, however, the data are characterized by a single value of $\phi_{max} = 0.81$, as determined from the intercept

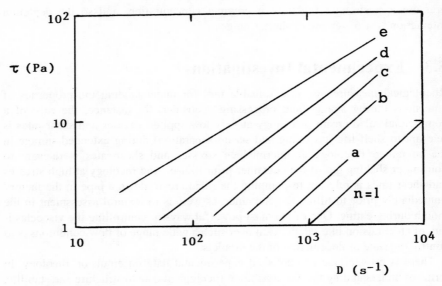

Figure 5.2 *Steady-state shear viscometry at 25 °C of coarse oil-in-water emulsions ($d_{32} \approx 10 \mu m$) of various concentrations made from mineral oil + 1 vol% solution of 1 vol% Triton X-100 in aqueous phase. Shear stress τ is plotted against shear-rate D on a log–log scale for different values of the oil volume fractions ϕ: (a) 0.2; (b) 0.4; (c) 0.5; (d) 0.6; (e) 0.7. The line labelled n = 1 [see eqn. (5.23)] has a slope of unity*
(Reproduced by permission of Elsevier Science Ltd. from ref. 71)

of an extrapolated plot of $\eta_r^{-0.5}$ against ϕ, as shown in Figure 5.3. This high value of ϕ_{max} is attributable[71] to the broad particle-size distribution and relatively high droplet deformability of coarse surfactant-stabilized emulsions.

Figure 5.4 shows a set of rheological data[72] for much finer oil-in-water emulsions (30–60 wt% mineral oil, 0.05 M aqueous phosphate buffer, pH 7, $d_{32} = 0.55 + 0.02 \mu m$) prepared with the nonionic surfactant Tween 20 (polyoxyethylene-sorbitan monolaurate) as sole emulsifier. Mutually consistent measurements of apparent viscosity as a function of shear stress were obtained[72] using controlled stress and controlled strain-rate rheometers with a double-gap concentric cylinder cell. The emulsions containing 30 or 40 wt% oil are close to Newtonian, whereas the 60 wt% one is strongly shear-thinning. Figure 5.5 shows a plot of the limiting zero-shear-rate relative viscosity η_r against oil volume fraction ϕ for the same Tween 20 emulsions in comparison with data for emulsions of slightly smaller average droplet diameter ($d_{32} = 0.44 \pm 0.02 \mu m$) made with the anionic surfactant SDS (sodium dodecyl sulfate). We can see that, whereas the data for the two sets of systems are identical up to $\phi \approx 0.45$, the limiting zero-shear-rate viscosities for the SDS emulsions are an order of magnitude larger at $\phi \approx 0.55$ and above. This is

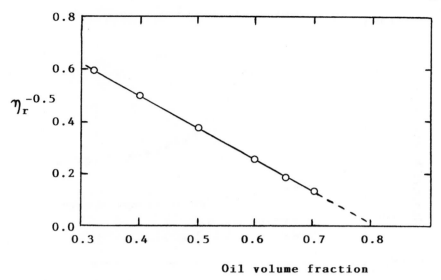

Figure 5.3 *Determination of theoretical maximum packing fraction $\phi_{max}(=K^{-1})$ in the Krieger–Dougherty equation [eqn. (5.9)] for the experimental data of Figure 5.2. The quantity $\eta_r^{-0.5}$, where η_r is the relative viscosity at stress $\tau = 6$ Pa, is plotted against the oil volume fraction ϕ*
(Reproduced by permission of Elsevier Science Ltd. from ref. 71)

attributable to a stronger interdroplet repulsion in the SDS emulsions (as well as a thicker adsorbed layer and/or lower droplet deformability), leading to a larger effective volume fraction ϕ_{eff} (a lower ϕ_{max}).

With very fine polymer-stabilized emulsions, the thickness of the adsorbed polymer layer can be of the same order of magnitude as the average particle size. This means that the space occupied by the polymer layer plays an extremely important role in determining the effective volume fraction of dispersed phase ϕ_{eff} [see eqn. (5.3)]. However, substantial deviations from hard-sphere behaviour also occur because the polymer layers can interpenetrate or compress, leading to a reduction in effective particle size.[73] By making measurements as a function of shear-rate, and assuming that there is no deformation of the particle core, it is feasible in principle to derive quantitative information about the nature of the interparticle forces between colliding droplets in the flow field. According to a recent report,[74] this approach appears to be valid for the case of photographic coupler emulsions, where the average droplet size is $d_{32} \approx 0.1$ μm, the oil phase is highly viscous, and the gelatin stabilizing layer is of similar thickness to the particle radius. Assuming that the effect of increasing applied shear forces is to compress further the adsorbed gelatin layer, changes in effective hard-sphere size of the dispersed particles derived from the Krieger–Dougherty flow curves can be

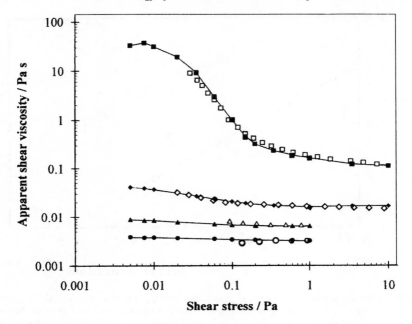

Figure 5.4 *Apparent shear viscosity as a function of shear stress for fine surfactant-stabilized mineral oil-in-water emulsions ($d_{32} = 0.55$ μm, 0.05 M phosphate buffer, pH 7) as determined at 25 °C in a controlled stress rheometer (filled symbols) and a controlled strain-rate rheometer (open symbols): ●, ○, 30 wt% oil, 1 wt% Tween 20; ▲, △, 40 wt% oil, 1.33 wt% Tween 20; ◆, ◇, 50 wt% oil, 1.67 wt% Tween 20; ■, □, 60 wt% oil, 2 wt% Tween 20*

used[75,76] to calculate an approximate particle–particle interaction potential $u(r)$, where r is the average distance between particle centres. For the case of the gelatin-stabilized photographic coupler emulsions, it was observed[74] that, upon increasing the stress up to *ca.* 500 Pa, the effective adsorbed layer thickness is reduced by between 15 and 30 nm, depending on the formulation.

5.3.1.2 Oil-in-water Emulsions with Added Polymer

High-molecular weight polysaccharides are widely used as thickeners and stabilizers in commercial emulsion formulations. The purpose of the added biopolymer is to enhance the texture and mouthfeel of the product, as well as to extend shelf-life by inhibiting or retarding destabilization processes such as creaming, coalescence and syneresis.[5] For maximum control of oil-in-water emulsion stability whilst maintaining liquid-like handling properties, the aqueous polymer solution should be extremely shear-thinning with a very high apparent viscosity at low shear stresses.[77,78] The extracellular microbial polysaccharide xanthan gum is an ideal water-soluble polymer from this rheological point of view.[79–81] Its extreme pseudo-plasticity is believed[82] to be due to the formation at low polymer concentrations of

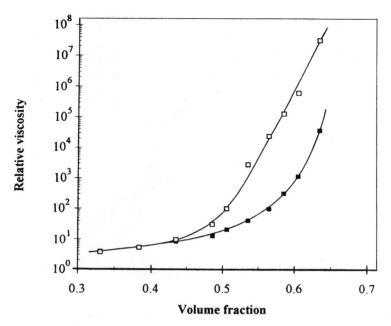

Figure 5.5 *Limiting zero-shear-rate relative viscosity as a function of volume fraction at 25 °C for fine emulsions stabilized by nonionic surfactant (Tween 20) or anionic surfactant (SDS) at surfactant:oil ratio 1:30 (wt basis):* ■, *Tween 20* $(d_{32} = 0.55 \; \mu m)$; □, *SDS* $(d_{32} = 0.44 \; \mu m)$

a complex loosely bound network of interacting rigid rod-like species composed of single or multi-stranded helices. Another biopolymer giving very strong shear-thinning behaviour in aqueous solution and in emulsions is the microbial poly-saccharide rhamsan.[83] This behaviour differs substantially from other hydrocolloids like carboxymethylcellulose (CMC) or guar gum, whose non-Newtonian solutions are only moderately shear-thinning and have much smaller low-stress limiting viscosities (at the same polymer concentrations). The non-Newtonian behaviour of solutions of CMC or guar gum is mainly due to the effect of entanglements of flexible non-interacting polymer chains.[84]

Solutions of mixed polysaccharide thickeners may have substantially enhanced viscoelastic properties over solutions of pure polysaccharides used alone.[85] For instance, a mixture of xanthan gum + guar gum exhibits remarkable rheology[86] due to a synergistic effect arising from specific interactions between unsubstituted regions of guar gum molecules and xanthan gum helices. This is illustrated by the data in Figure 5.6 for the influence of shear stress τ on the apparent viscosity η of 0.75 wt% polysaccharide solutions containing various proportions of xanthan and guar gums. (The solutions also contain 0.5 wt% nonionic surfactant to facilitate proper comparison with emulsion rheology data—see below.) Apparent viscosities were measured at 25 °C using a controlled stress rheometer with a cone-and-plate measuring system. We can see that the limiting low-stress apparent viscosity of the

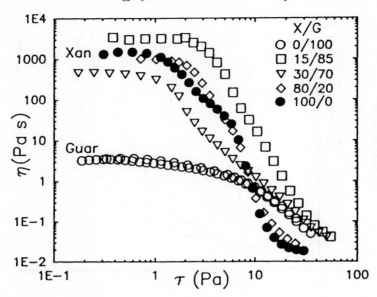

Figure 5.6 *Apparent shear viscosity η at 25 °C as a function of shear stress τ for mixed xanthan gum + guar gum solutions of total polysaccharide concentration 0.75 wt% and various xanthan/guar (X/G) ratios. Solutions also contain 0.5 wt% Triton X-100*
(Reproduced by permission of the American Institute of Chemical Engineers from ref. 86)

pure xanthan gum system is more than 200 times higher than that for pure guar gum, and that the xanthan solution is very much more shear-thinning, so that the two functions $\eta(\tau)$ cross at $\tau \approx 10$ Pa and $\eta \approx 1$ Pa s. The viscosity of the mixed polysaccharide solution is much higher than that expected on the basis of a linear relationship drawn through the pure component data, although the positive deviation from the 'mixing rule' decreases with the shear stress.[86] The mixed system with a xanthan:guar ratio of 15/85 is the composition of maximum positive synergistic interaction; it has a higher apparent viscosity than the pure xanthan system by a factor of between 2 and 10 over the whole shear stress range. It is noteworthy that, even though the mixed solution of biopolymer ratio 30/70 contains double the concentration of xanthan gum, its apparent viscosity over most of the stress range lies about an order of magnitude below than that for the 15/85 xanthan:guar system.

Figure 5.7 shows the rheology of concentrated oil-in-water emulsions (oil volume fraction $\phi = 0.65$) prepared with an aqueous phase containing 0.5 wt% Triton X-100 and 0.75 wt% polysaccharide.[86] As expected on the basis of the polysaccharide solution rheology, the emulsions containing pure xanthan gum have much higher limiting low-stress viscosities and are much more shear-thinning than

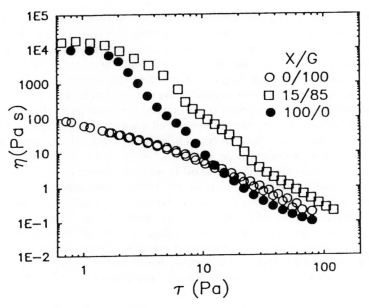

Figure 5.7 *Apparent shear viscosity η at 25 °C as a function of shear stress τ for surfactant stabilized emulsions (0.5 wt% Triton X-100, 0.75 wt% polysaccharide, 65 vol% petroleum oil) containing pure xanthan gum, pure guar gum or xanthan + guar gum (X/G = 15/85)*
(Reproduced by permission of the American Institute of Chemical Engineers from ref. 86)

those made with pure guar gum. Overall, the values of the emulsion viscosities are about an order of magnitude higher than those for the corresponding polysaccharide solutions (see Figure 5.6) at the same shear stress values. As with the mixed polysaccharide solutions, the viscosity values for the emulsion made with the xanthan:guar ratio of 15/85 are higher than those for the emulsions made with xanthan or guar gums alone in the suspending medium. The results demonstrate a good correlation between the non-Newtonian behaviour of the polysaccharide-containing emulsions and the corresponding polysaccharide solutions.

5.3.2 Flocculated Emulsions

5.3.2.1 *Depletion Flocculation by Non-adsorbing Polymer*

While high-molecular-weight polysaccharides are commonly added commercially to improve the stability of emulsions through rheological control, their presence at low concentrations has in many instances been demonstrated[72,83,87–91] to have actually the opposite effect because of the phenomenon of depletion flocculation.[47] An attractive interdroplet attraction occurs when two approaching droplet surfaces come closer together than the diameter of the non-adsorbing polymer molecule, resulting in the exclusion of the polymer from the intervening space. Reversible flocculation occurs above some critical polymer concentration c^*, but below the concentration at which a strong viscoelastic polymer network is formed in the aqueous medium between the droplets. While the most obvious manifestation of depletion flocculation is a greatly enhanced rate of emulsion creaming,[87–91] another important indicator is development of substantial non-Newtonian flow behaviour due to the formation of reversible aggregates and weak emulsion gel networks.[72,92]

Figure 5.8 shows the effect of various concentrations of the (non-gelling) nonionic microbial polysaccharide dextran (5×10^5 daltons) on the creaming kinetics of surfactant-stabilized mineral oil-in-water emulsions (30 wt% oil, 1 wt% Tween 20).[72] The data are based on simple visual observations of the rate of appearance of a distinct serum layer at the bottom of a small glass tube (height 60 mm). Samples without added dextran or containing 0.1 wt% dextran exhibit no discernible serum separation after 15 days storage at 25 °C. By contrast, those with 0.3 or 0.5 wt% dextran present exhibit rapid serum separation during the first few hours of storage. Further increases in polysaccharide to 1, 2, 5 and 10 wt% lead to a gradual lowering of the serum separation rate, although even at the highest

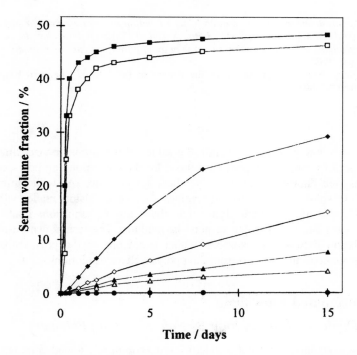

Figure 5.8 *Creaming kinetics of dextran-containing emulsions (30 wt% mineral oil, 1 wt% Tween 20, $d_{32} = 0.55 \mu m$, pH 7). Percentage of serum separated as a function of time at 25 °C is shown for various contents of added polysaccharide: ●, ≤0.1 wt%; ■, 0.3 wt%; □, 0.5 wt%; ◆, 1 wt%; ◇, 2 wt%; ▲, 5 wt%; △, 10 wt%*

dextran content the sample is less stable than the emulsions containing ≤0.1 wt% dextran. The influence of dextran (0.5, 1.0 or 10 wt%) on the rheology of the same emulsions (30 wt% oil, 1 wt% Tween 20) is shown in Figure 5.9. The polysaccharide-containing emulsions show strongly pseudoplastic behaviour, in contrast to the polysaccharide-free emulsion which is Newtonian (see Figure 5.4).[72] In the particular case of dextran, we can be sure that the shear-thinning behaviour of the polysaccharide-containing emulsion is *not* due to the direct effect of the added polymer on the aqueous phase rheology (since a 2 wt% aqueous dextran solution has a Newtonian viscosity only marginally above that of water). The origin of the pseudoplasticity must be attributed to the formation of a gel-like network of depletion flocculated emulsion droplets arising from the presence of non-adsorbing dextran at concentrations above c^*. The data in Figure 5.9 show an increase in limiting low-stress viscosity with increasing polysaccharide content. The very high degree of depletion flocculation at the highest dextran concentration (10 wt%) produces a relatively stable emulsion with respect to creaming (see Figure 5.8) because of the very slow rates of structural rearrangement and accompanying expulsion of aqueous serum from the strongly, but still reversibly, flocculated emulsion gel.

Depletion flocculation behaviour similar to that described above has also been observed in fine emulsions containing an anionic microbial polysaccharide,

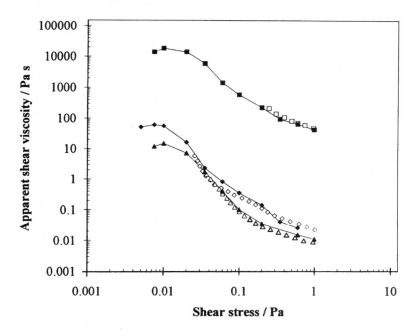

Figure 5.9 *Effect of dextran on rheology of surfactant-stabilized emulsions (30 wt% oil, 1 wt% Tween 20, d_{32} = 0.55 μm, 0.05 M phosphate buffer, pH 7). Apparent shear viscosity at 25 °C is plotted against shear stress as determined in a controlled stress rheometer (filled symbols) and a controlled strain-rate rheometer (open symbols):* ▲, △, *0.5 wt% dextran;* ◆, ◇, *1 wt% dextran;* ■, □, *10 wt% dextran*

xanthan[90] or rhamsan.[83] However, the situation is somewhat different with these hydrocolloids due to the formation of a weak gel-like structure in the aqueous phase at concentrations above c^*. Whereas a dilute dextran solution is a low-viscosity Newtonian liquid ($\sim 10^{-3}$ Pa s), a solution of $0.5–1.0$ wt% xanthan (or rhamsan) is highly shear-thinning with a limiting zero-shear-rate viscosity some 10^6 times larger (see Figure 5.5). This effect, together with stronger depletion flocculation induced by the xanthan polyelectrolyte than by the uncharged dextran polymer, can lead effectively to complete inhibition of serum separation during extended storage of an emulsion containing xanthan at a concentration as low as 0.3 wt%.[93]

In emulsions prepared with a common milk protein ingredient, sodium caseinate, as the sole emulsifier, it has been recently observed[94–96] that creaming and rheological behaviour are rather sensitive to the concentration of unadsorbed protein. Figure 5.10 shows sets of time-dependent creaming plots of oil volume fraction against height for fine oil-in-water emulsions (35 vol% n-tetradecane, pH 6.8, $d_{32} = 0.42 \pm 0.03$ μm) prepared with (a) 2 wt%, (b) 4 wt% and (c) 6 wt% sodium caseinate (based on the whole emulsion). Oil concentration profiles during quiescent storage in tubes of height 25 cm were determined at 30 °C using the ultrasound velocity scanning technique.[97,98] The creaming behaviour of the 2 wt% caseinate emulsion (Figure 5.10a) is typical of that of a stable (unflocculated) system. Within the first day or so, there is no change in profile; after about a week or so, there is the appearance of a thin cream layer right at the top of the sample; after about a month or so, the cream layer at the top has reached a thickness corresponding to around 10% of the whole sample height and there is slight depletion of oil droplets from near the bottom 10% of the sample height. The profiles in Figure 5.10a for the emulsion containing 2 wt% protein, which has a protein surface coverage of 2.4 mg m^{-2} and an unadsorbed protein concentration of below 1 wt% (based on the aqueous phase),[94,95] is characteristic of that expected for a stable fine emulsion composed of small individual droplets that cream in accordance with Stokes' Law.[8]

When the total concentration of protein in the fine caseinate-stabilized emulsion is increased to 4 wt% (corresponding to 3.4 wt% unadsorbed protein in the aqueous phase), it becomes much less stable to creaming, as illustrated in Figure 5.10b. After an initial lag period of a few hours, a serum layer of low oil content ($\phi \approx 0.03$) appears at the bottom of the sample tube and a sharp serum/emulsion interface moves steadily up the tube, reaching a height of $4–5$ cm after about 1 day.

Figure 5.10 *Influence of total protein content on the gravity creaming profiles of caseinate-stabilized emulsions (35 vol% n-tetradecane, $d_{32} = 0.39–0.45$ μm, 0.05 M phosphate buffer, pH 7). Oil volume fraction ϕ is plotted against height for various times of quiescent storage at 30 °C. (a) 2 wt% protein: □, 1 day; ▲, 9 days; △, 17 days; ●, 28 days; ○, 34 days. (b) 4 wt% protein: □, 6 h; ▲, 16.5 h; △, 18 h; ●, 21 h; ○, 24 h; ◆, 27 h; – – –, 31 days. (c) 6 wt% protein: □, 1 day; ▲, 8 days; △, 9 days; ●, 10 days; ○, 16 days; ◆, 30 days* (Reproduced by permission of Oxford University Press from ref. 95)

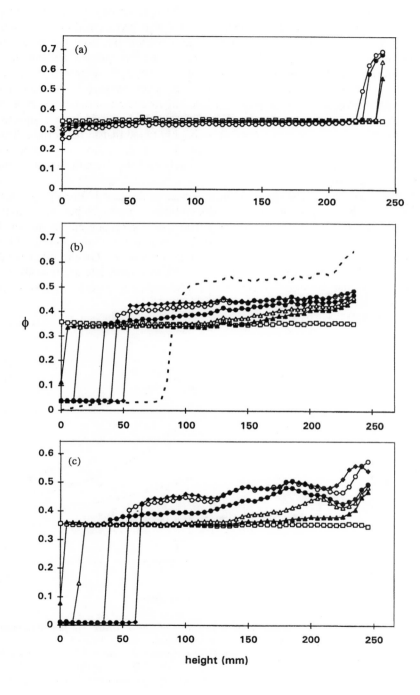

height (mm)

This behaviour is typical of an emulsion flocculated reversibly by non-adsorbing polymer. Increasing the protein content to 6 wt% also gives creaming behaviour characteristic of a flocculated emulsion (Figure 5.10c), but now the emulsion remains uniform for about 1 week before serum separation appears. The emulsion is still much less stable than the 2 wt% caseinate system (Figure 5.10a), but considerably more stable than the 4 wt% caseinate system (Figure 5.10b). Both the 4 and 6 wt% caseinate emulsions have a protein saturation surface coverage of 3.15 ± 15 mg m^{-2}, but the concentration of unadsorbed protein in the aqueous phase of the latter emulsion (~ 6.6 wt%) is roughly twice that in the former. This implies a theoretical interdroplet interaction due to depletion flocculation by spherical solute species [see eqn. (5.19)] of around twice the strength as that for the 4 wt% caseinate emulsion.

As with the polysaccharide-containing systems discussed already, a clear correlation can be observed for the caseinate-stabilized emulsions between gravity creaming behaviour[94] and shear rheology.[96] The 2 wt% caseinate emulsion, which has good creaming stability (Figure 5.10a), is a simple Newtonian liquid with a low viscosity ($\sim 10^{-2}$ Pa s) that does not change on storage. On the other hand, the depletion flocculated emulsions containing $\geqslant 4$ wt% caseinate are time-dependent non-Newtonian systems.[96] Figure 5.11 shows apparent viscosity against shear

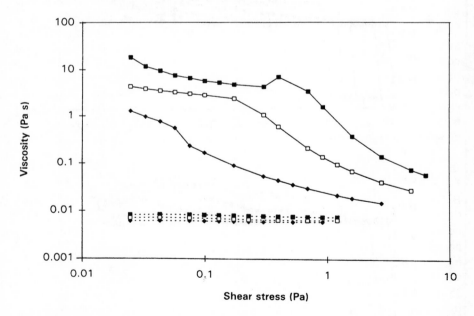

Figure 5.11 *Apparent shear viscosity as a function of shear stress for fine caseinate-stabilized emulsions (35 vol% n-tetradecane, $d_{32} = 0.39-0.45$ μm, 0.05 M phosphate buffer, pH 7, 30 °C). Data are shown for three different concentrations of sodium caseinate:* ◆, *4 wt%;* □, *5 wt%;* ■, *6 wt%. Dashed lines denote data for emulsions with the original continuous phase replaced by aqueous buffer*
(Reproduced by permission of Academic Press from ref. 96)

stress for emulsions containing 4–6 wt% protein (based on the whole emulsion) measured approximately 1 hour after preparation. The strong shear-thinning behaviour is characteristic of a flocculated structure. For these casein-rich emulsions, each containing a considerable excess of unadsorbed protein, the removal of the aqueous continuous phase (following centrifugation) and its replacement by aqueous buffer solution leads to a change to the same low-viscosity Newtonian behaviour as found with the 2 wt% caseinate emulsion. This clearly shows that the reversible flocculation causing the pseudoplasticity is due to the presence of unadsorbed protein in the aqueous phase between the droplets. Flocculation is attributed to a depletion interaction, not caused in this case by individual polymer molecules, but by spherical self-assembled protein particles (~10 nm) called caseinate sub-micelles. The postulated mechanism is therefore analogous to the reversible depletion flocculation of surfactant-stabilized emulsions induced by small-molecule surfactant micelles.[99–101] On this basis, one expects the depletion flocculation of the caseinate emulsion to be sensitive to any factor that disturbs the balance between adsorbed protein and unadsorbed protein or influences the state of casein self-association in the aqueous medium. This explains why addition of alcohol or calcium ions has a profound effect on the creaming stability and rheological behaviour.[102]

Depletion flocculated casein emulsions exhibit time-dependent rheological behaviour following high-speed shearing.[96,102] This can be attributed to some gradual restructuring of the transient gel network under the influence of reversible attractive interparticle interactions.[34] By analogy with results from particle gel computer simulations,[103] we can infer that the fractal-type network structure undergoes thermally driven reorganization and coarsening into locally more close-packed structures. It is the gradual expulsion of the serum phase during this emulsion gel restructuring process which is probably the main destabilizing mechanism during the creaming of emulsions containing a high concentration of unadsorbed protein.

5.3.2.2 Bridging Flocculation by Adsorbing Polymer

Bridging flocculation occurs in an emulsion made with a polymeric emulsifier if there is insufficient polymer present during emulsification to cover fully all the newly created oil–water interface. The presence of bridging flocs usually implies non-Newtonian rheology, and reduced long-term stability with respect to creaming and coalescence. In recent experiments with fine caseinate-stabilized emulsions prepared at a relatively low protein:oil ratio (i.e. 1 wt% sodium caseinate, 35 wt% oil), there was found[94,96] to be only slight bridging flocculation in the freshly prepared emulsion, but the degree of flocculation gradually became more extensive on storage over several days, as determined by measurements of changes in apparent droplet-size distribution and steady-state shear viscometry. The 1 wt% caseinate emulsion was found to have a low estimated surface coverage (1.4 mg m^{-2}) as compared with a saturated surface coverage of 2.5–3 mg m^{-2}. The tendency of the adsorbed protein to become shared between adjacent droplets leads to strong bridging flocculation that is not reversed on dilution;[94] this contrasts with the reversibility of depletion flocculation (see Figure 5.11).

An initially stable emulsion can exhibit a different kind of bridging flocculation if an adsorbing polymer, added later, interacts to form direct linkages between adjacent droplets. This situation occurs in protein-stabilized emulsions containing hydrocolloids where the protein–polysaccharide interaction at the surface of the droplets is net attractive.[104–106] The rheological implications of such a situation are illustrated in Figure 5.12. The plot shows the complex shear modulus G^* (1 Hz) of a BSA-stabilized emulsion (2.7 wt% protein, 40 vol% n-tetradecane, pH 6, $d_{32} = 0.55 \pm 0.01$ μm) containing various concentrations of ι-carrageenan added after emulsion formation.[107] There is a steep increase in modulus from $G^* \approx 1$ Pa in the absence of polysaccharide to $G^* > 10^2$ Pa at an added ι-carrageenan concentration of $C \approx 0.04$ wt%. Bridging flocculation on addition of polysaccharide is confirmed by a large increase in apparent average droplet size d_{32}^* which is not reversed on dilution.[107] A net attractive electrostatic interaction between BSA and ι-carrageenan in dilute aqueous solution (pH 6, 5 mM imidazole) has been confirmed by surface tension measurements; the interaction is strong at pH 6, but it is much weaker at pH 7, and it disappears in 0.1 M NaCl. Figure 5.12 shows that the complex shear modulus falls to much lower values as the polysaccharide content increases to $C > 0.1$ wt%. This is explained in terms of substantially reduced droplet bridging due to restabilization of the protein-coated emulsion droplets by a full secondary layer of adsorbed polysaccharide. Similar behaviour has also been observed in systems with ι-carrageenan replaced by dextran sulfate.[106]

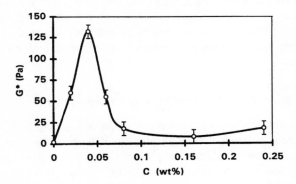

Figure 5.12 *Effect of ι-carrageenan added after emulsification on small-deformation rheology of fine BSA-stabilized emulsions (2.7 wt% protein, 40 vol% n-tetradecane, $d_{32} = 0.55$ μm, pH 6).[107] Complex shear modulus G* at 1 Hz and 30 °C is plotted against polysaccharide concentration c*
(Reproduced by permission from ref. 107)

Unlike some dispersions of small colloidal particles,[108] the bridging polymers in most emulsion systems are usually smaller in size than the emulsion droplets. Hence, the flocculation mechanisms of polymer bridging and polymer cross-linking are not so distinctive. In the case of a protein-stabilized emulsion, the droplets may become joined together during a separate processing step after emulsion formation through, say, heating[109] or treatment with a cross-linking enzyme.[46] The unflocculated low viscosity liquid-like emulsion is then converted into a viscoelastic emulsion gel. Improved understanding of the dependence of the small-deformation rheology of such a system on the interparticle interactions can be derived[62] from Brownian dynamics computer simulations.

5.4 References

1. P. Walstra, in 'Encyclopedia of Emulsion Technology', ed. P. Becher, Dekker, New York, 1996, vol. 4, p. 1.
2. G.A. Núñez, M. Briceño, C. Mata, H. Rivas and D.D. Joseph, *J. Rheol.*, 1996, **40**, 405.
3. S. Matsumoto and W.W. Kang, *J. Disp. Sci. Technol.*, 1989, **10**, 455.
4. R. Pal, *Langmuir*, 1996, **12**, 2220.
5. E. Dickinson, 'An Introduction to Food Colloids', Oxford University Press, Oxford, 1992, ch. 4.
6. E. Dickinson, *Annu. Rep. Prog. Chem., Sect. C*, 1986, **83**, 31.
7. B.P. Binks, *Annu. Rep. Prog. Chem., Sect. C*, 1996, **92**, 97.
8. E. Dickinson, in 'Food Structure—Its Creation and Evaluation', eds. J.M.V. Blanshard and J.R. Mitchell, Butterworths, London, 1988, p. 41.
9. D.H. Everett, 'Basic Principles of Colloid Science', Royal Society of Chemistry, London, 1988, ch. 8.
10. H.A. Barnes, J.F. Hutton and K. Walters, 'An Introduction to Rheology', Elsevier, Amsterdam, 1989.
11. S.B. Ross-Murphy, in 'New Physico-Chemical Techniques for the Characterization of Complex Food Systems', ed. E. Dickinson, Blackie, Glasgow, 1995, p. 139.
12. R.W. Whorlow, 'Rheological Techniques', Wiley, New York, 1980.
13. A. Einstein, *Ann. Phys.*, 1906, **19**, 289.
14. A. Einstein, *Ann. Phys.*, 1911, **24**, 591.
15. P. Sherman, 'Industrial Rheology', Academic Press, London, 1970, p. 164.
16. E. Dickinson and G. Stainsby, 'Colloids in Food', Applied Science, London, 1982, ch. 7.
17. G.I. Taylor, *Proc. R. Soc. (London)*, 1932, **A138**, 41.
18. J.G. Oldroyd, *Proc. R. Soc. (London)*, 1955, **A232**, 567.
19. G.K. Batchelor, *J. Fluid Mech.*, 1977, **83**, 97.
20. W.B. Russel, *J. Rheol.*, 1980, **24**, 287.
21. J.W. Goodwin, in 'Colloid Science', ed. D.H. Everett, Specialist Periodical Reports, Chemical Society, London, 1975, vol. 2, p. 246.
22. R.J. Phillips, J.F. Brady and G. Bossis, *Phys. Fluids*, 1988, **31**, 3462.
23. J.F. Brady and G. Bossis, *J. Chem. Phys.*, 1989, **91**, 4427.
24. J.F. Brady, *J. Chem. Phys.*, 1993, **99**, 567.
25. R.A. Lionberger and W.B. Russel, *J. Rheol.*, 1997, **41**, 399.
26. I.M. Krieger and T.J. Dougherty, *Trans. Rheol. Soc.*, 1959, **3**, 137.
27. M. Mooney, *J. Colloid Sci.*, 1951, **6**, 162.
28. D.S. Pearson and T. Shikata, *J. Rheol.*, 1994, **38**, 601.
29. D.G. Thomas, *J. Colloid Sci.*, 1965, **20**, 267.
30. B.E. Rodriguez, E.W. Kaler and M.S. Wolfe, *Langmuir*, 1992, **8**, 2382.
31. D. Bedeaux, *Physica A*, 1983, **121**, 345.

32. R. Pal, *AIChE J.*, 1996, **42**, 3181.
33. H. Barnes, *Colloids Surf. A*, 1994, **91**, 89.
34. E. Dickinson, in 'Food Colloids: Proteins, Lipids and Polysaccharides', eds. E. Dickinson and B. Bergenståhl, Royal Society of Chemistry, Cambridge, 1997, p. 107.
35. E. Dickinson and G. Stainsby, 'Colloids in Food', Applied Science, London, 1982, ch. 2.
36. E. Dickinson, R.H. Whyman and D.G. Dalgleish, in 'Food Emulsions and Foams', ed. E. Dickinson, Royal Society of Chemistry, London, 1987, p. 40.
37. E. Dickinson, J.A. Hunt and D.S. Horne, *Food Hydrocolloids*, 1992, **6**, 359.
38. E. Dickinson, *J. Chem. Soc., Faraday Trans.*, 1997, **93**, 2297.
39. E. Dickinson and G. Stainsby, 'Colloids in Food', Applied Science, London, 1982, ch. 3.
40. H. Mulder and P. Walstra, 'The Milk Fat Globule', Pudoc, Wageningen, Netherlands, 1974, ch. 8.
41. E. Dickinson, F.O. Flint and J.A. Hunt, *Food Hydrocolloids*, 1989, **3**, 389.
42. E. Dickinson and V.B. Galazka, *J. Chem. Soc., Faraday Trans.*, 1991, **87**, 963.
43. E. Dickinson, in 'Biopolymer Mixtures', eds. S.E. Harding, S.E. Hill and J.R. Mitchell, Nottingham University Press, Nottingham, 1995, p. 349.
44. J. Chen and E. Dickinson, *Colloids Surf. A*, 1995, **100**, 255.
45. D.J. McClements, F.J. Monahan and J.E. Kinsella, *J. Food Sci.*, 1993, **58**, 1036.
46. E. Dickinson and Y. Yamamoto, *J. Agric. Food Chem.*, 1996, **44**, 1371.
47. D.H. Napper, 'Polymeric Stabilization of Colloidal Dispersions', Academic Press, London, 1983, ch. 15.
48. W. Wolthers, D. van den Ende, M.H.G. Duits and J. Mellema, *J. Rheol.*, 1996, **40**, 55.
49. P. Wiltzius, *Phys. Rev. Lett.*, 1987, **58**, 710.
50. P.D. Patel and W.B. Russel, *Colloids Surf.*, 1988, **31**, 355.
51. H.M. Lee, J.W. Lee and O.O. Park, *J. Colloid Interface Sci.*, 1997, **185**, 297.
52. H.M. Lee and O.O. Park, *J. Rheol.*, 1994, **38**, 1405.
53. G.C. Ansell and E. Dickinson, *J. Colloid Interface Sci.*, 1986, **110**, 73.
54. J.S. Rowlinson and B. Widom, 'Molecular Theory of Capillarity', Oxford University Press, Oxford, 1982.
55. R. Pal, *J. Rheol.*, 1992, **36**, 1245.
56. R. Pal and E.J. Rhodes, *J. Rheol.*, 1989, **33**, 1021.
57. E. Dickinson and L. Eriksson, *Adv. Colloid Interface Sci.*, 1991, **34**, 1.
58. Y. Otsubo, *Adv. Colloid Interface Sci.*, 1994, **53**, 1.
59. E. Dickinson, 'Proceedings of 1st International Symposium on Food Rheology and Structure', eds. E. J. Windhab and B. Wolf, Vincentz Verlag, Hannover, 1997, p. 50.
60. T. van Vliet, J. Lyklema and M. van den Tempel, *J. Colloid Interface Sci.*, 1978, **65**, 505.
61. E. Dickinson and D.J. McClements, 'Advances in Food Colloids', Blackie, Glasgow, 1995, ch. 4.
62. M. Whittle and E. Dickinson, *Mol. Phys.*, 1997, **90**, 739.
63. E. Dickinson. *J. Colloid Interface Sci.,* 1987, **118**, 286.
64. C. Bondy, *Trans. Faraday Soc.*, 1939, **35**, 1093.
65. S. Asakura and F. Oosawa, *J. Chem. Phys.*, 1954, **22**, 1255.
66. J.Y. Walz and A. Sharma, *J. Colloid Interface Sci.*, 1994, **168**, 485.
67. Y. Mao, M.E. Cates and H.N.W. Lekkerkerker, *Physica A*, 1995, **222**, 10.
68. S. Asakura and F. Oosawa, *J. Polym. Sci.*, 1958, **33**, 183.

69. J.F. Joanny, J.F. Liebler and P.-G. de Gennes, *J. Polym. Sci., Polym. Phys. Edn.*, 1979, **17**, 1073.
70. B. Vincent, J. Edwards, S. Emmett and A. Jones, *Colloids Surf.*, 1986, **18**, 261.
71. Y. Yan, R. Pal and J. Masliyah, *Chem. Eng. Sci.*, 1991, **46**, 985.
72. E. Dickinson, M.I. Goller and D.J. Wedlock, *J. Colloid Interface Sci.*, 1995, **172**, 192.
73. J. Frith, W.B. Russel, T.A. Strivens and J. Mewis, *AIChE J.*, 1989, **35**, 415.
74. A.M. Howe, A. Clarke and T.H. Whitesides, *Langmuir*, 1997, **13**, 2617.
75. R. Buscall, *J. Chem. Soc., Faraday Trans.*, 1991, **87**, 1365.
76. R. Buscall, *Colloids Surf. A*, 1994, **83**, 33.
77. E. Dickinson, in 'Gums and Stabilisers for the Food Industry', eds. G.O. Phillips, D.J. Wedlock and P.A. Williams, IRL Press, Oxford, 1988, vol. 4, p. 249.
78. E. Dickinson, in 'Food Hydrocolloids: Structures, Properties and Functions', eds. K. Nishinari and E. Doi, Plenum, New York, 1993, p. 387.
79. W.E. Rochefort and S. Middleman, *J. Rheol.*, 1987, **31**, 337.
80. C.J. Carriere, E.J. Amis, J.L. Schrag and J.D. Ferry, *J. Rheol.*, 1993, **37**, 469.
81. R. Pal, *AIChE J.*, 1995, **41**, 783.
82. J.O. Carnali, *Rheol. Acta*, 1992, **31**, 339.
83. E. Dickinson, M.I. Goller and D.J. Wedlock, *Colloids Surf. A*, 1993, **75**, 195.
84. B. Launay, J.-L. Doublier and G. Cuvelier, in 'Functional Properties of Food Macromolecules', eds. J.R. Mitchell and D.A. Ledward, Elsevier Applied Science, London, 1986, p. 1.
85. J.-L. Doublier and G. Llamas, in 'Food Polymers, Gels and Colloids', ed. E. Dickinson, Royal Society of Chemistry, Cambridge, 1991, p. 349.
86. R. Pal, *AIChE J.*, 1996, **42**, 1824.
87. Y. Cao, E. Dickinson and D.J. Wedlock, *Food Hydrocolloids*, 1990, **4**, 185.
88. Y. Cao, E. Dickinson and D.J. Wedlock, *Food Hydrocolloids*, 1991, **5**, 443.
89. A.J. Fillery-Travis, P.A. Gunning, D.J. Hibberd and M.M. Robins, *J. Colloid Interface Sci.*, 1993, **159**, 1896.
90. E. Dickinson, J. Ma and M.J.W. Povey, *Food Hydrocolloids*, 1994, **8**, 481.
91. A. Parker, P.A. Gunning, K. Ng and M.M. Robins, *Food Hydrocolloids*, 1995, **9**, 333.
92. R. Pal, *Colloids Surf.*, 1992, **64**, 207.
93. E. Dickinson, J. Ma and M.J.W. Povey, *J. Chem. Soc., Faraday Trans.*, 1996, **92**, 1213.
94. E. Dickinson, M. Golding and M.J.W. Povey, *J. Colloid Interface Sci.*, 1997, **185**, 515.
95. E. Dickinson and M. Golding, *Food Hydrocolloids*, 1997, **11**, 13.
96. E. Dickinson and M. Golding, *J. Colloid Interface Sci.*, 1997, **191**, 166.
97. M.J.W. Povey, in 'New Physico-chemical Techniques for the Characterization of Complex Food Systems', ed. E. Dickinson, Blackie, Glasgow, 1995, p. 196.
98. M.J.W. Povey, 'Ultrasonic Techniques for Fluids Characterization', Academic Press, San Diego, 1997.
99. M.P. Aronson, *Langmuir*, 1989, **5**, 494.
100. M.P. Aronson, in 'Emulsions: A Fundamental and Practical Approach', ed. J. Sjöblom, Kluwer, Dordrecht, Netherlands, 1992, p. 75.
101. J. Bibette, D. Roux and B. Pouligny, *J. Phys. II*, 1992, **2**, 401.
102. M. Golding, Ph. D. thesis, University of Leeds, 1997.
103. B.H. Bijsterbosch, M.T.A. Bos, E. Dickinson, J.H.J. van Opheusden and P. Walstra, *Faraday Discuss.*, 1995, **101**, 51.
104. E. Dickinson and S.R. Euston, in 'Food Polymers, Gels and Colloids', ed. E. Dickinson, Royal Society of Chemistry, Cambridge, 1991, p. 132.
105. E. Dickinson and D.J. McClements, 'Advances in Food Colloids', Blackie, Glasgow, 1995, ch. 3.

106. E. Dickinson and K. Pawlowsky, in 'Gums and Stabilisers for the Food Industry', eds. G.O. Phillips, D.J. Wedlock and P.A. Williams, Oxford University Press, Oxford, 1996, vol. 8, p. 249.
107. E. Dickinson and K. Pawlowsky, *J. Agric. Food Chem.*, 1997, **45**, 3799.
108. B. Cabane, K. Wong, P. Lindler and F. Lafuma, *J. Rheol.*, 1997, **41**, 531.
109. E. Dickinson and S.T. Hong, *Colloids Surf. A*, 1997, **127**, 1.

CHAPTER 6

Phase Inversion and Drop Formation in Agitated Liquid–Liquid Dispersions in the Presence of Nonionic Surfactants

BRIAN W. BROOKS, HOWARD N. RICHMOND and MOHAMED ZERFA

Chemical Engineering Department, Loughborough University, Loughborough LE11 3TU, UK

6.1 Introduction

Over the last few decades, the production of emulsions has gained the interest of many scientists due to the wide range of technological products and processes involved. Stable emulsion systems are used in a large area of industry, such as the food industry, pharmacy, cosmetics and coatings. Although the procedures of direct emulsification of O/W and W/O systems are well established, the need for more efficient processes with a low energy input has become necessary. In recent years, the use of emulsification by phase inversion was seen as a possible solution.[1,2] Therefore, there are good reasons for developing the understanding of phase inversion processes.

When two immiscible liquids are mixed together, one of the phases can be dispersed as fine drops in the other phase (continuous phase). The dispersion produced is unstable because an increase in the interfacial area (due to mixing) results in an increase in the interfacial free energy, thus producing a thermo-dynamically unstable system. Consequently, the latter system cannot form an emulsion unless a third component is present to stabilise the system. This third component is often a surfactant (surface active agent) or a mixture of surfactants. The stability of such emulsions may last from a few minutes to many years, depending on the intended use. The nature of such an emulsion (O/W or W/O) depends mainly on the nature of the surfactant. The terms 'oil phase' and 'water phase' (aqueous phase) are used to differentiate between the two phases. If the surfactant is more soluble in the water phase than in the oil phase (hydrophilic), micelles are formed in the water phase and the water phase becomes continuous. On the other hand, if the surfactant is more soluble in the oil phase (lyophilic), micelles are formed in the oil phase and the oil phase becomes continuous. In some cases, the initial state of the emulsion (O/W or W/O) depends on the order of addition of the phases. When the conditions of the formed emulsion are changed, phase inversion may occur (the continuous phase becomes the dispersed phase and *vice versa*). This change can be brought about either by changing the volume fraction of one of the phases (catastrophic inversion) or by changing the affinity of the surfactant towards the two phases (transitional inversion).[3] The processes for the preparation of emulsions as well as the formation of multiple emulsions (O/W/O or W/O/W) by the latter method are not fully understood. Hence, it is necessary to develop an understanding of the essential features of phase inversion.

Relatively few published studies deal with the dynamics of phase inversion. The role of agitation has not been well established and the conditions necessary for the induction of phase inversion are usually difficult to define. Often the nature of the final emulsion cannot be predicted. Hence, a number of important points must be clarified if the production of emulsions by phase inversion on a large scale is to be a reliable and reproducible process.

6.2 Basic Principles of Phase Inversion

This section reviews the basic principles related to phase inversion. It also presents the relevant literature concerning nonionic surfactant–oil–water (nSOW) phase behaviour and emulsion phase inversion. It is always important to recall the IUPAC definition of an emulsion, which is: 'An emulsion is a dispersion of droplets of one liquid in another with which it is incompletely miscible. Emulsions of droplets of an organic liquid (an 'oil') in an aqueous solution are indicated by the symbol O/W and emulsions of aqueous droplets in an organic liquid as W/O. In emulsions the droplets often exceed the usual limits for colloids in size'. Although this definition is adequate, it is necessary to indicate the nature of the surfactant added to the liquid–liquid system as it is not possible to prepare an emulsion without this third component. The surface active agent (surfactant) has the property of accumulating at the oil–water interface. It helps drop break-up by reducing the system's interfacial tension and provides a mechanism for stabilising the emulsion drops against coalescence. Surfactants are available in many forms, usually either ionic type, nonionic, polymer or polyelectrolyte, finely divided solids or any combination of two or more of these materials.

Nonionic surfactant molecules contain a polar hydrophilic group (often a polyoxyethylene chain) and a non-polar (lipophilic) chain; this character is called amphipathy or amphilicity and results in a double affinity which can only be satisfied at a polar–non-polar interface. Commercial nonionic surfactants are polydisperse, having a distribution of hydrophilic chain lengths; they are sold as having an average chain length. Nonionic–oil–water systems are usually 'pseudo ternary', *i.e.* the surfactant may be a mixture of surfactants, the aqueous phase may be a salt solution and the oil phase may be a mixture of oils or contain dissolved polymers. The basic concepts related to the surfactant behaviour that need to be understood in order to explain the behaviour of emulsified systems are: solubility, formation of micelles and interfacial adsorption.

6.2.1   Dissolution State of Nonionic Surfactants

The most important properties of a surfactant are its solubility and its interfacial activity. When a nonionic surfactant is added to a two-phase liquid–liquid system, it preferentially adsorbs at the interface forming an adsorbed layer. At low surfactant concentrations an equilibrium exists between surfactant monomers dissolved in the oil phase, surfactant monomers dissolved in the water phase and the interfacial surfactant. In the case of a separated system (constant interfacial area), as the concentration of surfactant in the system increases, the amount of surfactant at the interface reaches a maximum possible concentration and, on further increase in surfactant concentration, excess surfactant will now form micelles in either the oil or aqueous phase, or form a surfactant phase depending on its affinity for the oil and water phases. The break point is termed the critical micelle concentration (CMC).

A surfactant micelle is generally pictured as a sphere of surfactant molecules with a liquid phase core, *e.g.* an aqueous micellar solution has continuous structure containing micelles with an oil phase core. This phase is thermodynamically stable and the oil within the micelle is termed solubilised. For conditions below the CMC, the effect of the adsorbed interfacial surfactant on the nSOW system's interfacial tension is governed by the Gibbs adsorption equation:

$$\Gamma = \frac{-1}{RT} \frac{\partial \sigma}{\partial \ln C} \tag{6.1}$$

where Γ = surface excess of surfactant (mol m^{-2}), R = gas constant, T = temperature, σ = interfacial tension and C = surfactant concentration. This equation is useful for calculating amounts of surfactant adsorbed or the surfactant molecule area of coverage at the interface from the variation of σ with $\ln C$.

To stabilise an emulsion, the surfactant must be present at a concentration above the CMC; hence, we shall be mainly concerned with systems of this sort. The phase in which the surfactant forms micelles is dependent on the surfactant's affinity for oil and water. The surfactant's affinity is controlled by a number of factors. Winsor[4,5] first addressed the problem of describing surfactant affinity. He introduced the concept of interaction energies between surfactant molecules adsorbed at the interface and the oil and water phases. Salager[3] identified different types of interactions. The ratio of the total interaction energies (per unit area of interface) of the surfactant for the oil and water phases is known as Winsor r (symbols are defined below):

$$r = \frac{A_{co}}{A_{cw}} = \frac{A_{lco} + A_{hco} - A_{oo} - A_{ll}}{A_{lcw} + A_{hcw} - A_{ww} - A_{hh}} \tag{6.2}$$

where A_{co} = surfactant–oil interaction energies, A_{cw} = surfactant–water interaction energies, A_{hcw} = interaction energies between the water and the hydrophilic part of the surfactant, A_{hco} = interaction energies between the oil and the hydrophilic part of the surfactant, A_{oo} = oil–oil interaction energies, A_{ll} = interaction energies between the lipophilic parts of the surfactant, A_{lcw} = interaction energies between the water and the lipophilic part of the surfactant, A_{lco} = interaction energies between the oil and the lipophilic part of the surfactant, A_{ww} = water–water interaction energies and A_{hh} = interaction energies between the hydrophilic parts of the surfactant. The three cases of $r < 1$, $r > 1$ and $r = 1$ correspond to type 1, type 2 and type 3 nSOW phase behaviour, respectively. Although Winsor's approach is purely quantitative, researchers have shown that it allows interpretation of all known effects, *e.g.* for a case $r < 1$, with a rise in temperature the hydration forces between the water phase and the hydrophilic group of a nonionic surfactant (A_{hcw}) decrease and therefore r increases; a progressive change from type 1 to type 3 to type 2 phase behaviour is seen with a rise in temperature.[1]

In Winsor's type 1 systems ($r < 1$) the affinity of the surfactant for the water phase exceeds its affinity for the oil phase. Thus, the interface will be convex towards water. Davis[6] used a triangular diagram to show that a type 1 nSOW system can be one or two phases. A system in the two-phase region will split into an oil phase containing dissolved surfactant monomers at CMC_o (critical micelle concentration in the oil phase) and an aqueous microemulsion—a water phase containing solubilised oil in normal surfactant micelles.

In Winsor's type 2 systems ($r > 1$), the affinity of the surfactant for the oil phase exceeds its affinity for the water phase and the interface will be convex towards oil. Again one- or two-phase nSOW systems are possible. A system in the two-phase region will split into a water phase containing surfactant monomers at CMC_w (critical micelle concentration in the water phase) and an oleic microemulsion—an oil phase containing solubilised water in reverse surfactant micelles.

In Winsor's type 3 systems ($r = 1$), the surfactant's affinity for the oil and the water phases is balanced. The interface will be flat. A type 3 nSOW system can have one, two or three phases depending on its composition. In the multiphase region the system can be: (a) two phase—a water phase and an oleic microemulsion; (b) two phase—an oil phase and an aqueous microemulsion; (c) three phase—a water phase containing surfactant monomers at CMC_w, an oil phase containing surfactant at CMC_o and a 'surfactant phase'. The surfactant phase may have a bicontinuous structure,[7] being composed of cosolubilised oil and water separated from each other by an interfacial layer of surfactant. The surfactant phase is sometimes called the middle phase because its intermediate density causes it to appear between the oil and the water phases in a phase-separated type 3 nSOW system.

6.2.2 Change of Phase Behaviour with Temperature

One way of altering the surfactant affinity in an nSOW system is by changing the temperature. This change in temperature will change the surfactant's affinity for the two phases. At high temperature the nonionic surfactant becomes more soluble in the oil phase while at low temperature it becomes more soluble in the water phase. Shinoda and Friberg[1] showed that, for surfactants with cyclohexane and water, the phase behaviour changed with temperature at a constant surfactant concentration. In that system, a rise in temperature increased the surfactant's affinity for the oil phase, hence the system moved from type 1 to type 3 to type 2 phase behaviour with increase in temperature. The water phase was continuous at low temperature.

With a similar system, Saito and Shinoda[8] showed that the solubilisation of oil into an aqueous micellar solution phase increased rapidly as the type 3 region is approached. Around the type 3 region, as the temperature rose the system moved from two phases (O and D, where D = surfactant phase), to three phases (O, W and D) and then back to two phases (W and D). With further rise in temperature, water solubilised in the oleic micellar phase was released to the water phase. Saito and Shinoda[8] also showed that the interfacial tension changed during the transition from type 1 to type 2 phase behaviour. In the three-phase region the interfacial tension is 'ultra low'; this has important consequences for the drop sizes and stability of emulsions of type 3 systems.

6.2.3 Factors Affecting Phase Inversion

It is apparent from the literature[1,2,3,9-11] that the phase inversion in nSOW systems has several controlling factors. Among these factors the following can be identified:

- Type of oil
- Surfactant type and concentration
- Temperature of the system
- Water to oil ratio
- Additives in the oil and water phases
- Mixing conditions
- Rate and order of additions of the different components

The first five factors affect the surfactant's affinity while the last two factors are dynamic variables. Various techniques and concepts have been used to correlate surfactant affinity variables and hence the emulsion type; these are described in chronological order below.

6.2.3.1 Hydrophile–Lipophile Balance (HLB)

Full descriptions of the hydrophile–lipophile balance (HLB) concept are given by Becher[12] and Becher and Schick.[13] Griffin[14] first defined the affinity of a nonionic surfactant in terms of an empirical quantity, the HLB. Surfactants are assigned an HLB number at 25 °C on a scale of 1 to 20, where low HLB numbers represent lipophilic surfactants and high HLB numbers represent hydrophilic surfactants. Generally, the application of a surfactant can be derived from its HLB number in accordance with Table 6.1.[15]

Table 6.1 *Application of surfactants based on their HLB number*

HLB number range	Application
3–6	W/O emulsifier
7–9	Wetting agent
8–18	O/W emulsifier
13–15	Detergent
15–18	Solubiliser

HLB numbers are calculated for a surfactant from simple formulae based either on analytical or composition data.[16] For polyoxyethylene nonylphenyl ethers (NPE), HLB $= E/5$ where $E =$ the weight % of polyoxyethylene in the surfactant. For example, for NPE12 (12 oxyethylene groups in the hydrophilic chain), HLB $= 14.2$. For a polyhydric fatty acid ester, HLB $= 20(1 - S/A)$, where $S =$ saponification number of the ester and $A =$ acid number of the fatty acid. For example, for polyoxyethylene(20) sodium monolaurate (trade name Tween 20), $S = 45.5$ and $A = 276$; hence HLB $= 16.7$.

Attempts have been made to assign HLB numbers to various oils to predict which surfactants will produce the most stable emulsion; these are called the required HLB of the oil (see Table 6.2).

Table 6.2 *Required HLB numbers to emulsify various oils*[17,18]

Oil	W/O emulsion	O/W emulsion
Paraffin oil	4	10
Beeswax	5	9
Lanolin, anhydrous	8	12
Cyclohexane	–	15
Toluene	–	15

However, it is well recognised that the HLB concept is limited; many researchers have found no correlation between emulsion type and HLB number; also, changes in emulsion type have been found to depend on the water to oil ratio, surfactant concentration and temperature.[2,10,11,19] The HLB concept's main failing stems from the fact that it does not allow surfactants to have different affinities for different oils. Graciaa *et al.*[20] showed that it was the HLB of the interfacial surfactant, rather than the overall surfactant HLB, that is the important affinity variable. This observation was made from the results of a model that describes the partitioning of surfactant between oil, water and interfacial surfactant phase.

6.2.3.2   *Phase Inversion Temperature (PIT)*

The sensitivity of emulsions to temperature was recognised by Shinoda *et al.*[21–23] They suggested the use of the PIT as a method of preparation of emulsions. The phase inversion temperature (PIT) was defined by Friberg *et al.*[24] as being the temperature at which the emulsifier shifts its preferential solubility from water to oil when the temperature is increased.

Effect of oil type.   Nonionic surfactants often become increasingly more lipophilic at elevated temperatures. The change in nSOW phase behaviour with temperature can bring about a phase inversion. The phase inversion takes place when the system is three phase (surfactant affinity balanced) and the temperature at inversion is known as the phase inversion temperature. Shinoda[25] has shown that, for a number of different oils, the HLB is a function of temperature (see Figure 6.1). HLB numbers have been assigned to surfactants by measuring the PIT of an emulsion containing the surfactant and checking this against a PIT *vs.* HLB calibration curve. Generally, HLB numbers derived from PIT measurements differ from formulae values by < 2 HLB numbers.

From Figure 6.1 it can be seen that, for any oil, the higher the HLB of a surfactant the higher the PIT. Shinoda and Friberg[1] also found that the solubility of a nonionic surfactant in a particular oil was inversely proportional to the PIT.

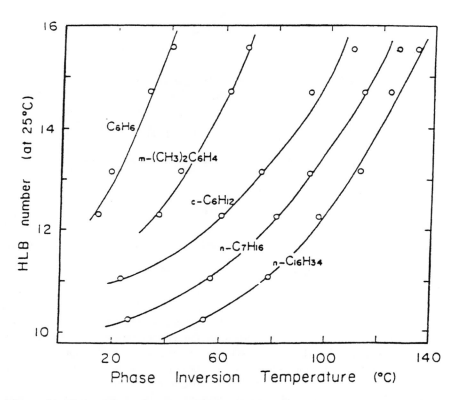

Figure 6.1 *The correlation between surfactant HLB number and the PIT in various oil/water (1:1) emulsions stabilised with nonionic surfactants (1.5 wt%)*
(Reproduced by permission of The Chemical Society of Japan from ref. 25)

Phase volume ratio.   The effect of the phase ratio of emulsions on the PIT depends on the kind of oil, the concentration of surfactant[1] and the distribution of the polyoxyethylene chain lengths in the surfactant.[26] In the case of nonpolar and saturated oils with surfactant concentrations >5%, Shinoda and Friberg[1] showed that the PIT is reasonably constant for a wide range of phase volume ratios. However, they describe the behaviour of aromatic hydrocarbons and polar oils as 'abnormal' and show that the phase volume ratio has a marked effect on the PIT (see Figure 6.2). The PIT was also affected by the 'way of shaking' in these cases.

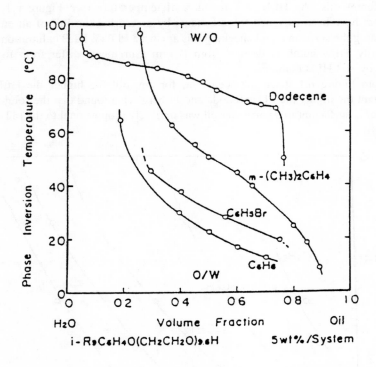

Figure 6.2 *The effect of phase volume on the PIT of emulsions stabilised with NPE(9.6) 5 wt%*
(Reproduced by permission of J. Wiley from ref. 1)

Shinoda and Friberg[1] showed that the PIT does not vary significantly with the phase ratio in nonpolar oil systems with surfactant concentration >5% wt, but does change in dilute solution (below 3% wt). Shinoda explains the variation of the PIT with surfactant concentration as follows. 'The saturation concentrations of nonionic homologues in water are all very small, but those in hydrocarbons are much larger and depend largely on the ethylene oxide chain length. Lipophilic homologues (shorter chain length) dissolve better than hydrophilic homologues in the oil phase. Hence, there will be a selection of more hydrophilic surfactant to be adsorbed at the oil/water interface. This effect is amplified in dilute solutions and when the volume fraction of oil is large'.

Surfactant mixtures. The efficiency of a surfactant at emulsifying and stabilising a liquid–liquid system is a function of the relative degrees of interactions of the various portions of the surfactant molecule with both the oil phase and the water phase. The HLB of an emulsion system, in which a mixture of surfactants is used (HLB mix), has been generally assumed to be an algebraic mean HLB of the individual surfactants (see eqn. 6.3). There is evidence that more stable emulsions are produced when a blend of surfactants is used.[1] PIT data indicate that when the difference of HLB of the surfactants in a blend is small, the surfactant mixture HLB is approximately that of the weight average HLB; however, this was not the case for blends of surfactants with large differences in HLB (see ref. 1).

$$HLB_{mix} = f_A HLB_A + (1 - f_A)HLB_B \qquad (6.3)$$

where f_A is the weight fraction of surfactant A and f_B is the weight fraction of surfactant B $(1 - f_A)$

The effect of additives in the water and oil phases. Although the exact mechanism by which various additives affect the phase inversion is not fully understood, their presence in nSOW systems has been shown to affect the PIT as well as the emulsion inversion point, EIP.[27-29] The phase inversion temperature varies with the amount and chemical type of additives in the water phase. Shinoda and Takeda[30] showed that inorganic salts can affect the PIT more strongly than their parent acids. Also, they showed that the effect of fatty acids and alcohols on the PIT for 1:1 volume ratio paraffin–water systems was independent of the chain length of the acid or alcohol.

The PIT of emulsions in which the oil phase is a mixture of oils (PIT$_{mix}$) was found to be expressed by the volume average of the PITs of the respective oils:[31]

$$PIT_{mix} = \sum f_i PIT_i \qquad (6.4)$$

where f_i = volume fraction of oil in the mixture and PIT_i = PIT of oil i (in °C).

Many studies of Shinoda *et al.* refer to oil–water systems containing NPE surfactants. However, other researchers have used different surfactant types and some have found different trends from those shown by Shinoda *et al.* Parkinson and Sherman,[32] who used Tween–Span blends, found that a maximum PIT occurred at an HLB corresponding to maximum stability.

6.2.3.3   Emulsion Inversion Point (EIP)

Much of the emulsion inversion point (EIP) work is due to Marzall.[18,33-35] Brooks *et al.*[2,10,11,36] also used this concept. The EIP is related to the inversion of W/O emulsions to O/W emulsions at constant temperature. The experimental method is described below.[33]

An aqueous phase is added (incrementally) to a finite amount of oil which contains a known amount of surfactant. The mixture is agitated by a turbine blender for 15 s on each addition and the emulsion type determined. A plot of EIP *vs.* HLB is made where:

$$\text{EIP} = \frac{\text{volume of aqueous phase at phase inversion point}}{\text{volume of oil phase}} \qquad (6.5)$$

The EIP tends to decrease with increasing HLB of the surfactant until a minimum is observed. The value of the HLB at the minimum is the required HLB of the oil to produce an emulsion with maximum stability.[37] However, Marzall[33] and Brooks and Richmond[2] have noted that the exact position of the EIP minimum can be affected by the agitation conditions. Marzall[18] discusses the events occurring in an EIP experiment in terms of the changes in the interface. It is argued that initially the interface is concave relative to water (W/O emulsion); hence Winsor $r > 1$. At the EIP it is proposed that the hydrophile–lipophile nature of the surfactant is balanced ($r = 1$) and once the EIP has been passed, $r < 1$, giving a concave interface relative to oil (O/W emulsion).

The findings of EIP experiments from various workers are summarised below:

- At the EIP minimum the inversion from W/O to O/W occurs and produces emulsions with very small drops
- The EIP increases with increasing concentration of lipophilic surfactant, whereas the EIP decreases with increasing concentration of hydrophilic surfactant
- In a series of alkanes, the higher the EIP the lower the required HLB
- Highest viscosity and minimum interfacial tension occur at the EIP
- For aromatic hydrocarbons, with increasing methyl group substitutions the EIP and the required HLB value decrease (the PIT increases)[35]
- The EIP has been found to show, accurately, changes in the required HLB of an oil brought about by the addition of additives, *e.g.* alcohols[27] and poly (ethylene glycol).[38]

6.2.3.4   *Equilibrium Water–Oil Ratio (WOR) Maps*

It is often convenient to plot the phase behaviour and the emulsion type on a single map in which the boundaries that separate the areas are referred to as inversion lines (or curves). Salager *et al.*[39–44] developed the work of Shinoda further. Salager reviews his work in ref. 3. Later, Brooks *et al.*[2,10,11,45] developed a more comprehensive map in which the volume fraction, the phase behaviour and the surfactant's HLB are represented. This type of map will be discussed in more detail in the next sections.

Formulation variables. Salager[3] points to the following formulation variables which are capable of changing the surfactant's affinity for oil and water:

- The salinity of the aqueous phase (expressed as wt% salt in aqueous solution). With increasing salt concentration the nSOW phase behaviour moves from type 1 to type 3 to type 2
- The surfactant HLB, where increasing the HLB results in the phase transitions type 2 to type 3 to type 1
- Temperature, where increasing the temperature (for nonionic surfactants) results in the phase transitions type 1 to type 3 to type 2 (Shinoda *et al.*'s[1] PIT work)
- The addition of alcohols, which cause the transition type 1 to type 3 to type 2 with increasing concentration
- The nature of the oil. This was expressed as ACN (alkane carbon number for alkanes) and EACN (equivalent alkane number for non-alkanes)

Many researchers have shown there to be a direct relationship for increasing carbon number in a homologous series of hydrocarbons with the PIT.[1] By measuring the interfacial tension between an aqueous phase and alkanes, Cash *et al.*[46] developed a calibration curve (interfacial tension *vs.* alkane number); then by mixing various hydrocarbons with alkanes and measuring the interfacial tension, an EACN could be given to the mixture and, hence, to the hydrocarbon. They showed that by using molar mixing rules a range of EACN numbers could be produced. It was also shown that parts of the oil molecule had a specific EACN: e.g. phenyl group, EACN = 0 (each CH_3 addition = 1); cyclohexyl group, EACN = 4. An increase in EACN results in the phase behaviour transition type 2 to type 3 to type 1.

Bi-dimensional scanning procedure. To scan the effect of each of the above variables an equilibrium technique was used by Salager *et al.*[41] in which nSOW samples, with a set water volume fraction and surfactant concentration, were made progressively by changing one of the formulation variables. The samples were allowed to come to equilibrium over a 48 hour period without emulsifying. The phase behaviour was then noted (two-phase, three-phase). The sample was emulsified with a turbine blender and the emulsion type determined. Figure 6.3 shows the results of a bi-dimensional scan using the average ethylene oxide number (EON) as the formulation variable. The diagram is a map of emulsion type, nSOW phase behaviour and an inversion locus.

Optimum formulation. It has been found that ultra-low interfacial tension is associated with three-phase behaviour. This condition enables the production of very fine emulsions with low energy input. It is also of interest for enhanced oil recovery processes because ultra-low interfacial tension offsets the capillary forces which maintain the residual oil trapped in the porous matrix of a reservoir.[3] Maximum oil recovery is obtained in the three-phase region; hence, three-phase systems have been termed 'optimum formulation'.

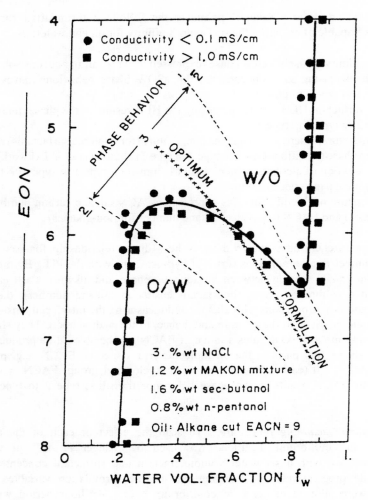

Figure 6.3 *Bidimensional formulation/WOR map showing the nSOW phase behaviour, the inversion locus and the emulsion types in each region (MAKON mixtures are ethoxylated nonylphenol nonionic surfactants with different average numbers of ethylene oxide groups)*
(Reproduced by permission of Marcel Dekker Inc. from ref. 41)

Attempts have been made to correlate variables with the optimum formulation, *e.g.* Bourrel and Salager[47] for ethoxylated nonionic surfactants:

$$\alpha - \text{EON} - k\text{EACN} + mA_i + bS + C_T(T - 28) = 0 \qquad (6.6)$$

where α depends on the lipophilic group of the surfactant, EON = average number of ethylene oxide groups per surfactant molecule, A_i = alcohol concentration, $S =$

salinity of the aqueous phase, T = temperature, m and b are parameters depending on the type of alcohol and electrolyte, and k and C_T are constants.

Therefore, the optimum formulation may be written as a linear relationship between formulation variables:

$$SAD = \sum D_i E_i \qquad (6.7)$$

where D_i = formulation coefficient, E_i = formulation variable and SAD = surfactant affinity difference, which relates to the nSOW phase behaviour: SAD < 0, type 1 phase behaviour (termed SAD$-$); SAD > 0, type 2 phase behaviour (termed SAD$+$); SAD $= 0$, type 3 phase behaviour. SAD represents the same concept as the Winsor r, but it is expressed in terms which are experimentally obtainable and is, therefore, more useful for practical applications. Salager[3] uses SAD plus optimum formulation in a schematic diagram which describes all equilibrium emulsion types.

Schematic WOR map. Salager *et al.*[41] present a bi-dimensional map, shown in Figure 6.4, that is divided into six regions by the optimum formulation (SAD = 0) line and the inversion locus. For positive SAD, phase behaviour at equilibrium is type 2 and according to Bancroft's rule a W/O emulsion is expected. This is true in regions B$^+$ and A$^+$; however, in region C$^+$ an 'abnormal' O/W emulsion is produced because the volume of oil is too small to make it the continuous phase.

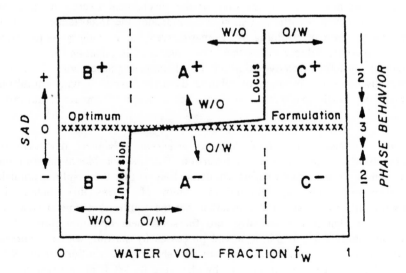

Figure 6.4 *Schematic plot of SAD versus. WOR showing the inversion locus and the different emulsion type regions*
(Reproduced by permission of Marcel Dekker Inc. from ref. 41)

The abnormal C^+ emulsions are often W/O/W type, which is a way in which Bancroft's rule is partially satisfied. For negative SAD the nSOW phase behaviour is type 1. A^- and C^- are O/W emulsion regions and B^- is an 'abnormal' W/O emulsion region.

Emulsion stability is closely related to the region boundaries; normal A^+, B^+ (W/O emulsions) and A^-, C^- (O/W emulsions) regions are found to be relatively stable, with increasing stability (at constant SAD) approaching the A^+/C^+ or A^-/B^- limit, *i.e.* with increasing disperse phase fraction. Stability decreases from both sides as the A^+/A^- boundary is approached, *i.e.* near the three-phase region. It is found that abnormal (B^-, C^+) emulsions break readily.

The viscosity of emulsions in the A^+, A^- regions far from SAD = 0 can be high with respect to their external phase. However, close to SAD = 0 the emulsion viscosity can be extremely low, probably because of the low interfacial tension, which allows easy deformation of droplets near the A^+/A^- boundary. Abnormal emulsions have low internal phase ratio and exhibit viscosities similar to their external phase. However, real systems can show large deviations from the schematic WOR map.

Factors affecting the inversion locus.   The central A^+/A^- boundary depends on the position of the optimum formulation transition. However, it does not always correspond to a straight line crossing the three-phase region. In systems which exhibit wide three-phase regions (vertically) and a narrow A region (horizontally) there is no neat plateau.[16] The position of the lateral branches of the inversion locus may depend on surfactant type and alcohol formulation and the oil viscosity may also alter and shift the locus.[41]

Salager also notes that an increase in the mechanical energy input during emulsification can widen the A region. This indicates the importance of using a set procedure when determining WOR maps, even when using pre-equilibrated samples. However, Salager states that most industrial applications involve non-equilibrium systems (steady addition of dispersed phase until inversion). Salager[3] discusses the problem of relating dynamic inversion with equilibrium systems using WOR maps and introduces the concept of apparent equilibrium time.

Apparent equilibrium time and dynamic inversion.   Salager[3] discussed two experiments which concern dynamic inversion. The first considered the minimum contact time between phases before emulsification so that the resulting emulsion was indiscernible from an equilibrium system. He terms this contact time 'apparent equilibrium time' and concluded that the equilibrium time decreases as SAD → 0 and that it is essentially zero for some near optimum formulations. The second study tried to mimic actual processes by starting with an emulsion produced from a pre-equilibrated mixture and shifting its position on the WOR map, *e.g.* by changing temperature, or by changing its WOR at constant SAD. A shift across the inversion locus is a dynamic inversion. The results are shown in Figure 6.5.

The arrows in Figure 6.5 show the direction of change of WOR or SAD for

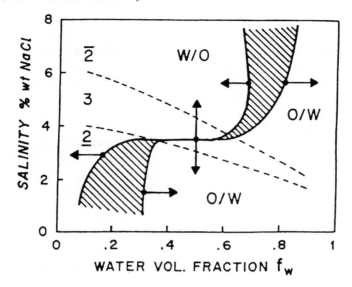

Figure 6.5 *Dynamic inversion locus showing the hysteresis zones (shaded). The arrows indicate the direction of change*
(Reproduced by permission of Marcel Dekker Inc. from ref. 3)

each inversion. The dynamic inversion due to SAD change through the A^+/A^- boundary occurs at the same point regardless of the direction of change, *i.e.* this inversion is reversible. On the other hand, the dynamic inversions due to change in WOR induced by addition of the disperse phase under agitation show a hysteresis effect from W/O to O/W. Therefore, the emulsion type in the shaded zone is dependent on its history. Note the hysteresis zone widens as SAD moves further from SAD = 0. Hysteresis effects have also been noted by Becher[12] for nSOW systems.

This leads to the observation that there are two inversion types:

- *Transitional inversion.* The inversion across SAD = 0 is continuous from W/O to MOW (microemulsion–oil–water) to O/W and is induced by altering the surfactant's affinity, which alters the nSOW system phase behaviour.
- *Catastrophic inversion.* Inversions across A^+/C^+ and A^-/B^- boundaries involve a catastrophic change from W/O to O/W and from O/W to W/O, respectively. Inversions of this type are induced by altering the system's WOR.

6.3 Dynamic Inversion Maps

There are many apparent contradictions to be found in the phase inversion literature. Some of these contradictions have been explained by Salager's observation that there are two types of phase inversion: transitional inversion induced by changing the nSOW phase behaviour and catastrophic inversion induced by changing the system's WOR. In Salager's SAD–WOR equilibrium map technique, surfactant affinity (and nSOW phase behaviour) is plotted against the volume fraction of water. Surfactant affinity difference (SAD) was used to describe phase behaviour types. Therefore, a SAD *vs.* WOR map provides a good framework for studying dynamic phase inversion because the variables connecting both inversion types are included in the map.

Shinoda *et al.*'s[1] PIT work and Marzall's[27,33,38] EIP work are studies of dynamic inversion. PIT inversions (induced by change in temperature altering the surfactant affinity) can now be seen to be transitional inversions, while EIP inversions (induced by adding a dispersed water phase to a continuous oil phase) are catastrophic inversions.

WOR maps are limited to examining one surfactant concentration per map. Phase behaviour changes with composition in these systems are sometimes represented on triangular diagrams (these have been used for surfactant–oil–water systems by Smith *et al.*[48–50]). However, triangular diagrams can only apply to nSOW systems when the surfactant is a single species. When commercially distributed nonionic surfactants are used (or the surfactant is made up from a mixture of surfactants), triangular diagrams are inappropriate because the surfactant's affinity depends on the HLB of the interfacial surfactant, not on the overall weight average HLB. In the case of a distributed surfactant, the HLB of the interfacial surfactant varies with WOR and with surfactant concentration. Hence, the construction of a meaningful triangular diagram (which can only show one surfactant affinity condition) is impossible with mixed surfactants.

In order to construct a WOR map, Salager *et al.*[40,41] checked the phase behaviour of pre-equilibrated mixtures (at known WOR and SAD) before emulsification. After emulsification, emulsion type was determined by electrical conductivity. Salager noted that large deviations from the generalised WOR map could occur. A more detailed inversion map was developed by Brooks and Richmond.[2] In their work, a standardised procedure was used to construct nSOW inversion maps. For each nSOW system a suitable pair of lipophilic and hydrophilic surfactants (different grades from the same family) was chosen to prepare surfactant blends with a range of HLB values. The aqueous phase was distilled water containing 0.5% of potassium chloride (to enhance electrical conductivity). The oil phase was either cyclohexane, *n*-heptane or toluene. A titration process was used to locate the inversion line (curve). The inversion points were induced either by varying the surfactant composition (changing the HLB), thus producing a transitional inversion, or by changing the phase ratio of oil to water which produces a catastrophic inversion. The results of changing the volume ratio or the HLB in a particular isothermic system can be predicted from the path of the inversion map.[2] A typical phase inversion map for NPE systems is shown in Figure 6.6. It should be noted that all the other parameters (such as agitation speed, temperature, surfactant concentration and rate of addition of components) should be kept constant.

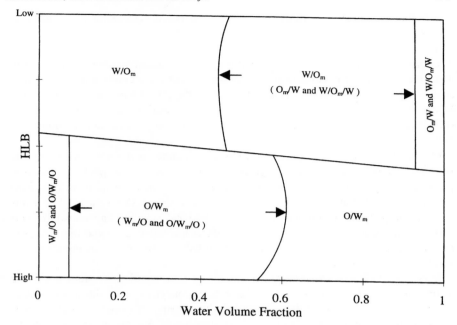

Figure 6.6 *Generalised phase inversion map for an oil–water–nonionic surfactant system (symbols explained in text below; arrows show direction of change)*

6.4 Catastrophic Inversion

This section is concerned with the factors affecting the drop sizes in nSOW systems before and after catastrophic inversions. There are a number of different drop types present at different stages of a catastrophic inversion. Therefore, in order to help the reader's understanding of this section, the drop types and drop notations used are described below.

Drops present before catastrophic inversion. When the catastrophic inversion is brought about by the addition of the aqueous phase to the oil phase (high HLB), two drop types can be present before phase inversion:[10]

- Unstable water drops containing surfactant micelles, in a continuous oil phase (*i.e.* W_m/O).
- Stable oil drops within water drops, in a continuous oil phase (*i.e.* $O/W_m/O$).

When the catastrophic inversion is brought about by adding the oil phase to the water phase (low HLB), the drop types present before inversion can be one of:

- Unstable oil drops containing surfactant micelles, in a continuous aqueous phase (*i.e.* O_m/W).

- Stable water drops within oil drops, in a continuous aqueous phase (*i.e.* $W/O_m/W$).

Drops present after catastrophic inversion. After catastrophic inversion has taken place the resulting emulsion consists of stable oil drops in a continuous water phase containing surfactant micelles (*i.e.* O/W_m) when the initial continuous phase is oil. When the initial continuous phase is aqueous the resulting emulsion consists of stable water drops in a continuous oil phase containing surfactant micelles (*i.e.* W/O_m).

Ostwald[51,52] first modelled catastrophic inversions as being caused by the complete coalescence of the dispersed phase at the close packed condition (corresponding to a dispersed phase fraction of 0.74 in Ostwald's uniform hard sphere model). Other studies, *e.g.* Marzall,[33] have shown that catastrophic inversions (though these inversions were not called catastrophic inversions by that author) can occur over a wide range of WOR. It has been suggested that this may be due to the formation of double emulsion drops[12] ($O/W_m/O$), boosting the actual volume of the dispersed phase.

Very few studies of dynamic factors affecting catastrophic inversions are found in the literature. Virtually all studies have been concerned with the movement of inversion boundaries with either changes in the system composition, or changes in the system's dynamics. EIP studies of catastrophic inversions in nSOW systems, however, have not been concerned with the system dynamics. Those studies that have looked at the effect of agitation conditions on catastrophic inversion boundaries have been concerned with oil–water systems with no surfactant present.[53–55]

In the case of systems that do not contain stabilising surfactant, inversion hysteresis has been shown to occur.[56] Some studies have looked at the effect of the system composition. Clarke and Sawistowski[56] studied the effect of additives in the oil and water phases. Selker and Sleicher[57] noted that as the viscosity of the oil phase increased, the more likely it was to become the dispersed phase. Many studies have examined the effect of stirrer speed on catastrophic inversion boundaries. Quinn and Sigloh[58] and Selker and Sleicher[57] showed that inversion was shifted to a higher dispersed phase fraction as the stirrer speed increased. This was also noted by Arashmid and Jeffreys[59] and Guilinger *et al.*[53] In batch mode the location of the impeller can be critical.[57,58,60,61] Generally, it was found that for a certain range of WOR, the phase in which the impeller was initially immersed became continuous. Guilinger *et al.*[53] examined catastrophic inversions in a continuous mixer system. They found that the height of the agitator above the vessel bottom had no effect on inversion and that the rate of addition of feed to the mixer had little effect on the inversion point. Catastrophic inversion in batch systems has also been found to occur with an increase in agitation.[56,61] Hence, it has been concluded that the difference between the coalescing rates of the oil drops and those of the water drops is the controlling factor which dominates the dispersion types.[61]

In oil–water systems with no surfactant present the formation of O/W/O and W/O/W drops may be limited.[62] However, some authors have noted the presence of these drop types.[61] Delayed catastrophic inversion can occur in oil–water disper-

sions over a narrow range of volume fractions.[62] During the delay time there is a
rise in the drop sizes of the dispersion; the growth is controlled by the relative rates
of drop break-up and coalescence.[62] Hence, catastrophic inversion, although it may
be controlled by the presence of a surfactant, is ultimately dependent on the drop
types and size distribution of the emulsion. Therefore, its understanding requires a
detailed knowledge of the drop sizes produced at the catastrophic inversion point
and how these are related to the drop sizes prior to inversion. Recently, Brooks and
Richmond[2,11] presented detailed work on catastrophic inversion using nSOW
systems. They showed that the location of the catastrophic inversion boundaries
depended markedly on dynamic conditions. Figure 6.7 shows that the location of a
phase inversion boundary depends on stirrer speed for cyclohexane–water–NPE
systems when the initial continuous phase is the oil (high HLB). At high stirrer
speeds the inversion boundary approached a constant position. When the addition
rate of the aqueous phase was reduced the value of f_w for the phase inversion point
was also reduced. Brooks and Richmond[11] also found that the drop size distribution
during catastrophic inversion depends on the stirring speed and the rate of addition
of the aqueous phase.

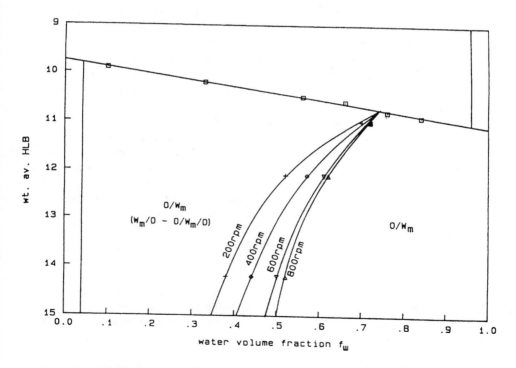

Figure 6.7 *Effect of stirrer speed on the phase inversion boundary for cyclohexane and
NPE surfactants (2 wt%)*
(Reproduced by permission of Elsevier Science from ref. 2)

6.5 Transitional Inversion

It was shown earlier that transitional inversion is directly related to the surfactant–oil–water phase behaviour. Shinoda and Friberg[1] investigated how the drop sizes of cyclohexane/NPE 9.7 (3% wt) O/W_m emulsions vary with temperature, for temperatures up to the phase inversion temperature (PIT). This work can be regarded as an investigation of the effect of a SAD change from SAD$^-$ to SAD $= 0$ on the drop sizes of O/W_m emulsions, where SAD $= 0$ is located at the PIT. In the study, O/W_m emulsions were made up at varying temperatures up to the PIT and their drop sizes were measured immediately after agitation was stopped and also 5 hours after agitation was stopped. From the resulting size distributions, number average diameters were calculated (at constant temperature for each experiment). Shinoda *et al.*[1] showed that the initial mean drop diameter of the emulsions decreased as the temperature approached the PIT. They stated that this reflects the change of the oil–water interfacial tension and the fraction of surfactant phase being present in this system as a function of temperature. Five hours after agitation, the drop diameter of emulsions close to the PIT rose rapidly with time because coalescence was facilitated due to the ultra-low interfacial tension.

Measurements of the change in the interfacial tension across the transitional inversion have been made by several workers. Kunieda and Shinoda[63] used a sessile drop technique to show that Antonoff's relationship for the interfacial tensions between three phases held for conditions near the PIT. Cayais *et al.*[64] measured the interfacial tension across a transitional inversion directly using a spinning drop apparatus. Their results are shown in Figure 6.8 (EACN $= 8$ is equivalent to SAD $= 0$, EACN < 8 is equivalent to SAD$^+$ and EACN > 8 is equivalent to SAD$^-$ conditions). It is to be noted that some authors[3,65] have quoted values as low as 10^{-6} N m^{-1}.

The results of Shinoda *et al.* for emulsion stability across the phase transition show that for O/W_m emulsions the drainage rate was low at temperatures about 20–40 °C below the PIT and also low for W/O_m emulsions at temperatures about 20–40 °C above the PIT. This led to Shinoda's observation that 'the HLB numbers of surfactants whose HLB temperatures in an oil–water system are 25–70 °C higher than the emulsion's storage temperature are the required HLB numbers for emulsification of that system'. Shinoda *et al.*[1] also showed that the stability of emulsions falls to a minimum in type 3 systems. Many other studies have shown stability maxima either side of a stability minimum in the three-phase region,[3,66,67] although there has been much debate as to the relevance of the measurement of stability used in some studies.[68]

Shinoda also studied emulsions that were produced at temperatures below the PIT, at the PIT and above the PIT with rapid cooling to aid stability. Drop sizes of emulsions produced at the PIT were retained in the final cooled emulsion. Shinoda noted that emulsions with the finest drops were produced by emulsifying 2–4 °C below the PIT and then cooling. Shinoda termed this emulsification method 'emulsification by the PIT method'. The study also showed that 'emulsification by the inversion method', *i.e.* emulsification above the PIT as a W/O_m emulsion and then cooling, did not result in such small drops.

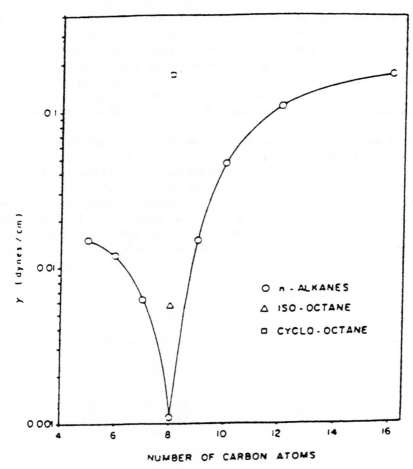

Figure 6.8 *The variation of interfacial tension across a SAD transition, measured using spinning drop apparatus (hydrocarbons* vs. *0.2% petronate/10% NaCl at 27 °C after 24 hours of stirring)*
(Reproduced by permission of the American Chemical Society from ref. 64)

Friberg *et al.*[69,70] performed similar experiments to Shinoda and found that, at temperatures above the PIT, there was no surfactant phase present and there was no reduction in drop sizes. However, at temperatures below the PIT when the system was three phase and the emulsion volume contained 20% and 50% surfactant phase, there was a large reduction in the drop sizes. Hence, it was concluded that as the surfactant phase separate on cooling it will produce extremely small drops. These extremely small drops skew the average droplet size to smaller values.[1]

Parkinson and Sherman[32] looked at the drop sizes at the PIT of the systems stabilised by Tween–Span mixtures and found only small differences in the drop sizes of emulsions made at the PIT with those produced at other temperatures. The results of these workers may again be explained by noting that conditions for transitional inversion points, in systems stabilised by these surfactants, may not exist if the disperse phase fraction is <80% as the surfactant phase may not become continuous. Hence, the inversion observed by these workers may have been a catastrophic inversion.

Brooks and Richmond[2,10] studied the dynamics of transitional phase inversion for various combinations of nSOW systems. They found that drop sizes decreased as the transitional inversion point was approached. Figure 6.9 shows the change in drop diameter across the phase transition for the oil–water–NPE system. This change in drop diameter reflects the interfacial tension change through the phase transition. For all the individual mixtures that were studied by Brooks and Richmond,[2] the locus for the transitional inversion was found to be a straight line. It was also found that, contrary to what was suggested by other workers,[1] when

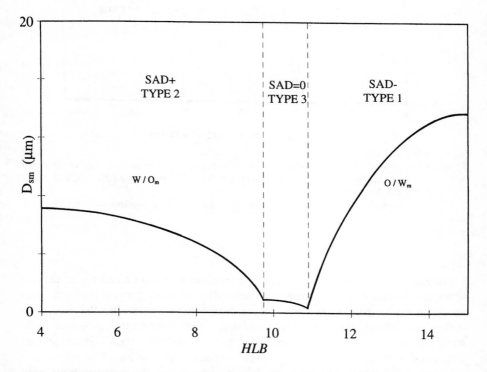

Figure 6.9 *Change in the steady state drop diameter* (D_{sm}) *across the phase transition (cyclohexane–water–NPE system at 25 °C)*

surfactant mixtures are used the boundary corresponding to SAD = 0 was not horizontal but sloped downwards from left to right. Furthermore, the slope of the transitional inversion line changes with surfactant concentration, as seen in Figure 6.10.

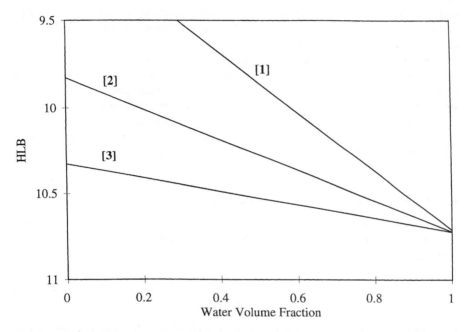

Figure 6.10 *Transitional inversion lines: cyclohexane/NPE(5)–NPE(12) system at total surfactant concentrations: [1] 0.0277 M, [2] 0.0454 M, [3] 0.1135 M*

6.6   Partitioning of Surfactant

In this section, the partitioning of surfactant between the oil phase, water phase and the interface is considered, with particular reference to transitional inversion. A phase separation model has been used to describe the partitioning of surfactant between oil–water–surfactant phases.[71,72] Harusawa and Tanaka[73] investigated the partitioning of NPE surfactants between water and cyclohexane. Graciaa *et al.*[74] and Allan *et al.*[75] studied the partitioning of surfactants having a distribution of chain lengths using octylphenol ethoxylates (OPE) in water–isooctane and NPE in water–hexane systems, respectively. These models were based on the assumption that there is ideal mixing, thermodynamic equilibrium and that the nSOW system can be split into three constituent parts:

(*i*)   a water phase containing surfactant monomers at a mixed CMC_w
(*ii*) an oil phase containing surfactant monomers at a mixed CMC_o

(*iii*) a pseudo surfactant phase, which represents the surfactant in micelle form and surfactant at the oil–water interface.

Brooks and Richmond[71,72] developed a simple surfactant partitioning model that can be applied to isothermal transitional inversions. The linearity and gradient variation of transitional inversion lines in systems containing distributed nonionic surfactant was also explained. The derived model used mixed surfactant theory to predict the slope of the SAD = 0 line with surfactant concentration, for transitional inversion induced by varying the amount of a homogeneous lipophilic and a homogeneous hydrophilic surfactant in an oil–water system.

Graciaa *et al.*[20] assumed that microemulsion phases (O_m, W_m, surfactant phase microemulsion (M_s), type 3 oleic microemulsion (M_o), type 3 aqueous microemulsion (M_w)), despite being single thermodynamic phases, are composed of submicroscopic regions of oil and water separated by an interfacial layer of surfactant. Consequently, it is possible for each of the different microemulsion phases to be constructed from the three constituents parts described earlier.

6.6.1   Basis of Phase Separation Model

If surfactant concentrations in the water and oil phases are below their respective critical micelle concentration (CMC), the partition coefficient for surfactant monomer *i* between oil and water can be written as:[73]

$$K_i = \frac{CMC_{oi}}{CMC_{wi}} \tag{6.8}$$

For conditions above the CMCs, using an analogy to Raoult's law, it is shown that for an ideal solution:

$$C_{oi} = x_i\, CMC_{oi} \tag{6.9}$$

$$C_{wi} = x_i\, CMC_{wi} \tag{6.10}$$

where x_i is mole fraction of surfactant *i* in the pseudo surfactant phase and C_{oi} and C_{wi} are the concentrations of monomeric surfactant dissolved in the oil and water phases, respectively. For details of the derivation of eqns. (6.9) and (6.10), see ref. 71.

6.6.2   Theory Applied to Transitional Inversion

Graciaa *et al.*,[20] in a series of experiments using the isooctane–OPE system found that, using CMC data, isothermal transitional inversion occurs at a specific value of the HLB of the pseudo surfactant phase. They calculated the HLB of the pseudo surfactant phase from the mole average of each chain length in the phase. Phase separation models are known to require a large amount of CMC data. Brooks and

Richmond[71] developed a simple method for determining the HLB of the pseudo surfactant phase and the slope of the SAD $= 0$ line that required only a minimum of inversion points. This method is shown below.

$$x_i = \frac{\begin{array}{l}(\text{total number of moles of surfactant } i \text{ present in all phases}) - \\ (\text{number of moles of } i \text{ dissolved as monomer in the oil and water phases})\end{array}}{(\text{total number of moles of surfactant in the pseudo phase})}$$

$$x_i = \frac{C_i V - C_{oi} V_o - C_{wi} V_w}{\sum_{i=1}^{N}(C_i V - C_{oi} V_o - C_{wi} V_w)} \tag{6.11}$$

where V = total volume of the system $(V_o + V_w)$, V_o = volume of the oil phase, V_w = volume of the water phase, C_i = effective overall concentration of surfactant i, C_{oi} = concentration of monomeric surfactant in oil phase and C_{wi} = concentration of monomeric surfactant in water phase.
Thus

$$x_i = \frac{C_i - (1 - f_w)C_{oi} - f_w C_{wi}}{\sum_{i=1}^{N}(C_i - (1 - f_w)C_{oi} - f_w C_{wi})} \tag{6.12}$$

For transitional inversion they found that there are advantages in splitting the surfactant into hydrophilic and lipophilic groups. They used the simple case of isothermal transitional inversion brought about by varying the ratio of a homogeneous lipophilic and a homogeneous hydrophilic surfactant in an oil–water system.

For a two surfactant system, the mole average HLB of the pseudo surfactant (HLB$_{act}$) can be calculated using the following equation:

$$\text{HLB}_{act} = \sum_{i=1}^{N} x_i \text{HLB}_i \tag{6.13}$$

The pseudo surfactant phase composition is the important variable because, for a two surfactant system, the value of HLB$_{act}$ is entirely dependent on x_h and x_l (subscripts l and h denote lipophilic and hydrophilic surfactants, respectively). Knowing that $x_l + x_h = 1$ and that:

$$C_{ol} = x_l \text{CMC}_{ol}$$

$$C_{wl} = x_l \text{CMC}_{wl}$$

$$C_{oh} = x_h \text{CMC}_{oh}$$

$$C_{wh} = x_h \text{CMC}_{wh}$$

$$C_t = C_h + C_l$$

where C_{ol} and C_{wl} = concentration of lipophilic surfactant in the oil and water phase, respectively; C_{oh} and C_{wh} = concentration of hydrophilic surfactant in the oil and water phase, respectively; CMC_{ol} and CMC_{wl} = lumped critical micelle concentration of the lipophilic surfactant chain lengths in the oil and water phase, respectively; CMC_{oh} and CMC_{wh} = lumped critical micelle concentration of the hydrophilic surfactant chain lengths in the oil and water phase, respectively; and C_t = total concentration of surfactant in the system.

By substituting these equations into eqn. (6.12) we arrive at the following quadratic equation:

$$-x_h^2[(1 - f_w)(CMC_{ol} - CMC_{oh}) + f_w(CMC_{wl} - CMC_{wh})] +$$

$$x_h[(1 - f_w)(CMC_{ol} - CMC_{oh} + f_w(CMC_{wl} - CMC_{wh}) - (C_h + C_l)] + C_h = 0$$

(6.14)

Although commercial nonionic surfactants have a distribution of chain lengths and are sold as having an average chain length, there are certain cases when this equation could be applied to isothermal transitional inversion in systems containing distributed surfactants.[71]

6.6.3 Lumping of Distributed Surfactant Terms

Allan *et al.*[75] found that, to a first approximation, the CMC of a Poisson distributed surfactant is the same as that of a homogeneous surfactant of the same chain length. The required average ethylene oxide chain length of the surfactant molecules, in the pseudo surfactant phase of a system containing a distributed surfactant, was shown to approach a constant above a certain overall surfactant concentration.[74] From the above statements, the distributed surfactant above a certain overall concentration in an oil–water system could be regarded as a single component. Brooks and Richmond[71] stated that, in the case of transitional inversion in systems containing distributed surfactants, if the overall surfactant concentration is such that each of the surfactant pairs may be regarded as single components, then eqn. (6.14) can also be applied to these systems.

Crook *et al.*[76] assumed that, for dissolved surfactant monomers in the oil and aqueous phase, there are weak interactions between molecules of varying ethylene oxide chain length. This assumption does not look unreasonable as the CMCs of surfactant monomers are normally of the order 10^{-5} to 0.1 M and the molecules are nonionic.

The partition coefficient of a distributed surfactant between oil and water can be defined as:

$$K_{olw} = \frac{\sum_{i=1}^{N} C_{oi}}{\sum_{i=1}^{N} C_{wi}}$$

(6.15)

Crook *et al.*[76] found that eqn. (6.15) was in good agreement with experimental results. Brooks and Richmond[71] used this technique of lumping CMC values to derive expressions for each of the terms in eqn. (6.14). For each of the surfactant pairs they were concerned only with the number of moles of that surfactant dissolved in each phase; hence:

$$x_l = \sum_{i=1}^{N} x_{li} \qquad (6.16)$$

$$CMC_{ol} = \frac{\sum_{i=1}^{N} x_{li} \, CMC_{oi}}{x_i} \qquad (6.17)$$

where x_{li} is the lipophilic surfactant's contribution to the overall mole fraction of chain length i in the pseudo phase and CMC_{oi} is the CMC of chain length i in the oil phase. Similar expressions can be derived for x_h, CMC_{oh}, CMC_{wh} and CMC_{wl}.

Brooks and Richmond[71] applied eqn. (6.14) to transitional inversion data and used it to predict the variation of the slope of the SAD = 0 inversion line with surfactant concentration at constant temperature. More analytical details on the use of this method are given in refs. 16 and 71.

6.7 Concluding Remarks

During the last few decades, considerable work has been done in trying to understand the parameters affecting nonionic surfactant–water–oil systems. Recently, more interest has been shown in the phase inversion process as a means of producing stable and cheaper emulsions.[36,77] Different techniques have been developed for this purpose. A link between phase inversion types and nSOW phase behaviour has been established. The controlling factors of the nSOW system have been investigated qualitatively and quantitatively. As a result, maps relating the nSOW phase behaviour to other parameters affecting the system were developed. Also, owing to a better understanding of these systems, models predicting certain nSOW behaviours were proposed. Despite all this, there is much scope for improvement if the phase inversion process is to be used more widely in industry. The need for further work in this area is more pressing than ever as new technologies and processes in the chemical industry are being developed very rapidly. Hence, it is important to define some of the areas where there is scope for improvement. These are presented below:

- Development of a wide range of phase inversion maps for a wider range of surfactants and oils. This might help produce a generalised map predicting the phase behaviour of different types of nSOW systems.
- There is need for a better understanding of the multiple emulsions produced during phase inversion. Their quantitative and qualitative prediction could be useful for certain industries such as the drug industry.

- The mixed surfactant model could be generalised if a better definition of the surfactants present in the nSOW system could be found. The location of the nonionic surfactant during the different stages of phase inversion is also important.

6.8   References

1. K. Shinoda and S. Friberg, 'Emulsions and Solubilisation', Wiley, New York, 1986.
2. B.W. Brooks and H.N. Richmond, *Colloids Surf. A*, 1991, **58**, 131.
3. J.L. Salager, in 'Encyclopedia of Emulsion Technology', ed. P. Becher, Dekker, New York, 1988, vol. 3, p. 79.
4. P.A. Winsor, *Trans. Faraday Soc.*, 1948, **44**, 376.
5. P.A. Winsor, 'Solvent Properties of Amphiphilic Compounds', Butterworth, London, 1954.
6. H.T. Davis, *Colloids Surf. A*, 1994, **91**, 9.
7. K. Shinoda, *Prog. Colloid Polym. Sci.*, 1983, **68**, 1.
8. H. Saito and K. Shinoda, *J. Colloid Interface Sci.*, 1970, **32**, 647.
9. V. Yu. Lobanova, T.F. Svitova, T.S. Rogova and L.V. Kolpakov, *Colloid J. USSR*, 1990, **52**, 852.
10. B.W. Brooks and H.N. Richmond, *Chem. Eng. Sci.*, 1994, **49**, 1053.
11. B.W. Brooks and H.N. Richmond, *Chem. Eng. Sci.*, 1994, **49**, 1065.
12. P. Becher, 'Emulsion: Theory and Practice', 2nd edn., Reinhold, New York, 1966.
13. P. Becher and M.J. Schick, in 'Nonionic Surfactants: Physical Chemistry', ed. M.J. Schick, Surfactant Science Series, vol. 23, Dekker, New York, 1987, p. 435.
14. W.C. Griffin, *J. Soc. Cosmet. Chem.*, 1949, **1**, 311.
15. W.C. Griffin, *J. Soc. Cosmet. Chem.*, 1954, **5**, 249.
16. H.N. Richmond, 'Phase Inversion in Nonionic Surfactant–Oil–Water Systems', Ph.D. Thesis, Loughborough University, UK, 1992.
17. P. Becher, in 'Nonionic Surfactants', ed. M.J. Schick, Surfactant Science Series, vol. 1, Dekker, New York, 1967, p. 604.
18. L. Marzall, in 'Nonionic Surfactants: Physical Chemistry', ed. M.J. Schick, Surfactant Science Series, vol. 23, Dekker, New York, 1987, p. 493.
19. S. Lehnert, H. Tarabishi and H. Leuenberger, *Colloids Surf. A*, 1994, **91**, 227.
20. A. Graciaa, J. Lachaise and G. Marion, *Langmuir*, 1989, **5**, 1315.
21. K. Shinoda and H. Arai, *J. Phys. Chem.*, 1964, **68**, 3485.
22. K. Shinoda and H. Arai, *J. Colloid Sci.*, 1965, **20**, 93.
23. K. Shinoda and H. Saito, *J. Colloid Interface Sci.*, 1969, **30**, 258.
24. S. Friberg, I. Lapczynska and G. Gillberg, *J. Colloid Interface Sci.*, 1976, **56**, 19.
25. K. Shinoda, *J. Chem. Soc. Jpn.*, 1968, **89**, 435.
26. T. Mitsui, Y. Machida and F. Harusawa, *Bull. Chem. Soc. Jpn.*, 1970, **43**, 3044.
27. L. Marzall, *J. Colloid Interface Sci.*, 1977, **60**, 570.
28. T. Forster, F. Schambil and W. von Rybinski, *J. Disp. Sci. Technol.*, 1992, **13**, 183.
29. T. Forster, W. von Rybinski and A. Wadle, *Adv. Colloid Interface Sci.*, 1995, **58**, 119.
30. K. Shinoda and H. Takeda, *J. Colloid Interface Sci.*, 1970, **32**, 642.
31. H. Arai and K. Shinoda, *J. Colloid Interface Sci.*, 1967, **25**, 396.
32. C. Parkinson and P. Sherman, *J. Colloid Interface Sci.*, 1972, **41**, 328.
33. L. Marzall, *Cosmet. Perf.*, 1975, **90**, 37.
34. L. Marzall, *Colloid Polymer Sci.*, 1976, **254**, 674.

35. L. Marzall, *J. Colloid Interface Sci.*, 1985, **107**, 572.
36. M. Zerfa and B.W. Brooks, 'IChemE Jubilee Research Event', Nottingham, 1997, vol. 2, p. 1213.
37. T. Chand, *Colloid Polymer Sci.*, 1980, **258**, 1204.
38. L. Marzall, *J. Colloid Interface Sci.*, 1977, **59**, 376.
39. J.L. Salager, I. Louiza-Maldonado, M. Minana-Perez and F. Silva, *J. Disp. Sci. Technol.*, 1982, **3**, 279.
40. J.L. Salager, M. Minana-Perez, J.M. Andrez, J.L. Grossa, C.I. Rojas and I. Layrisse, *J. Disp. Sci. Technol.*, 1983, **4**, 161.
41. J.L. Salager, M. Minana-Perez, M. Perez-Sanchez, M. Ramirez-Gouveia and C. I. Rojas, *J. Disp. Sci. Technol.*, 1983, **4**, 313.
42. J.L. Salager, G. Lopez-Castellanos, M. Minana-Perez, C. Parra, C. Cucuphat, A. Graciaa and J. Lachaise, *J. Disp. Sci. Technol.*, 1991, **12**, 59.
43. R.E. Anton, P. Castillo and J.L. Salager, *J. Disp. Sci. Technol.*, 1986, **7**, 319.
44. M. Minana-Perez, P. Jarr, M. Perez-Sanchez, M. Ramirez-Gouveia and J.L. Salager, *J. Disp. Sci. Technol.*, 1986, **7**, 331.
45. B.W. Brooks, H.N. Richmond and M. Zerfa, 'Proc. First European Congress on Chemical Engineering', Florence, May 1997, vol. 1, p. 109.
46. I. Cash, J.L. Cayais, G. Fournier, D. Macallister, T. Schares, R.S. Schecter and W.H. Wade, *J. Colloid Interface Sci.*, 1977, **59**, 39.
47. M. Bourrel and J.L. Salager, *J. Colloid Interface Sci.*, 1980, **75**, 451.
48. D.H. Smith and P.D. Fleming, *J. Colloid Interface Sci.*, 1985, **105**, 80.
49. D.H. Smith and L. Kyung-Hee, *J. Phys. Chem.*, 1990, **94**, 3746.
50. D.H. Smith and G.K. Johnson, *J. Phys. Chem.*, 1995, **99**, 10853.
51. W. Ostwald, *Kolloid Z.*, 1910, **7**, 64.
52. W. Ostwald, *Kolloid Z.*, 1910, **7**, 103.
53. T.R. Guilinger, A.K. Grislingas and O. Erga, *Ind. Eng. Chem. Res.*, 1988, **27**, 298.
54. A.W. Pacek, I.P.T. Moore, R.V. Calabrese and A.W. Nienow, *Trans. Inst. Chem. Eng. A*, 1993, **71**, 340.
55. A.W. Pacek, I.P.T. Moore, A.W. Nienow and R.V. Calabrese, AIChE J., 1994, **40**, 1940.
56. S.I. Clarke and H. Sawistowski, *Trans. Inst. Chem. Eng.*, 1978, **56**, 50.
57. A.H. Selker and C.A. Sleicher, *Can. J. Chem. Eng.*, 1965, **43**, 298.
58. J.A. Quinn and D.B. Sigloh, *Can. J. Chem. Eng.*, 1963, **41**, 15.
59. M. Arashmid and G.V. Jeffreys, *AIChE J.*, 1980, **26**, 51.
60. W.A. Rodger, V.G. Trice and J. H. Rushton, *Chem. Eng. Prog.*, 1956, **52**, 515.
61. S. Kato, E. Nakayama and J. Kawasaki, *Can. J. Chem. Eng.*, 1991, **69**, 222.
62. A. Gilchrist, K.N. Dyster, I.P.T. Moore and A.W. Nienow, *Chem. Eng. Sci.*, 1989, **44**, 2381.
63. H. Kunieda and K. Shinoda, *Bull. Chem. Soc. Jpn.*, 1982, **55**, 1777.
64. J. L. Cayais, R.S. Schecter and W.H. Wade, in 'Adsorption at Interfaces', ed. K.L. Mittal, ACS Symp. Series, Washington,1975, vol. 8, p. 234.
65. R. Strey, *Colloid Polym. Sci.*, 1994, **272**, 1005.
66. A. Graciaa, Y. Barakat, R.S. Schecter, W.H. Wade and S. Yiv, *J. Colloid Interface Sci.*, 1982, **89**, 217.
67. M. Bourrel, A. Graciaa, R.S. Schecter and W.H. Wade, *J. Colloid Interface Sci.*, 1979, **72**, 161.
68. F.S. Milos and D.T. Wasan, *Colloids Surf.*, 1982, **4**, 91.
69. S. Friberg, I. Lapczynska and G. Gillberg, *J. Colloid Interface Sci.*, 1976, **56**, 19.
70. S. Friberg and C. Solans, *J. Colloid Interface Sci.*, 1978, **66**, 367.
71. B.W. Brooks and H.N. Richmond, *J. Colloid Interface Sci.*, 1994, **162**, 59.

72. B.W. Brooks and H.N. Richmond, *J. Colloid Interface Sci.*, 1994, **162**, 67.
73. F. Harusawa and M. Tanaka, *J. Phys. Chem.*, 1981, **85**, 882.
74. A. Graciaa, J. Lachaise, J.G. Sayous, P. Grenier, S. Yiv, R.S. Schecter and W.H. Wade, *J. Colloid Interface Sci.*, 1983, **93**, 474.
75. G.C. Allan, J.R. Aston, F. Grieser and T.W. Healy, *J. Colloid Interface Sci.*, 1989, **128**, 258.
76. E.H. Crook, D.B. Fordyce and G.F. Trebbi, *J. Colloid Sci.*, 1965, **20**, 191.
77. A. Kabalnov and J. Weers, *Langmuir*, 1996, **12**, 1931.

CHAPTER 7
Coalescence in Emulsions

ALEXEY S. KABALNOV

Hewlett Packard, 1000 NE Circle Boulevard, Corvallis, OR 97330-4239, USA

7.1 Introduction

Coalescence is the primary mechanism for the loss of emulsion stability. It is due to coalescence that a mixture of oil and water, without surfactant present, resolves into oil and water within several seconds. If a proper surfactant is added, however, a stable emulsion forms, which can be infinitely stable to coalescence.

How does the nature of a surfactant molecule affect the emulsion stability? Which molecular parameters directly affect the emulsion stability and which are less important? Why do some surfactants tend to stabilize O/W emulsions and others W/O? What surfactant concentration is sufficient to prevent coalescence? Why does the addition of some surfactants to a stable emulsion cause rapid coalescence and demulsification? These questions have been addressed since the beginning of this century. Already in these very early studies, correct experimental trends were established. However, only relatively recently, with the advances in the physics of surfactant monolayers, has a mechanistic picture of emulsion coalescence started to emerge.

The objective of this chapter is to provide a brief introduction to the field of emulsion coalescence. The literature on this subject is voluminous and controversial. To aggravate the situation, there is a substantial barrier dividing theoreticians and experimentalists. In this chapter, an attempt is made to diminish this barrier by making the theoreticians aware of the recent, as well as very old, experimental developments. On the other hand, the author hopes that some theoretical aspects of the film rupture kinetics can be of interest for practical emulsion technologists.

Because of the limitations of this chapter, not all aspects of the coalescence phenomenon will be covered and the discussion will be substantially biased to the interests of the author. Firstly, only the systems stabilized by reversibly adsorbing micelle-forming surfactants are discussed. Secondly, the surfactant concentration is assumed to be above the critical micelle concentration, *i.e.* all the emulsions discussed are stabilized by saturated surfactant monolayers. Thirdly, the volume fraction of the disperse phase is assumed to be high, so that the coalescence is substantially film rupture controlled. The coupling between flocculation and coalescence is not considered, because the emulsion is either already fully flocculated or is very concentrated to start with. Fourthly, the subject of this chapter is relatively stable emulsions, with lifetimes from several minutes to years. On this timescale, the contribution of film thinning to the overall emulsion lifetime is typically negligible. Finally, this chapter does not address in any way emulsion formation, but only the stability of already formed emulsions.

This chapter has the following outline. First, an introduction to the behavior of thermodynamically stable equilibrium oil–water–surfactant (O–W–S) systems is given. The main discussion topics are the phase equilibria, surfactant monolayer elasticity, and the interfacial tensions between the co-existing phases. This section is necessary to introduce the terminology and theoretical formalism which will then be used throughout the chapter. In the next section, an overview of experimental trends in macroemulsion stability is made. Next, the theory of hole nucleation in emulsion films is presented, which is followed by comparison of the theory with experiment.

To complete this introductory section, several references to the aspects not covered in this review are given. The theory of emulsion film thinning prior to rupture has been considered by many authors, and a monograph[1] and review[2] are recommended as an introduction. The role of surfactant surface viscosity in emulsion film thinning and stability was reviewed by Malhotra and Wasan.[3] The effect of the Gibbs elasticity on emulsion stability is discussed by Lucassen and Lucassen-Reynders.[4,5] The theory of film rupture by varicose fluctuation growth has been reviewed[6] and its critical comparison with experiment has been given by Sonntag and Strenge.[7] All these theories concern the stability of emulsions on timescales of, typically, several minutes. The stabilization of emulsions by irreversibly adsorbed polymers has been reviewed by Dickinson in this book. The theory of emulsification has been discussed by Walstra.[8,9] The forced coalescence of emulsions by osmotic pressure is discussed by Binks in this book.

7.2 Thermodynamic Properties of Surfactant Monolayers: Phase Behavior, Monolayer Elasticity and Interfacial Tension

Why must a 'chapter on macroemulsions consider the phase behavior and thermo-dynamics of oil–water–surfactant systems at all? Indeed, macroemulsions are non-equilibrium systems, whereas the phase behavior concerns thermodynamically stable equilibrium systems — microemulsions and liquid crystals. The reason for it is that microemulsions and macroemulsions normally co-exist in the same system and their properties are interrelated. The surfactant monolayers covering micelles and macroemulsion droplets are in thermodynamic equilibrium. This equilibrium leads to many peculiar effects.

7.2.1 Surfactant Adsorption and Micellization

Surfactant molecules consist of two blocks, one with affinity to water and the other to oil. When added to the mixture of oil and water, they self-assemble at the oil–water interface, so that the hydrophilic block stays in water and the hydrophobic one remains in oil. As the surfactant concentration increases, the monolayer at the oil–water interface becomes more densely populated and the interfacial tension decreases, as predicted by the Gibbs adsorption equation:

$$\mathrm{d}\sigma = -\Gamma_s\, \mathrm{d}\mu_s \tag{7.1}$$

where σ is the interfacial tension, Γ_s is the surface concentration of surfactant and μ_s is the surfactant chemical potential. Both Γ_s and μ_s increase with surfactant concentration, c.

At a certain concentration (or rather in a very narrow concentration range), surfactant molecules start to self-associate in the bulk of one of the phases, forming supramolecular aggregates, $e.g.$ spherical micelles. After reaching this point, which is called the critical micelle concentration (CMC) regardless of the particular form of the aggregates, the surfactant chemical potential stays essentially constant, because all the surfactant added is consumed by the micelles, $cf.$ eqn. (7.1) and Figure 7.1. The interfacial tension at the CMC is the lowest possible under given experimental conditions. It plays a very important role in emulsion formulation and will be discussed in more detail at the end of Section 7.2.

Consider now the microscopic structure of micelles. In three-component oil–water–surfactant systems, the core of a micelle contains a solubilizate, which is the oil, if the micelles are formed in water, and water if the micelles are formed in oil. The oil–water interface is therefore present not only between macroscopic phases, but also on the microscopic level within the micelles. From this standpoint, the surfactant monolayer can be considered as a basic construction block of both the macroscopic oil–water interface and the microscopic micellar interface (Figure 7.2).

7.2.2 Elasticity of Monolayers

Because of their peculiar molecular architecture, surfactants are adsorbed at the interface between oil and water. As a result, when the surfactant monolayer is deformed, it shows not only a viscous response to deformation, but also an elastic response. There are several deformation modes of the monolayer to be considered: stretching, bending, shearing, and tilting. Each of the modes can be characterized by its own elastic and viscous moduli. The main interests of this chapter are the stretching and bending deformation modes in the elastic régime.

7.2.2.1 Stretching Mode

Consider a patch of a surfactant monolayer at the interface between oil and water. The free energy of the monolayer F is proportional to the surface area of the patch.

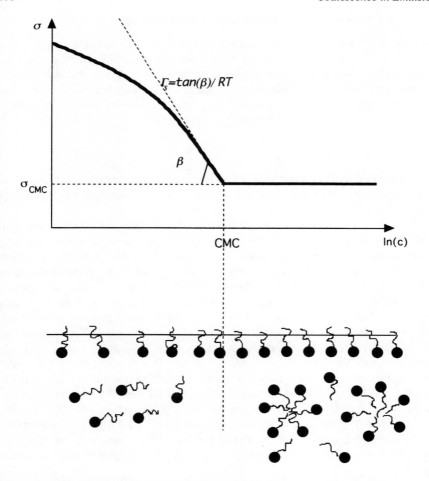

Figure 7.1 *Interfacial tension isotherm of a micelle-forming nonionic surfactant*

The most familiar contribution is the interfacial tension term, which can be presented as a difference:

$$F = A \times \sigma = A \times [\sigma_{O-w} - \Pi(A_s)] \tag{7.2}$$

where A is the surface area of the patch, σ_{O-w} is the bare oil–water interfacial tension, Π is the surfactant surface pressure, which is a function of the area per surfactant molecule, $A_s = 1/\Gamma_s$. As the area per surfactant molecule decreases, the surface pressure increases and the interfacial tension falls. However, the surface pressure never reaches the bare σ_{O-w} value and the interfacial tension remains positive. This happens because of surfactant micellization. Once the chemical potential of the surfactant at the macroscopic oil–water monolayer reaches that of the surfactant in the micelles, the surface pressure stays constant and the interfacial

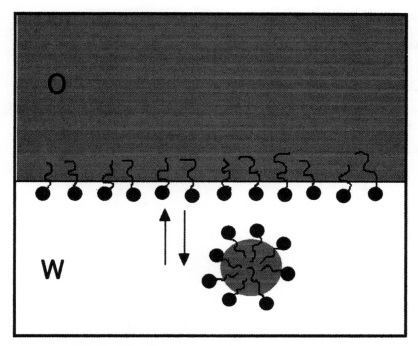

Figure 7.2 *Equilibrium between macroscopic and microscopic surfactant monolayers*

tension does not decrease. The micellar environment is always more attractive for the surfactant molecules than the macroscopic oil–water monolayer, which makes the tension positive. In some cases, however, the standard chemical potentials of these two states are very close and very low σ values, down to 10^{-4} mN m^{-1}, can be reached.

On the other hand, *microscopic* monolayers covering the micellar interface are under *zero* tension. In a closed system, the molecules at the micellar interface adjust their area per molecule in such a way that the free energy is minimized. In this state, the bare oil–water interfacial tension is completely counterbalanced by the interfacial pressure:[10,11]

$$\sigma_{O-w} = \Pi(A_s^*) \tag{7.3}$$

This balance is reached at the 'optimum' area per surfactant molecule, A_s^*. The state in which the bare oil–water interfacial tension is completely counterbalanced by the surface pressure is called the Schulman limit. The free energy has a minimum at A_s^* and deviations from this value cost energy, as determined by the following Hookean form:[12]

$$F = A \times \frac{\lambda}{2} \frac{(A_s - A_s^*)^2}{(A_s^*)^2} \tag{7.4}$$

where λ is the *compression modulus*. The compression modulus of monolayers is similar to Young's elastic modulus for solids and values range from 100 to 1000 mN m^{-1} for typical surfactant systems.

7.2.2.2   Bending Mode

The free energy of a monolayer under bend can be presented by another Hookean equation:[13]

$$dF = \frac{\kappa}{2}\left(\frac{1}{R_1} + \frac{1}{R_2} - \frac{2}{R_0}\right)^2 dA + \overline{\kappa}\,\frac{1}{R_1 R_2}\,dA \qquad (7.5)$$

Here R_1 and R_2 are the local principal radii of curvature, $H_0 = 1/R_0$ is the spontaneous curvature, κ is the bending modulus, and $\overline{\kappa}$ is the saddle splay modulus. The half-sum of the local curvatures is called the mean curvature H:

$$H = \frac{H_1 + H_2}{2} = \frac{1/R_1 + 1/R_2}{2} \qquad (7.6)$$

Surfactant monolayers have two sides which are not identical. Therefore, all the curvatures have a sign, and the states with positive and negative curvatures are physically different. According to the sign convention, the curvature is counted as positive if the monolayer is curved towards oil (*e.g.* in O/W micelles). Thus, for an O/W spherical micelle with radius R, $H = H_1 = H_2 = 1/R$. For an infinite O/W cylinder with the radius R, $H_1 = 1/R$ and $H_2 = 0$, and $H = 1/2R$. For a plane, $H_1 = H_2 = H = 0$. For saddle-shaped surfaces, the principal curvatures have opposite signs and the mean curvature can be equal to zero, even though both H_1 and H_2 are non-zero (*i.e.* when $H_1 = -H_2$, see Figure 7.3).

The bending elasticity of a monolayer is therefore characterized by three parameters: bending and saddle splay moduli which have dimensions of energy, and spontaneous curvature, which has a dimension of reciprocal length. In essence, eqn. (7.5) introduces the concept of the preferred shape of surfactant monolayers, stating that deviations of the mean curvature from the spontaneous curvature are costly in energy, with the energy penalty being proportional to κ. Whereas the physical meaning of the bending modulus of the spontaneous curvature is quite intuitive along these lines, it is less so for the saddle splay modulus. This parameter is coupled substantially with the topology of the monolayers. Thus, when two particles merge, the saddle splay energy changes by $\Delta F = -4\pi\overline{\kappa}$, independent of the size and shape of the particles.

Bending elasticity moduli can be measured, or at least estimated experimentally. The spontaneous curvature is close to the reciprocal radius of spherical micelles in equilibrium with solubilizate,[14] and therefore is rather straightforward to measure. The bending modulus is more difficult to measure experimentally.[15,16] Both κ and $\overline{\kappa}$ are usually presented in units of thermal energy, $k_B T$. Typical values of the bending modulus vary from 0.1 ('flexible' monolayers) to 10 $k_B T$ ('rigid' monolayers).[16–18] There is still very little data concerning the experimental value of the saddle splay

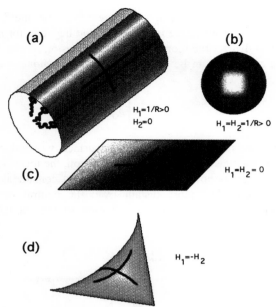

Figure 7.3 *Principal curvatures of various surfaces: cylinder* (a), *sphere* (b), *plane* (c) *and saddle surface* (d)

modulus. While the bending modulus is always positive, the saddle splay modulus can (or rather must be[19]) negative. The spontaneous curvature can vary in a continuous way from large and negative ($\sim -10^{-9}$ m^{-1}) through zero to large and positive ($\sim 10^{9}$ m^{-1}) values.

7.2.3 Spontaneous Curvature

The spontaneous curvature plays a very important role in surfactant science. For molecules with a large polar head and small tail, the molecules have a distinct wedge shape. This means that the monolayer, if allowed to do whatever it wants, tends to curve towards oil ($H_0 > 0$). Similarly, for molecules with a bulky tail and a small polar head, the monolayer curves towards water ($H_0 < 0$). Of course, the nature of the spontaneous curvature goes beyond this simple cartoon. Not only is the architecture of the polar head and alkyl chain important, but the nature of the oil and aqueous solvents, the presence of electrolytes, and temperature also play a role. These effects can be more clearly seen when one considers the distribution of the pressure in the interfacial region. It is well known that the interfacial tension between two phases can be represented as the zeroth moment of the excess pressure in the interfacial region:[20]

$$\sigma = \int \Delta p(z) \, dz \tag{7.7}$$

where $\Delta p(z)$ is the excess pressure at a distance z from the interface. The pressure is counted as positive for attractive interactions, and negative for repulsive ones. It is possible to show that for the monolayers at zero tension (in the Schulman limit), the spontaneous curvature can be evaluated as the *first moment* of the transverse pressure profile:[21]

$$H_0 = -\frac{1}{2\kappa} \int \Delta p(z) z \, dz \qquad (7.8)$$

The origin of the z axis is placed in the *neutral* surface, which does not change the surface area under the bend. For the interfacial tension, the location of the repulsive and attractive pressure is not significant, but it has a considerable effect for the spontaneous curvature. The more distant from the neutral surface the excess pressure is located, the larger is its contribution to the spontaneous curvature (Figure 7.4).

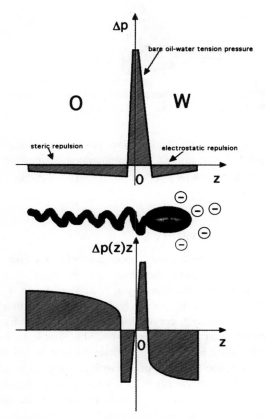

Figure 7.4 *Cartoon showing the transverse pressure profile in a monolayer of an ionic surfactant, showing the regions of steric chain repulsion, electrostatic repulsion, and bare oil–water attraction. The area under the Δp vs. z curve represents the interfacidl tension, and the area under the Δp(z)z vs. z curve is equal to $-2\kappa H_o$, i.e. proportional to the spontaneous curvature*

The spontaneous curvature of surfactant layers can be controlled in many ways; see Table 7.1. For ionic surfactants, one can control the contribution coming from the counter-ions by adjusting the concentration of salt. As the electrolyte concentration increases, the ionic atmosphere approaches the oil–water interface (Debye length decreases), the transverse pressure moment decreases and the spontaneous curvature falls.[22,23] Increasing the temperature does the opposite, because the osmotic pressure (π) of the counter-ions is proportional to temperature due to the osmotic pressure 'ideal gas' law: $\pi = cRT$.

Table 7.1 *Effects of various factors on the spontaneous curvature*

Factor	Effect on H_0 (taking the sign into account)	
	Nonionic surfactants of ethylene oxide type	Ionic surfactants
Size of surfactant chain	↓	↓
Branching of surfactant chain at a given number of carbon atoms	↓	↓
Size of polar group	↑	n/a
Increase in temperature	↓	↑
Electrolytes		
NaF, NaCl, NaBr, Na_2SO_4, Na_3PO_4, Na_2CO_3	↓	↓
NaI, $NaClO_4$, NaSCN	↑	normally, ↓
Desorbing neutral species, *e.g.* sugars	↓	↓
Nature of hydrocarbon oil		
Chain length	↑	↑
Polarity	↓	↓

For nonionic surfactants of the ethylene oxide type, the temperature effect is opposite to that of ionic surfactants, because of the peculiar temperature dependence of the hydration of ethylene oxide groups. As the temperature increases, the ethylene oxide chain loses its hydration water and the spontaneous curvature decreases. To a good approximation, the spontaneous curvature can be approximated by the first term of expansion in series *versus* temperature:[24,25]

$$H_0 = -\alpha(T - \overline{T}) \tag{7.9}$$

where $\overline{T}$ is called the balanced temperature (or the balanced point), at which the spontaneous curvature becomes equal to zero, and α is an empirical coefficient.

The chain length and polarity of the oil also influence the spontaneous curvature. Shorter chain and polar oils tend to penetrate to a larger degree into the chains of the surfactant monolayer, creating a wedge effect on the oil side of the monolayer and decreasing H_0.[26,27] Finally, adding surface-inactive species to the water makes the spontaneous curvature smaller.[28] The depleted solutes stay away from the adsorption layer and dehydrate the surfactant polar heads, which creates an additional bending moment in the negative direction.

7.2.4   Implications for the Phase Behavior

The significance of the surfactant shape for the phase behavior was recognized first by Winsor,[29] who introduced the 'R-parameter', which is analogous to the spontaneous curvature. The behavior of surfactant systems is most easily illustrated with the example of an oil–water–surfactant system, where the surfactant is an oligo(ethylene oxide) ether of a fatty alcohol, $C_iH_{2i+1}(OCH_2CH_2)_jOH$, abbreviated below as C_iE_j. The spontaneous curvature in these systems can be easily controlled by varying the temperature. When surfactant, oil, and water are mixed together, the surfactant assembles in such a way that the bending energy is minimized. Thus, when the spontaneous curvature is large and positive, micelles are formed in water. The oil is solubilized by the micelles until saturated with the solubilizate. This saturation happens when the micellar radius becomes close to the reciprocal monolayer spontaneous curvature $1/H_0$, or in primitive terms, when the surfactant wedge shape fits the curvature of the micellar interface. When more oil is added, it separates from the system as a second phase. The equilibrium of an aqueous micellar solution with excess oil is called the Winsor I equilibrium (Figure 7.5a). Similarly, at large (in absolute value) and negative spontaneous curvatures, the micelles are formed in oil and solubilize water as solubilizate; the excess water separates as a second phase and the Winsor II equilibrium is formed. Sometimes, the Winsor I equilibrium is designated as $\underline{2}$, and Winsor II as $\overline{2}$, to show the location of the surfactant in the two-phase equilibrium (the oil is tacitly assumed to have a lower density than water). In between the Winsor I and Winsor II regions the phase behavior is quite complicated. Normally, a three-phase equilibrium of 'oil' (upper), 'water' (lower), and 'surfactant' (middle) phases is observed. This state is called the Winsor III equilibrium. The three-phase region is separated from the Winsor I and Winsor II equilibria by two end-points, in which two of the Winsor III phases become critical with respect to each other. At the lower end point, the lower and middle phase are critical to each other, while the upper phase is non-critical to them. At the upper end-point, the upper and middle phases are critical and the lower phase is the non-critical 'spectator'.

In the Winsor III state, the 'oil' and 'water' phases represent weak molecular solutions of the surfactant in the solvents. On the other hand, the structure of the surfactant phase is quite intriguing. Because the spontaneous curvature is small, the surfactants associate into structures of zero, or low, curvature, which are either lamellar liquid crystals with $H_1 = H_2 = 0$, or a sponge-like bicontinuous microemulsion phase, where the surface has a locally saddle shape: $H_1 \approx -H_2$. It is believed that lamellar phases are favored for rigid surfactant monolayers, with $\kappa \approx 10\ k_B T$, while bicontinuous microemulsion phases are favored for flexible monolayers, with $\kappa \approx 1\ k_B T$.[11]

The whole sequence Winsor I–Winsor III–Winsor II in alkane–water–C_iE_j systems can be observed on increasing the temperature, which can be more easily visualized with the help of a prism diagram (Figure 7.5b). Essentially the same

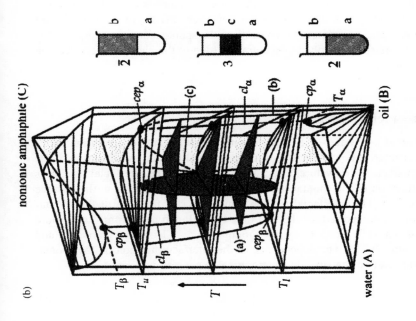

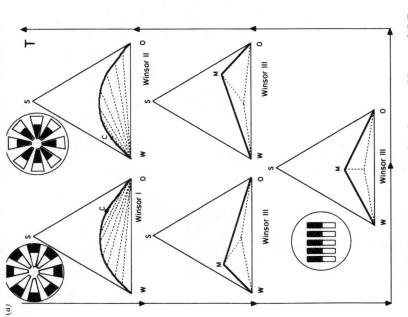

Figure 7.5 (a) *Schematic changes in the phase equilibrium of C_iE_j–alkane–water system on raising the temperature. The inserts show the preferred molecular packing of the surfactant molecules, black areas corresponding to the polar groups and white to surfactant tails. As temperature increases, the spontaneous curvature changes from positive to negative values and the phase equilibrium changes from Winsor I to Winsor III to Winsor II. The tie-lines are dashed. The locations of the critical points (C) and the surfactant middle phase (M) are shown on the plot.*
(b) *Triangular diagrams of (a) arranged into a prism diagram*
(Reproduced by permission of the American Institute of Physics from ref. 30)

sequence can be observed when the spontaneous curvature is controlled by other means, *e.g.* by adding a salt (due to a salt depletion effect). In ionic surfactant systems, the effective parameters for controlling the spontaneous curvature and the phase behavior are the electrolyte concentration (due to the electrostatic screening effect) and temperature.

7.2.5   Interfacial Tension

Consider again an oil–water–C_iE_j system with the surfactant concentration above the CMC. Our interest now is the oil–water interfacial tension as a function of the monolayer spontaneous curvature. Experiment shows that the interfacial tension is a parabolic function of the spontaneous curvature, with a minimum at the balance point (Figure 7.6). The interfacial tension minimum as a function of the sponta- neous curvature was reported for the first time by Wellman and Tartar,[31] and, more recently, systematically studied by the Yokohama,[32-34] Hull,[35-39] Paris,[17,40] and Göttingen groups.[24,25,30,41] The dependence of the interfacial tension on the sponta- neous curvature can be described by an empirical expansion-in-series:[42,43]

$$\sigma = \sigma_0 + 2\varepsilon H_0^{\,2} \qquad (7.10)$$

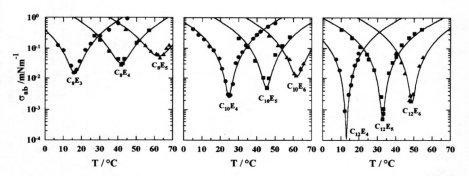

Figure 7.6 *Interfacial tensions of* n-*octane against water in the presence of various* C_iE_j *surfactants above the CMC as a function of temperature*
(Reproduced by permission of the American Institute of Physics from ref. 30)

where the first term, σ_0, is the interfacial tension in the balanced point, and the expansion coefficient ε has a dimension of energy. There have been several attempts to explain this dependence theoretically. By considering the equilibrium of surfactant monolayers covering the micellar surface and the macroscopic oil– water interface, several authors came to the following result:[11,44,45]

$$\sigma = \frac{2\kappa^2}{\kappa + \overline{\kappa}/2} H_0^2 \qquad (7.11)$$

This equation is based on neglecting the entropy of mixing term for the micelles, and is applicable only to spherical micelles (*i.e.* not over the Winsor III region). Although it is close in form to the empirical expansion, eqn. (7.10), it fails to account for the finite interfacial tension at the balance point, *i.e.* the σ_0 term. Experiment shows that the coefficient ε is indeed close to $k_B T$ for typical micro-emulsion systems, while the values of σ_0 vary considerably. Normally, σ_0 decreases with the surfactant chain length from 0.1 to 10^{-4} mN m^{-1} (see Figure 7.6).

Note that in the Winsor III equilibrium there are three phases in equilibrium and, therefore, three interfacial tensions. The σ value we have been discussing so far in this section is the interfacial tension between the upper (oil) and lower (water) phases, denoted σ_{ab} in Figure 7.7. The interfacial tensions between the middle (surfactant) phase and upper and lower phases also strongly depend on the sponta-neous curvature because of the critical phenomena at the three-phase-body end points (Figure 7.7). The wetting phenomena over the Winsor III region are very peculiar. In the balanced point of long-chain surfactants, the middle phase does not

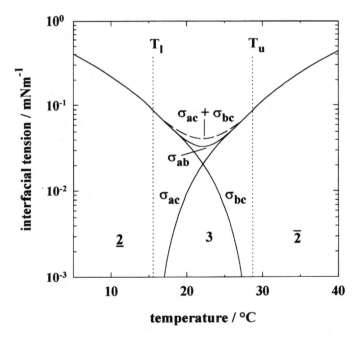

Figure 7.7 *Interfacial tensions in an $H_2O-oil-C_iE_j$ system over the Winsor III region as a function of temperature. Here σ_{ab} is the interfacial tension between upper and lower phases, σ_{ac} is the interfacial tension between the middle and lower phases and σ_{cb} is the interfacial tension between the middle and upper phases. σ_{cb} vanishes at the upper end-point T_u and σ_{ac} vanishes at the lower end-point T_l because of the critical phenomena. Except for the very vicinity of the end-points, $\sigma_{ac} + \sigma_{cb} > \sigma_{ab}$ and the middle phase does not wet the interface between the upper and lower phases*
(Reproduced by permission of the American Institute of Physics from ref. 30)

wet the upper and lower phases and forms a lens at the interface. The wetting is restored close to the end points. For short-chain surfactants, the middle phase wets the upper and lower phases over the whole Winsor III region.[41]

So far, we have been concerned with the interfacial tension ($\sigma \equiv \sigma^{planar}$) of a planar emulsion film in equilibrium with a micellar solution. If the emulsion film is curved to the mean curvature H, the interfacial tension ($\sigma^{curved}(H)$) is expected to be equal to:[43]

$$\sigma^{curved}(H) = \sigma^{planar} + 2\kappa(H - H_0)^2 - 2\kappa H_0^{\,2} \qquad (7.12)$$

The interfacial tension of a curved film has a minimum at $H = H_0$, where the spontaneous curvature is equal to the curvature of the film. Note that this 'monolayer frustration' correction applies only to the saturated surfactant monolayers and is different from the well-known Tolman curvature correction to the surface tension,[46] which applies to much higher curvatures.

7.3   Experimental Trends in Macroemulsion Stability

Since the basic ideas in surfactant science have been introduced, one can now proceed with the main topic of this chapter: macroemulsion stability. Macroemulsions are formed by mechanical mixing of oil and water in the presence of surfactants, *e.g.* by mixing the phases of the Winsor I equilibrium in each other. As a result of mixing, one of the phases breaks into macroscopic droplets, while the other stays continuous. Macroemulsions are thermodynamically unstable and gradually resolve with time into two distinct layers. However, in some cases they show a remarkable kinetic stability. Most experimental trends in macroemulsion stability were established a long time ago and will be outlined below.

7.3.1   Surfactant Nature: Oriented Wedge Trend

At the beginning of this century it was discovered that long-chain carboxylates of monovalent metals tend to stabilize O/W emulsions, and those of polyvalent metals, W/O emulsions. Moreover, when salts of polyvalent metals were added to O/W emulsions stabilized by soaps of monovalent metals, macroemulsions either broke or inverted into W/O systems. Since double-tail surfactants tend to stabilize W/O emulsions, and single-tail surfactants stabilize O/W emulsions, Harkins *et al.*[47] and Langmuir[48] independently concluded that the steric constraint in the monolayer is important for emulsion stability (Figure 7.8). Harkins *et al.* wrote:

> 'The stability of emulsoid particles seems to be brought about by orientation
> of molecules at the interface with the medium of dispersion... For the
> emulsoid particle to be stable... the molecules of the 'film' should fit the
> curvature of the drop. From this standpoint, the surface tension of very small
> drops is a function of the curvature of the surface.'

This concept received the name of the 'oriented wedge theory' and, in modern terms, can be reformulated as a relationship between the macroemulsion type and

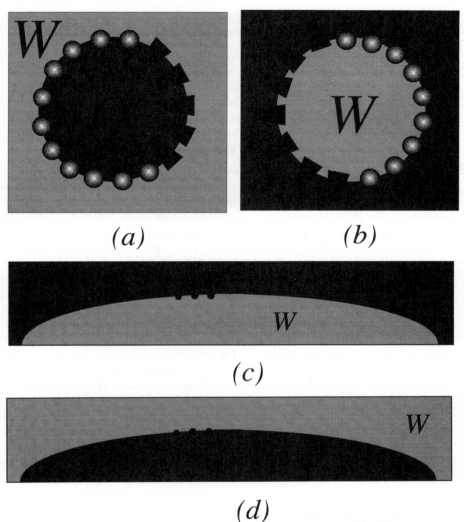

Figure 7.8 (a *and* b): *Cartoon representing the oriented wedge theory, as presented by Harkins* et al.[47] *and Langmuir.*[48] *The monolayers covering emulsion droplets have different frustration energies, which favor one emulsion type over another. Note that the picture shows the macroscopic emulsion droplets and not the surfactant micelles. The theory is wrong because the monolayers are essentially planar on the molecular scale* (c *and* d)

the monolayer spontaneous curvature. It should be noted that essentially the same idea was proposed earlier by Bancroft,[49] who introduced the spontaneous curvature not as a simple steric argument, but in terms of the transverse pressure moment in the surfactant monolayer. Bancroft considered the mixture of two liquids: water

(A), oil (B) and models the surfactant film as a third liquid (C). Liquid C spreads between liquids A and B. He then compared the interfacial tensions σ_{AC} and σ_{BC}.

> 'If $\sigma_{AC} > \sigma_{BC}$, the film C will tend to become convex towards the B phase because... this shortens the surface AC relatively to the surface BC. In other words, water will tend to be emulsified in oil. If $\sigma_{AC} < \sigma_{BC}$, the film C will tend to become convex to phase A because this shortens the surface BC relatively to the surface AC. In other words, oil will tend to be emulsified in water. We thus see that the sole factor determining the type of emulsion is the sign of the difference between σ_{AC} and σ_{BC}.'

By 1940, it was realized that the oriented wedge theory, whether formulated in Langmuir–Harkins or in Bancroft terms, was wrong.[50] The radius of spontaneous curvature of typical unbalanced surfactants amounts to $\sim 10^{-7}$ cm, several orders of magnitude less than the radius of macroemulsion droplets. For these surfactant molecules the interface of O/W and W/O emulsion droplets is essentially planar on the molecular scale and the monolayer frustration in both cases is nearly identical (Figure 7.8). The oriented wedge theory should rather be applied not to macro-emulsion droplets, but to surfactant micelles, as it was indeed re-discovered many years later.[14,29,51] After this flaw in the oriented wedge theory was found, it was gradually forgotten.

Despite the problems with the oriented wedge theory, a closer look reveals that the empirical rule-of-thumb, relating the spontaneous curvature and macroemulsion type and stability, is very powerful. For positive spontaneous curvature, O/W emulsions tend to be stable, while for negative spontaneous curvature, W/O emulsions tend to be stable. Below we will call this correlation the *oriented wedge trend*, to distinguish it from the incorrect oriented wedge *theory* of Bancroft–Harkins–Langmuir. The most successful empirical correlations—Phase Inversion Temperature (PIT) and Optimal Surfactant Formulation models, can be reduced to the oriented wedge trend. Two other empirical correlations, Bancroft's rule and the Hydrophile–Lipophile Balance (HLB) scale, also usually (but not always) follow it.

7.3.2 Bancroft's Rule

In spite of the fact that the mechanistic interpretation of emulsion stability by Bancroft[49] was essentially the oriented wedge theory, this paper became acclaimed because of another empirical correlation. Bancroft states:

> 'The simplest way to emulsify oil in water is to add a water-soluble colloid which is adsorbed strongly at the interface, and the simplest way to emulsify water in oil is to add an oil-soluble colloid which is adsorbed strongly in the interface. As a matter of fact, this is the way in which almost all emulsions are made.'

In modern terms, this correlation can be reformulated—water-soluble surfactants tend to stabilize O/W emulsions and oil-soluble surfactants tend to stabilize W/O emulsions. Bancroft realized that this was just an empirical correlation and not a fundamental law of nature. In a later paper, he gave several examples where the Bancroft rule was violated.

In the formulation of the rule, the author did not distinguish between the molecular and micellar solubilities of the surfactant, simply because micelles were not known at that time. If the surfactant concentration is above the CMC, the net 'solubility' has micellar and molecular components. In the case when the critical micelle concentration is negligibly small, Bancroft's rule is reduced to the oriented wedge trend. Indeed, for positive spontaneous curvature, micelles are formed in water and the surfactant is 'water-soluble'. Conversely, for negative spontaneous curvature, the surfactant forms micelles in oil and is 'oil-soluble'. The situation is more complicated when the critical micelle concentration is comparable to the net micellar concentration. This type of solution behavior, for example, is shown by most C_iE_j surfactants, which show a very high CMC value in alkanes, ~ 1 wt%. Experiment shows that in these cases Bancroft's rule is violated and nature follows the oriented wedge trend.[52,53]

There were several attempts to interpret Bancroft's rule in terms of the damping of surface corrugations by the Gibbs elasticity[50] or in terms of the rate of thinning of emulsion films prior to coalescence.[54-56] Another interpretation is concerned not with the stability of an already formed emulsion, but with the process of emulsification.[8]

7.3.3   HLB Scale

In the late 1940s, Griffin[57] proposed to characterize surfactants with a Hydrophile–Lipophile Balance (HLB) value, which measures the ability of surfactants to stabilize O/W and W/O emulsions, as well as to act as wetting agents, detergents, and solubilizers (Table 7.2). According to Griffin, surfactants having HLB values of 4–6 tend to stabilize W/O emulsions, while surfactants with HLB values of 8–18 tend to stabilize O/W emulsions. Griffin also included the nature of the oil component into the HLB model. For each oil, an experimental procedure to evaluate the 'required HLB value' was proposed. Within the ranges shown in Table 7.2, more polar oils required higher-HLB surfactants for making O/W emulsions and lower-HLB surfactants for W/O emulsions. In addition, a simple rule of thumb to evaluate the HLB value of surfactant mixtures was proposed: HLB values of individual components were averaged with the coefficients proportional to their mass fractions in the mixture.

Table 7.2   *A summary of HLB ranges and their application*[57]

HLB range	Use
4–6	W/O emulsifiers
7–9	Wetting agents
8–18	O/W emulsifiers
13–15	Detergents
15–18	Solubilizers

The method of HLB evaluation originally included a number of emulsion stability tests, and was quite tedious. Later, Griffin proposed empirical equations, which allowed the evaluation of the HLB value from the molecular formula of some surfactants.[58] Thus, for polyoxyethylene stearates, the HLB value was estimated as HLB $= E/5$, where E is the weight percent of ethylene oxide in the molecule. A further development occurred when Davies[59-61] suggested an additive scheme to evaluate HLB values for a much broader range of surfactants. In Davies' method, a group number is assigned to various emulsifier component groups and the HLB is then calculated by an empirical relation (see also Tables 7.3 and 7.4):

$$\text{HLB} = 7 + \sum (\text{hydrophilic group numbers}) - \sum (\text{hydrophobic group numbers}) \tag{7.13}$$

Table 7.3 *HLB group numbers*[60]

Groups	Group number
Hydrophilic	
$-SO_3^-Na^+$	38.7
$-COO^-Na^+$	19.1
$-COO^-K^+$	21.1
Sulfonate	~ 11
Ester (sorbitan ring)	6.8
Ester (free)	2.4
$-OH$ (sorbitan ring)	0.5
$-OH$ (free)	1.9
$-COOH$	2.1
Ether oxygen, $-O-$	1.3
$-(CH_2CH_2O)-$	0.33
$-(CH(CH_3)CH_2O)-$	-0.15
Lipophilic	
$-CH-$	0.475
$-CH_2-$	0.475
$-CH_3-$	0.475

Table 7.4 *HLB values of some surfactants*[60]

Surfactant	Experimental HLB value
Na lauryl sulfate	40
K oleate	20
Na oleate	18
Tween 80 (sorbitan monooleate + 20 $-(CH_2CH_2O)-$ groups)	15
Span 80 (sorbitan monooleate)	4.3
Cetyl alcohol	1
Oleic acid	1

The HLB scale enjoys great popularity among emulsion technologists and is generally accepted in practice. Indeed, when a new surfactant is synthesized, the first rough estimate of its properties can be readily made with the HLB model. An extensive list of HLB values of commercial surfactants can be found in Becher's monograph.[62] Several updates of bibliography on HLB have been published.[63] However, the HLB scale has severe limitations. Firstly, if one uses the Davies additive scheme for HLB evaluation, the structure of the hydrocarbon backbone and the polar head is not included in the model, because it is only the number of atoms (groups) that counts. For example, a linear chain C_{16} surfactant has the same HLB value as the double chain C_8 surfactant with the same polar head. This is an oversimplification: double chain surfactants have a much stronger tendency to stabilize inverse emulsions than the single chain surfactant with the same carbon number. Similarly, branching in the polar head also has a significant effect on the surfactant properties. (This problem was apparently evident to Griffin, who insisted on experimental evaluation of HLB numbers.) Secondly, the temperature effects are not incorporated into the model. For some surfactants, the temperature dependence is quite drastic and the same surfactant can be an O/W and W/O emulsion stabilizer (see below). Thirdly, for nearly balanced surfactants with HLB of $\approx 6-10$, the HLB method often predicts the wrong emulsion type. More discrepancies between the HLB model and experiment are discussed by Shinoda and Friberg[64] and by Davis.[65]

Several empirical correlations between the HLB value and physicochemical parameters of the surfactants were proposed. Griffin suggested a correlation between the HLB value and the nonionic surfactant cloud point.[57] Becher related the HLB value to the free energy of surfactant micellization, and to the Israelachvili–Mitchell–Ninham packing parameter (*i.e.* spontaneous curvature).[66] There was also an attempt to provide a mechanistic interpretation of the HLB scale by Davies *et al.*[60,61] The authors examined the rates of coalescence of O/W and W/O emulsions. For high-HLB surfactants, the coalescence barrier of O/W emulsions is predicted to be high, while that of W/O emulsions is low. Eventually, O/W emulsions will dominate over the time, which explains the HLB. However, the way the coalescence barriers were estimated is rather primitive.[43]

7.3.4 'Molecular Complexes'

In the early 1940s, Schulman and co-workers[67,68] discovered that using two surfactants of an opposite packing type (*i.e.* a mixture of 'oil-soluble' and 'water-soluble' surfactants) could drastically improve emulsion stability. The authors attributed it to the formation of 'molecular complexes' between the surfactants at the oil–water interface. It was discovered that two surfactants together reduced the interfacial tension to very low values (~ 0.1 mN m^{-1}), much lower than each of them separately.[69] Many years later the studies of the phase behavior of similar systems revealed that the idea of 'molecular complexes' was incorrect. The surfactants do not form a stoichiometric complex, but associate into a mixed bilayer and form a lamellar phase, which co-exists with oil and water in a three-phase equilibrium.

The idea of mixing of two surfactants with an opposite packing type for making emulsions is widely used in practice. After a system containing two surfactants with an opposite packing type is selected, their ratio in the mixture is optimized by trial and error. In most cases, the emulsion stability, as judged by the disperse phase separation, passes through a maximum at some surfactant ratio.[70] At high concentrations the emulsion solidifies because of the lamellar phase present, which is called the 'self-bodying action' of the surfactant mixtures. The systems of that sort are very common in many commercial pharmaceutical and cosmetic preparations.[71]

7.3.5   PIT Concept

This empirical correlation was discovered by Shinoda and Saito in 1968[72] with surfactants of the ethylene oxide type. The authors demonstrated that the same surfactant can act as an O/W or a W/O emulsion stabilizer. At low temperatures, over the Winsor I region, O/W macroemulsions can be easily formed and are quite stable. On raising temperature, the O/W macroemulsion stability decreases, and the macroemulsion finally resolves when the system reaches the Winsor III state. Within this region, both O/W and W/O macroemulsions are unstable, with the minimum stability in the balanced point. At higher temperatures, over the Winsor II region, W/O emulsions become stable (Figure 7.9). This behavior is always observed in nonionic systems if the surfactant concentration is above the CMC and the volume fractions of the components are not extreme. The macroemulsion stability pattern is essentially symmetrical with respect to the balanced point, just as the phase behavior is. At positive spontaneous curvature, O/W emulsions are stable, while at negative spontaneous curvature, W/O emulsions are stable, *i.e.* the PIT concept can be reduced to the oriented wedge trend.

Figure 7.10 represents the most clear-cut image of the macroemulsion inversion as a function of temperature. Equal volumes of oil and water were emulsified at various temperatures. Five hours after preparation, macroemulsions completely sediment. Below the balanced temperature, a stable O/W cream layer is formed, which does not show any visible coalescence. Similarly, above the balanced point, a stable sediment of a W/O macroemulsion is formed. (Note that the W-phase layer below the balanced point and O-layer above the balanced point are formed due to creaming and not coalescence.) Close to the balanced point, in the narrow temperature range between 66 and 68 °C, where the three-phase equilibrium is observed, neither O/W nor W/O macroemulsions are stable.

There are two more subtle features of the nonionic macroemulsion stability to be discussed. Firstly, within the Winsor III region, the stability of macroemulsions is very temperature sensitive. Although exactly in the balanced state, the macroemulsions are very unstable and break within minutes, the system becomes stable only several tenths of a degree away from the balanced point, while still being within the Winsor III region. Secondly, the macroemulsion stability pattern is not completely symmetric. W/O emulsions reach maximum stability at *ca.* 20 °C above the balanced point, after which the stability starts to decrease. On the other

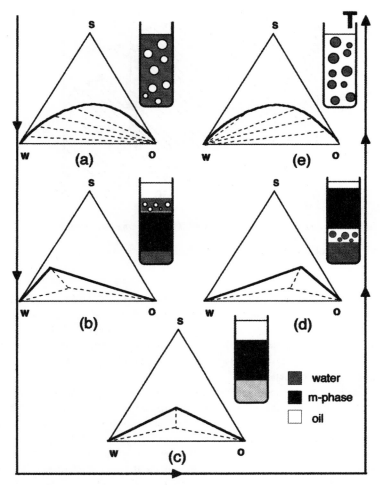

Figure 7.9 *The PIT concept. As the temperature increases, the macroemulsion type changes from O/W over the Winsor I region to W/O over the Winsor II region, through the emulsion breakage at the balanced point*

hand, there is no similar maximum in the O/W emulsion stability at very low temperatures.

Up to this point, we were concerned with three-component nonionic O–W–S systems. It is known that the microemulsion phase behavior can be 'tuned' not only by changes in temperature, but also by adding different 'co-solvents' or 'co-surfactants'. Some of these additives merely produce temperature shifts in the phase diagrams, leaving the general picture of the phase behavior unaltered. For instance, the balanced point of the n-C_8H_{18}–$C_{10}E_5$–water system is at $\sim$45 °C, while that of the n-C_8H_{18}–$C_{10}E_5$–10% NaCl brine is at $\sim$28 °C.[41] The changes in microemulsion phase behavior induced by additives lead to a similar shift in the

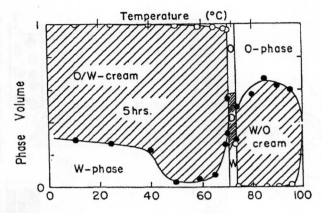

Figure 7.10 *Macroemulsion stability diagram of cyclohexane–water–polyoxyethylene (9.7)*
nonylphenyl ether system
(Reproduced by permission of Academic Press from ref. 73)

macroemulsion stability profile.[64,74] Thus, when 10% of NaCl is added to the
system, the new balanced point is established at 28 °C and now *macro*emulsions
prepared below 28 °C will have an O/W type, and those prepared above 28 °C will
have a W/O type. The same effect is found when the location of the balanced point
is controlled by adding co-solvents to oil and water, changing the chain length of
the oil, and adding co-surfactants.[64]

PIT behavior can be readily observed in test-tube experiments. Close to the
balanced state, the samples of macroemulsions can be readily prepared by hand
because of the low interfacial tension between oil and water. Moreover, the
surfactant purity is not very important. Impurities just shift the position of the
macroemulsion inversion point, leaving the general inversion trend unchanged. In
fact, the PIT trend itself was established not with pure surfactants but with a
polydisperse surfactant product. On the other hand, good temperature control is
necessary, in particular close to the balanced point.

7.3.6 'Optimal Surfactant Formulation'

The 'Optimal Surfactant Formulation' model can be considered as an extension of
the PIT concept for complex surfactant mixtures. In 1973, the oil embargo caused a
considerable amount of research aimed at enhanced oil recovery. Among the
proposed methods were low-tension surfactant flooding processes, in which a
surfactant solution was injected into the oil reservoir to produce a low interfacial
tension between the crude oil and water, in order to reduce the capillary pressure
resistance. The researchers were aiming at an 'optimal surfactant formulation', at
which the interfacial tension has a minimum, *i.e.* at the balanced surfactant
composition. The systems studied represented very complex mixtures, containing
polydisperse surfactants, alcohols, salts, hydrocarbons, and water. Elaborate em-

pirical equations were proposed to evaluate the location of the balanced point of these compositions. Salager *et al.*[75] proposed an empirical formula for determining the 'optimal surfactant formulation' (*i.e.* the balanced point) for anionic surfactant systems:

$$\ln S - K \times \text{ACN} - f(A) + \sigma - a_T \Delta T = 0 \qquad (7.14)$$

Here S is the NaCl concentration in wt%, ACN, or Alkane Carbon Number, is a characteristic parameter of the oil phase, ΔT is the temperature deviation from a certain reference (25 °C), $f(A)$ is a function of the alcohol type and concentration, and K, σ, and a_T are empirical parameters characterizing the surfactant. A similar empirical equation was proposed for nonionic surfactants:[76,77]

$$\alpha - \text{EON} + bS - k\text{ACN} - \phi(A) + c_T \Delta T = 0 \qquad (7.15)$$

where EON is the average number of ethylene oxide groups per surfactant molecule, $\phi(A)$ is another empirical function of the alcohol type and concentration, and α, c_T, and k are the empirical constants characterizing the surfactant. It is *aposteriori* clear that the left-hand sides of these equations are proportional to the monolayer spontaneous curvature, taken with the opposite sign.

As the by-product of the search for the optimal surfactant formulation, an interesting relationship between the macroemulsion composition and stability was discovered.[78-81] The macroemulsions invert as any of the composition parameters is continuously varied in such a way that the system passes through the optimal formulation (balanced state), and the left-hand side of eqns. (7.14) or (7.15) changes sign. Figure 7.11 shows the macroemulsion stability trends for complex multicomponent mixtures, containing anionic surfactants, alcohols, oil (or mixture of oils), and inorganic salts. As the spontaneous curvature is varied by changing the mole fraction of one of the surfactants in mixture, composition of the oil phase, or the mole fraction of alcohol, systems pass through the Winsor I–Winsor III–Winsor II sequence. In all cases the macro-emulsions invert at the balanced point with a sharp change in their conductivity. The macroemulsion lifetime has a deep minimum of several minutes in the balanced point, which increases sharply away from the balanced point to several hours or even to several days. Again, the experimental data fit the 'oriented wedge trend'.

7.3.7 Other Empirical Correlations

Several other half-empirical scales of surfactant emulsifying action have been proposed. Beerbower and Hill[82] introduced the Cohesive Energy Ratio (CER), defined as the ratio of the surfactant adhesion energy to the oil phase to the surfactant adhesion to the water phase. The calculation of adhesion energies is done in terms of the Hildebrand solubility parameters. Kruglyakov *et al.*[83,84] proposed the Hydrophile–Lipophile Ratio (HLR), which is the ratio of the energy of adsorption of the surfactant molecule from the water phase to its energy of

228

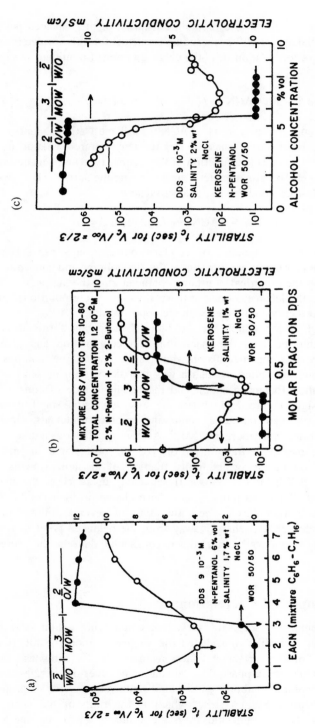

Figure 7.11 *Macroemulsion inversion as a function of monolayer spontaneous curvature. The spontaneous curvature is varied by changing the ratio of two surfactants in mixture (a), by varying the 'equivalent alkane carbon number' (EACN), i.e. by mixing two oils in different ratios (b), and by the addition of alcohol (c). All the system are studied over the $\underline{2}$–3–$\underline{2}$ transition range; see the bars at the top of the plots. The left-hand axis shows the macroemulsion lifetime, denoted as the time for the separation of two-thirds of the disperse phase by volume. The right-hand axis shows the macroemulsion conductivity. All emulsions were prepared at a 1:1 oil-to-water ratio; other details of the compositions are shown on the plots. DDS and WITCO TRS 10-80 are commercial surfactants (Reproduced by permission of Marcel Dekker Inc. from ref. 78)*

adsorption from the oil phase. Salager[77,85] proposed the Surfactant Affinity Difference (SAD) parameter which is the difference in the standard chemical potentials of the surfactant in the oil and in the water. The parameters of Kruglyakov *et al.* and Salager are essentially identical. Both HLR and SAD are believed to depend on the surfactant partition coefficient between the oil and water:

$$\text{HLR} \approx \text{SAD} \approx \mu_0^{\text{W}} - \mu_0^{\text{O}} = RT \ln K \tag{7.16}$$

where K is the partition coefficient of the surfactant between the oil and water and μ_0^{O} and μ_0^{W} are the standard chemical potentials of the surfactant in the oil and water, respectively. On a phenomenological level, the HLR and SAD models are identical to Bancroft's rule. On a thermodynamic level, the standard chemical potentials and the partition coefficient have meanings only when the surfactant forms a molecular solution both in oil and water, *i.e.* the system is below the CMC.

The Equivalent Alkane Carbon Number (EACN) is an empirical correlation which helps to evaluate the location of a balanced point for different emulsified oils and oil mixtures.[77,85,86] The 'Calculation of Phase Inversion in Concentrates' (CAPICO) method[71] is similar to the EACN.

7.3.8 Multiple Emulsions

Most often, emulsions are either O/W or W/O. In some cases, however, emulsions have a more complex structure and the disperse phase drops contain inclusions of still smaller droplets of continuous phase. These emulsions were discovered in 1925 by Seifritz[87] and are called *multiple emulsions*. These systems are typically observed as transient structures close to the emulsion inversion point and are rather unstable.[62]

Multiple emulsions can be prepared with two surfactants of an opposite packing type.[88] Thus, to produce an O/W/O emulsion, a surfactant with negative spontaneous curvature is first dissolved in oil. Then water is added and a W/O emulsion is prepared. The system is then emulsified again as a continuous phase of its own, in an aqueous solution of surfactant with positive spontaneous curvature, to produce a multiple O/W/O emulsion. The systems prepared this way, however, often resolve into O/W or W/O types on standing. Only recently, kinetically stable multiple emulsions were prepared.[89]

7.3.9 Surfactant Concentration: Tight Monolayer Packing

A review of early studies is presented in the Harkins monograph.[90] The author writes:

> 'Emulsions are stabilized by the presence of a film between the oil droplets and the water. According to McBain,[91] this film is many molecules thick, but it has been shown to be only monomolecular... The stability of an emulsion is highly dependent upon the tightness of packing of the molecules in this monolayer.'

This conclusion was based on a careful study of O/W emulsions stabilized by sodium oleate.[92] The particle size distributions of the emulsions were determined by optical microscopy, and the surfactant concentration in the bulk solution was measured prior to and after emulsification. This allowed the evaluation of the area per surfactant molecule in the adsorbed layer of the emulsion droplets. The authors showed that the area per surfactant molecule at the interface of stable (>1 h) emulsion droplets falls within the range 20–45 Å². Fischer and Harkins note that this range of areas per molecule corresponds to a dense molecular packing, but is not sufficiently small to allow more than one molecule for the thickness of the film, and conclude that one densely packed surfactant monolayer is necessary and sufficient for stabilization.

When a micelle-forming surfactant is used as an emulsion stabilizer, the surface coverage reaches a maximum at the CMC and stays constant as the surfactant concentration increases further. As a simple empirical rule of thumb generalizing available experimental data,[7,84,93] one can state that the emulsion stability sharply increases in a rather narrow surfactant concentration region between $\sim 0.2 \times$ CMC and $1 \times$ CMC. After reaching the CMC, the stability typically (but not always[94]) levels off. The emulsion stability is indeed very sensitive to the extent of surface coverage, since over the surfactant concentration range between 0.2 and $1 \times$ CMC, the coverage of typical surfactants only changes slightly. Not only does the macroemulsion stability steeply increase close to monolayer saturation, but the whole pattern of the macroemulsion inversion as a function of spontaneous curvature can only be seen for nearly saturated monolayers. In recent careful experiments[53] it was shown that in $C_{12}E_5$–heptane–water mixtures, the oriented wedge trend is observed only above a surfactant concentration of ~ 0.5 CMC, where the surface coverage is essentially that of a saturated monolayer.

The tight monolayer packing concept applies not only to emulsions, but also to foams. The stability of foam films formed by a homologous series of potassium carboxylates (C_8-C_{14}) shows a sharp increase in the vicinity of the CMC, after which it levels off or slightly decreases.[95] The data covering several orders of magnitude in surfactant concentration are very convincing.

7.3.10   Concentration for Formation of Black Films, C_{black}

This parameter was introduced by Scheludko and Exerowa.[96,97] According to the authors, above some surfactant concentration in the bulk, called the concentration for the formation of black films, C_{black}, liquid films can transform into 10–100 nm 'black' films, stabilized by disjoining pressure. The films are called black because they have a thickness which is less than the wavelength of light and look black in reflected light. At this surfactant concentration, emulsion and foam stability shows a sharp increase. The values of C_{black} were experimentally measured for a number of surfactants both in emulsion and foam systems.[7,84,97] The data indicate that C_{black} of foam films can be several orders of magnitude lower than the CMC.[97] Most literature data on emulsions show that C_{black} is close to the CMC, and support the 'tight monolayer packing' idea of Harkins.

7.3.11 Stabilization by a Lamellar Phase

Although many emulsions are stabilized by only one nearly saturated monolayer, this is not always the case. If the surfactant is balanced, it can form a lamellar phase (L_α), co-existing with oil and water. If emulsification is performed in such a system, the emulsion droplets are often covered by several monolayers back to back, which drastically improves the emulsion stability.[98-101] It should be noted that the original idea that emulsion drops must be stabilized by several monolayers back-to-back in order to be stable belongs to McBain.[91]

Among researchers involved in surfactant phase diagram studies, it is well known that whenever the lamellar phase region is reached, emulsification becomes very noticeable and the construction of the phase diagram becomes a headache. This can be illustrated with the example of the phase diagram of a balanced nonionic microemulsion. Let us adjust the temperature to the balanced state, mix oil and water and gradually increase the surfactant concentration (see Figure 7.12). At very low concentrations of surfactant the system is below the CMC and the two-phase equilibrium of oil and water is observed. Over this region, emulsions are very unstable. As the system enters the Winsor III three-phase equilibrium, the

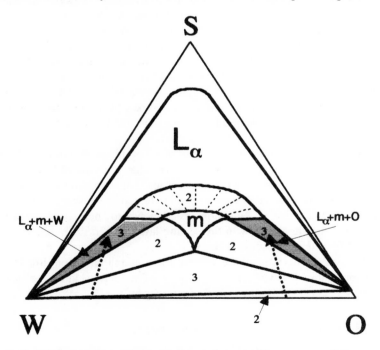

Figure 7.12 *Schematic phase diagram of a balanced alkane–water–C_iE_j system (see Strey[102] for an experimentally determined phase diagram of the H_2O–$C_{10}E_5$–n-octane system). Here L_α is the lamellar phase and m is the bicontinuous microemulsion phase. The dark three-phase triangles are characterized by a very strong emulsification, by contrast with the Winsor III triangle (unshaded), in which emulsions are very unstable. Dashed arrows show the composition trajectories*

macroemulsions are still very unstable. A spectacular change happens, however, when the system enters the three-phase triangles oil$-$m$-$L$_a$ or water$-$m$-$L$_a$, where m refers to the bicontinuous microemulsion phase (Figure 7.12). In these regions, emulsification is very evident even in the balanced state.

Friberg *et al.*[103] attributed the stabilization effect of the nested lamellar layers to the change in dispersion attraction between emulsion droplets. Another interpretation is based on the hole nucleation in multilamellar films, originally proposed by de Gennes and Prost,[104] and discussed below in more detail.

7.3.12   Effect of Volume Fraction of Disperse Phase and Droplet Size

Macroemulsions can be prepared in a very broad range of disperse phase volume fraction ϕ, which is not limited by the close sphere packing value $\phi = 0.74$. Indeed, one can prepare a macroemulsion which contains only 1 wt% of continuous phase! This can be done because liquid drops need not remain spherical and can pack into a structure, called a biliquid foam.

When the effect of ϕ on the emulsion stability is considered, one must distinguish between the effect of the volume fraction on the ease of emulsification and on the subsequent emulsion stability. The volume fractions of oil and water have a considerable effect on the ease of emulsification. For example, it is very difficult to prepare an O/W emulsion at an oil-to-water volume ratio of 9:1, even if an appropriate surfactant is selected. As a rule, the resulting system has an O/W/O type, which quickly converts into O/W plus excess oil after the mixer has been turned off.[105,106] However, if the mixing cycle is repeated several times, or the oil is added not as a single batch but gradually, one can obtain a very concentrated and stable O/W macroemulsion.

A common anecdotal observation is that small droplet size emulsions are more stable to coalescence than the coarse ones. This has been recently confirmed by more systematic studies on single emulsion drops[107] and concentrated emulsions in regimes of spontaneous coalescence[108] and forced coalescence in an ultracentrifuge.[93] The lifetime of emulsions τ is shown to increase drastically with decreasing the particle size d, with the scaling law $\tau \sim 1/d^2$,[108] in agreement with the model of film rupture discussed below.

The effects of the disperse phase volume fraction and emulsion droplet size on the emulsion stability have been the subject of numerous kinetic models; see Tadros and Vincent[70] and Danov *et al.*[109] for reviews.

7.4   Theory of Emulsion Coalescence

7.4.1   Rupture of Single Films and the Emulsion Lifetime

After considering experimental trends, let us proceed with a simple model*. Consider the coalescence in a creamed O/W emulsion layer. We model the system

---

*See Deminiere[108] for a similar approach

as a stack of monodisperse cubic cells (Figure 7.13). Each cell has six facets separating it from its neighbors. If rupture is a completely random process, one can expect that the probability of rupture is proportional to the surface area of the film:

$$\tau_{\text{film}} \approx \frac{1}{Af} \tag{7.17}$$

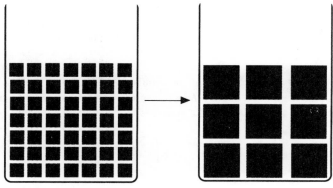

Figure 7.13 *Cartoon illustrating coalescence in concentrated emulsions*

where A is the surface area of the film, τ_{film} is the film lifetime, and f is the frequency of rupture per unit area $[\text{m}^{-2}\,\text{s}^{-1}]$. It is very difficult to measure the lifetime of one emulsion film. However, one can experimentally evaluate f by measuring the mesh size d of the emulsion as a function of time t. Let us assume that the thickness of films separating the oil droplets is very low, so that the total volume of the system V is equal to:

$$V = N(t)d^3(t) = \text{const} \tag{7.18}$$

where N is the number of cells in the system. The total number of coalescence acts per unit time can be evaluated as:

$$-\frac{\mathrm{d}N}{\mathrm{d}t} = f \times A_{\text{total}} = \frac{Z}{2} N(t)\, d^2(t) f \tag{7.19}$$

where $Z = 6$ is the coordination number of the cubic cell. From eqns. (7.18) and (7.19), one straightforwardly concludes that the mesh size in the emulsion increases with time according to the following law:

$$\frac{1}{d_0^2} - \frac{1}{d^2(t)} = \frac{Z}{3}\, ft \tag{7.20}$$

where d_0 is the initial droplet diameter. The time to complete coalescence ($d = \infty$) therefore equals:

$$\tau = \frac{3}{Z d_0^2 f} \approx \frac{1}{2 d_0^2 f} \tag{7.21}$$

According to this equation, the emulsion lifetime decreases with an increase in the frequency of the hole nucleation and in the emulsion droplet size. Sometimes it is difficult to follow the changes in the mesh size of the concentrated emulsion experimentally. In this case, one can monitor the rate at which the layer of free oil is formed at the top of the vessel. It is easy to show that the time for separation of half of the volume of the oil, $\tau_{1/2}$, is close to τ predicted by eqn. (7.21), if the height of the cream layer (h) is much larger than the size of the emulsion droplets: $h \gg d_0$.

The problem of the evaluation of the emulsion lifetime is therefore reduced to determining the rupture frequency f. In the spirit of the physical kinetics approach, it can be regarded as the product of a pre-exponent f_0 and an exponent:

$$f = f_0 \exp\left(-\frac{F^*}{k_B T}\right) \tag{7.22}$$

where F^* is the activation energy. The exponent is the essential rate-controlling term, as will be discussed in detail below. The pre-exponent term is less dependent on the external conditions. As a first rough estimate of the pre-exponent f_0, one can use the 'natural frequency' of the system equal to the speed of sound reduced to molecular dimensions and molecular surface area: $f_0 \approx 10^{30}$ m^{-2} s^{-1}.[104] On a more qualitative level, factors controlling the pre-exponent will be discussed later.

7.4.2 Activation Energy: de Vries Theory

The somewhat mysterious behavior of macroemulsions is related to the nucleation behavior of the emulsion film rupture event: for the coalescence act to occur, a nucleation hole must be formed, which involves overcoming a free energy penalty.[110] Consider a flat-parallel liquid film of half-thickness b. The liquid film is under interfacial tension σ. Assume a hole of radius a has been formed; the edge of the hole is round (Figure 7.14a). The emulsion film rupture is driven by reducing the surface area of the planar part of the emulsion film. On the other hand, the edge of the nucleation hole creates an extra surface area, and, therefore, a free energy penalty. The energy barrier of nucleation comes from the interplay of the free energy penalty at the edge of the hole and the free energy gain at the planar part. The surface area loss at the planar part is equal to:

$$A_1 = -2\pi(a + b)^2 \tag{7.23}$$

The created surface area of the edge, which can be modeled as a surface of revolution, formed by revolving a semicircle around the symmetry axis, equals:

(a)

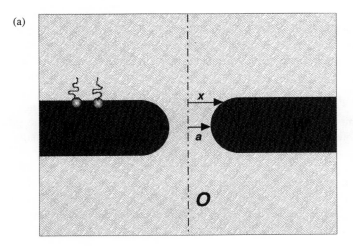

(b)

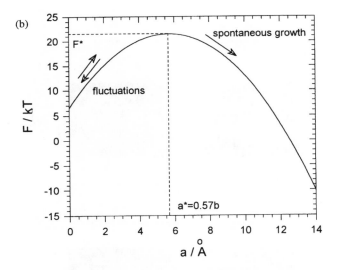

Figure 7.14 (a) *Geometry of the nucleation hole.* (b) *de Vries model of the hole nucleation and growth. After reaching the critical size,* $a^* = 0.57b$, *the growth becomes spontaneous*

$$A_2 = 2\pi[\pi b(a + b) - 2b^2] \tag{7.24}$$

Assume first that the interfacial tensions of the curved and planar parts are the same. The free energy balance of the nucleation pore in a liquid film states that:

$$F = \sigma(A_1 + A_2) = -2\pi a^2\sigma + 2\pi a[\sigma(\pi - 2)b] + (2\pi^2 - 6\pi)b^2\sigma \tag{7.25}$$

At constant film thickness, the excess surface area as a function of the hole radius a passes through the maximum $A^* = 2.94b^2$ at $a^* = 0.57b$, after which it decreases to negative values (Figure 7.14b). After passing through the maximum, the hole growth becomes spontaneous; the maximum, however, can only be reached by a thermal fluctuation. Naturally, the activation barrier for the film to break is equal to the free energy at the maximum, $F^* = \sigma A^* = 2.94\sigma b^2$. Assuming $\sigma = 30 \, \text{mN m}^{-1}$ and $f_0 \approx 10^{30} \, \text{m}^{-2} \, \text{s}^{-1}$, one concludes that the 1 cm^2, 2 nm thick film breaks almost instantly [with $\tau_{\text{film}} \approx 10^{-16}$ s, $F^* = 22 \, k_B T$]. On the other hand, twice as thick a film has an activation energy of 87 $k_B T$ and is virtually infinitely stable ($\tau_{\text{film}} \approx 10^{12}$ s).

Consider now the hole nucleation in a stack of N liquid films (say, in a lamellar liquid crystalline film composed of N bilayers). If the hole pierces all the N bilayers, the unfavorable edge is formed on all the N bilayers. On the other hand, the favorable reduction of the total surface area is the same as in the case of only one film (Figure 7.15). The total balance of the surface area states that

$$\Delta A = A_1 + NA_2 \tag{7.26}$$

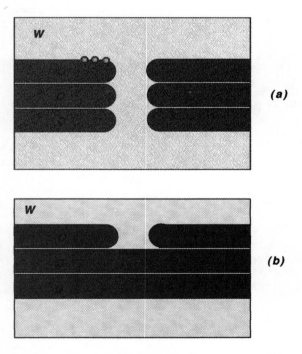

Figure 7.15 *Hole nucleation in a multilamellar-stabilized film. (a) When a hole in three bilayers back-to-back is formed, the free energy penalty is three times larger than in the case of one bilayer, but the free energy gain is the same. (b) When only one bilayer of the three is ruptured, there is no free energy gain to oppose the penalty, and this process is not favorable*[104]

The excess free energy has a maximum at $a^* = b(N\pi/2 - 1)$ equal to $F^* = \sigma b^2(N^2\pi^3/2 - 4\pi N)$. One can see that the coalescence energy barrier drastically increases with the number of films to be pierced. At $b = 1$ nm, one bilayer is unstable to rupture, whereas two bilayers back-to-back are infinitely stable ($F^* = 277$ $k_B T$). Note that the bilayers cannot just pop one after another: there is no free energy gain in this process, only the free energy penalty (see Figure 7.15b).[104]

Consider again eqn. (7.25). The first term on the right-hand side is proportional to the hole surface area and is the *interfacial tension* contribution. The second term is proportional to the hole perimeter and is therefore the *line tension* contribution. (The third term is constant for constant film thickness b and can be incorporated into the pre-exponent.) The alternative way to model the film rupture is to present the free energy of the nucleation pore as the sum of the line and surface tension terms:

$$F = 2\pi a\gamma - 2\pi a^2\sigma \qquad (7.27)$$

Here $\gamma \equiv \sigma(\pi - 2)b$ is the line tension; the coefficient 2 on the second term accounts for the two sides of the film. The free energy has a maximum at the critical hole radius a^*:

$$a^* = \gamma/2\sigma \qquad (7.28a)$$

equal to

$$F^* = \pi\gamma^2/2\sigma \qquad (7.28b)$$

which is the thermal activation barrier.

The treatment of the hole nucleation in terms of the interfacial and line tensions is equivalent to the surface area analysis presented above. It has an advantage, however, that the line tension can be considered as a phenomenological parameter which is not necessarily equal to $\sigma(\pi - 2)b$. The analysis of the film rupture in terms of surface and line tensions was developed by several groups.[111–113]

7.4.3   The Kabalnov–Wennerström Theory

The analysis of de Vries is based on the assumption that the interfacial tension of the edge of the film is the same as that of the planar part of the film. It is a good approximation for the case when the film is formed by a pure liquid. The situation changes, however, when the film is covered by a saturated surfactant monolayer. In this case, there is considerable monolayer frustration at the edge of the nucleation hole which must be accounted for;[43] see Figure 7.16.

Consider a flat-parallel O/W/O emulsion film with a hole in it. By contrast with the previous section, the film is assumed to be covered with a surfactant monolayer, being in thermodynamic equilibrium with the micellar solution in the

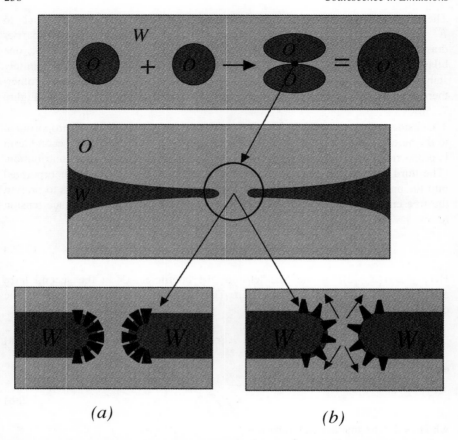

Figure 7.16 *By contrast to the classical oriented wedge theory, in the Kabalnov–Wenner-ström theory it is assumed that the spontaneous curvature affects not the free energy of the emulsion droplets but the free energy of the coalescence transition state: the hole in the film. If the spontaneous curvature of the surfactant molecule fits the neck, the hole propagates without a significant barrier (a). In the opposite case, the nucleation is suppressed and the emulsions are stable (b)*

bulk (the surfactant concentration is assumed to be above the CMC). The approach is analogous to that of de Vries, with two major differences. Firstly, one accounts for the fact that the surface tension at the edge is different from that of the planar film, because the monolayer 'frustration' at the edge is not the same as that in a planar monolayer; see eqn. (7.12). To additionally elucidate this equation, consider an emulsion film patch, having a mean curvature H and being in thermodynamic equilibrium with the micellar solution. Compare the state of monolayers covering the interface of the micelles and the emulsion film patch. In the micellar solution, the monolayer has a third dimension to its disposal and can acquire the curvature which is close to the spontaneous curvature H_0 (for a moment, we assume the saddle splay modulus equals zero). On the other hand, the monolayer covering the emulsion film patch is constrained to a certain curvature, and is in general frustrated, the more so the larger the difference between H and H_0. Therefore, the 'curvature component' of the surfactant chemical potential is bigger in the emulsion film than in the micelle. Since both monolayers are in thermodynamic equilibrium with each other, this increment of the chemical potential must be canceled by the reduced lateral pressure of the monolayer covering the emulsion film. This results in the dependence of the interfacial tension of emulsion films on the monolayer spontaneous curvature H_0 and the actual (geometrical) mean curvature of the film H: the closer these values, the lower is the interfacial tension.

The second difference of this model from de Vries's theory is that both the hole radius and the film thickness are allowed to vary. At an arbitrary hole radius a, the film adjusts its thickness b to a value at which the hole free energy has a minimum. At this stage, the disjoining pressure penalty which might be involved in this adjustment is disregarded. This may be a reasonable approximation, because, as will be shown later, the optimal thickness may be of the order of hundreds of Å, a distance at which the interactions are weak.

The free energy of the hole F can be represented as the sum of four terms. The first term is equal to the increment of the surface area for the planar part of the film multiplied by the interfacial tension of the planar monolayer:

$$F_1 = -2\pi\sigma^{\text{planar}}(a + b)^2 \tag{7.29}$$

The second term refers to the surface of revolution formed by revolving a semicircle around the symmetry axis multiplied by the interfacial tension of the *planar* monolayer:

$$F_2 = 2\pi\sigma^{\text{planar}}[\pi b(a + b) - 2b^2] \tag{7.30}$$

The first two terms are identical to those of de Vries. The third 'bending energy' term accounts for the extra (positive or negative) interfacial tension of the surface of revolution with respect to the planar state:

$$F_3 = \frac{\kappa}{2}\left[\int (2H_0 - H_1 - H_2)^2 - 4H_0^2\right] dA \tag{7.31}$$

The principal radii of curvature at a given point of the surface of revolution are equal to $H_1 = -1/b$ (the 'meridional' curvature) and $H_2 = \sin\phi/x$ (the 'parallel' curvature), where x is the distance from a given point on the surface of revolution to the symmetry axis, and ϕ is the angle between the vertical and the vector connecting the center of the semicircle and the point on the surface of revolution (Figure 7.14a). The signs of H_1 and H_2 refer to the passage in an O/W/O film; for a W/O/W film, the signs must be reversed. The $4H_0^2$ term accounts for the bending energy of the planar monolayer. The integral (in eqn. 7.31) can be evaluated in explicit form:

$$F_3 = 2\pi\kappa\left\{2\pi H_0 a + \frac{2(a+b)^2}{b\sqrt{a(a+2b)}}\arctan\sqrt{\frac{a+2b}{a}} + 2(\pi-4)bH_0 - 4\right\} \quad (7.32)$$

Again, for W/O/W films, the sign of the terms proportional to H_0 must be reversed. The fourth term accounts for the fact that the fusion of two emulsion droplets reduces the number of separate droplets in the system:

$$F_4 = -4\pi\bar{\kappa} \quad (7.33)$$

Note that this contribution does not depend on the film surface area, nor on the film orientation, *i.e.* is the same for O/W/O and W/O/W films.

The total free energy is equal to the sum of eqns. (7.29), (7.30), (7.32), and (7.33):

$$F = F_1 + F_2 + F_3 + F_4 \quad (7.34)$$

As has been discussed above, the interfacial tension of a planar emulsion film shows a deep minimum in the balanced state, the dependence being described by the expansion in series (eqn. 7.10). At any fixed spontaneous curvature, the σ^{planar} value is determined by eqn. (7.10), and the free energy (eqn. 7.34) becomes a unique function of the hole radius a and the film half-thickness b: $F = F(a, b)$. The problem is dependent on the three parameters having a dimension of energy (κ, $\bar{\kappa}$ and ε), one parameter having a dimension of reciprocal length (H_0), and one of the energy per unit area (σ_0). Note that, rigorously speaking, one cannot discriminate the linear- and quadratic-in-a terms in the free energy, and one cannot, therefore, treat the hole nucleation problem in terms of the line and surface tensions, as was done in the previous section.

The equations describing the problem are not very transparent. Before proceeding with numerical calculations, we outline the physical scenario of hole nucleation. Consider an O/W/O film with $H_0 \ll 0$, or a W/O/W film with $H_0 \gg 0$ (we define these films as 'inside-out' films). At any arbitrary (in absolute value) spontaneous curvature, the film can adjust its thickness in such a manner that the spontaneous curvature of the monolayer H_0 will fit the mean curvature of the hole edge, *viz.* $H = (H_1 + H_2)/2 \approx H_0$. At this optimal thickness the surface tension of the hole edge is very low, and stays low for any arbitrary hole radius a. This means that the film breaks without a significant energy barrier (Figure 7.17a).

Consider now an O/W/O film with $H_0 \gg 0$, or a W/O/W film with $H_0 \ll 0$ (we call these films 'normal') (Figure 7.17b). In this case, the mean curvature of the monolayer at the edge of the hole H and the spontaneous curvature H_0 have opposite signs. Thus, there is always a monolayer frustration energy involved in the hole formation. The frustration can be reduced by increasing the film thickness: in this case, $H \to 0$ and the difference between H and H_0 decreases. The decrease in the Hookean ($H - H_0)^2$ term is, however, traded for an increase in the total surface area of the edge. As a result, at any given hole radius, the free

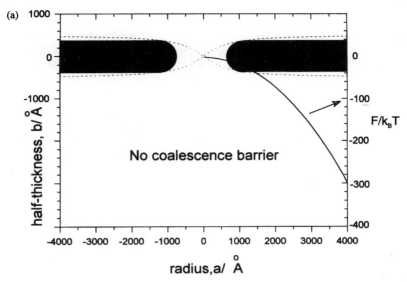

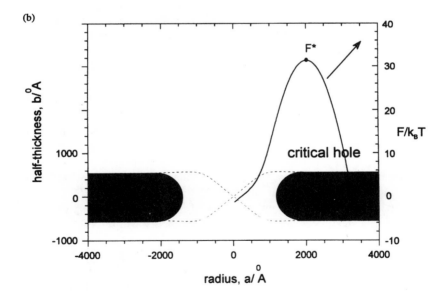

Figure 7.17 *Cross-sectional representation of the hole propagation between two oil droplets separated by a water film for the* n-*octane*–$C_{12}E_5$–*water system. The thickness of the film is adjusted to minimize the free energy of the hole. The dashed lines trace the hole propagation path.* (a) *Temperature is 1 K above the balanced point. The free energy of the neck uniformly decreases with the hole radius and there is no coalescence barrier to overcome; see the solid curve, representing the dependence of the free energy on the hole radius.* (b) *Temperature is 1 K below the balanced point. The free energy of the hole passes through the maximum, which is the coalescence activation barrier*

energy must have a *minimum* at some finite ('optimal') b^* value. Now, allow the radius of the hole to increase and keep the film thickness variable at the 'optimal' value $b = b^*(a)$. Increasing the hole radius increases the interfacial free energy by reducing the surface area of the planar part, which is, loosely speaking, $\sim a^2$. This is offset somewhat by an increase in the surface area of the edge, $\sim a$. As in traditional 2-D nucleation theories, the squared-in-radius term overweighs the linear-in-radius term and the $F(a, b^*(a))$ curve must have a maximum at $a = a^*$. Hence, the free energy surface $F(a, b)$ of the 'normal' film must have a *saddle point*: a maximum *versus* a and a minimum *versus* b. The problem of finding the energy barrier against coalescence is then reduced to determining the saddle point of this surface:

$$\frac{\partial F}{\partial a} = 0; \quad \frac{\partial^2 F}{\partial a^2} > 0$$

$$\frac{\partial F}{\partial b} = 0; \quad \frac{\partial^2 F}{\partial b^2} < 0 \tag{7.35}$$

One can numerically examine the behavior of the $F(a, b)$ function by using the parameter set of the n-octane$-C_{12}E_5-$water system: $\kappa = 0.6\ k_B T$; $\overline{\kappa} = 0.3\ k_B T$; $\sigma_0 = 4 \times 10^{-4}\ \text{mN m}^{-1}$; $\varepsilon = \kappa^2/(\kappa + \overline{\kappa}/2)$; $H_0 = -1.525 \times 10^3\ \Delta T\ [\text{m}^{-1}]$, $\Delta T = T - \overline{T}$, and the balanced temperature $\overline{T} = 32.65\ °\text{C}$.[24] Figure 7.18 represents projections of the optimal trajectories on the $F-a$ plane at different temperatures. Note that these plots show a very simple behavior. At $\Delta T < 0$ (the 'normal' film case), they have a single maximum corresponding to the saddle point on the $F(a, b)$ surface (see the inset). We refer to the energy at the saddle point as the energy barrier. The energy barrier steeply decreases with increasing temperature in a $1-2\ °\text{C}$ vicinity of the balanced state. A small energy barrier exists in the balanced state; it disappears at slightly positive ΔT values.

The dependence of the energy barrier on temperature is represented in more detail in Figure 7.19. Note that far from the balanced state ($\Delta T < -5\ °\text{C}$) the energy barrier is virtually constant at a value of $\sim 40\ k_B T$. Increasing the temperature results in a steep decrease in the activation energy within $\Delta T = -1$ to $2\ °\text{C}$. There is some non-zero barrier in the balanced state, which disappears at $\Delta T \approx 0.05\ °\text{C}$. Figure 7.20 shows the 'critical' hole dimensions. Note that the values cover the range of hundreds to thousands of Å. The values of a^* and b^* increase as the temperature gets closer to the balanced point. Both values show a maximum at slightly negative ΔT and then decrease. Far from the balanced state, a^* is greater than b^* by a factor of ~ 3. At the balanced state the values are almost equal.

Although the system of eqns. (7.35) cannot be solved analytically, asymptotic solutions may be obtained for the cases of large and small (in absolute value) spontaneous curvatures: $\sigma_0 \gg 2\varepsilon H_0^2$ or $\sigma_0 \ll 2\varepsilon H_0^2$. The first case corresponds to the region of steep change in the coalescence barrier with temperature in the vicinity of the balanced state; the second, to the region of high (in absolute value) ΔT, where the activation barrier levels off to a constant value.

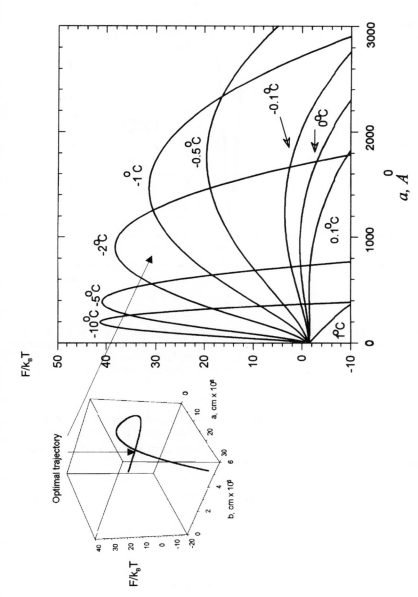

Figure 7.18 *Results of hole free energy simulations for the n-octane–$C_{12}E_5$–water system (O/W/O films). Projections of the 'optimal' hole trajectories on the F–a plane. Each trajectory joins the minima F(a, b) vs. b points of the F(a, b) surface (see inset). The maximum of the curve corresponds to the saddle point of the F(a, b) surface. Each curve corresponds to a specific temperature from the balanced point, ΔT (Reproduced by permission of the American Chemical Society from ref. 43)*

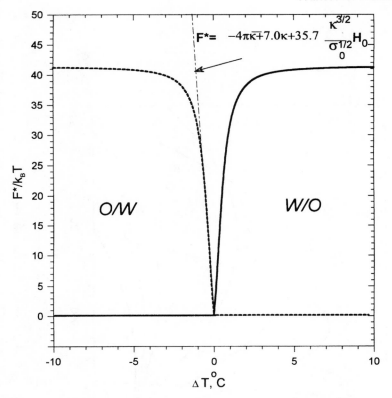

Figure 7.19 *Hole activation barrier* F^* *vs. temperature from the balanced point* ΔT *for the n-octane$-C_{12}E_5-$water system. The dashed line shows the analytical approximation of the dependence in the vicinity of the balanced point* (Reproduced by permission of the American Chemical Society from ref. 43)

(i) Consider the vicinity of the balanced state. The spontaneous curvature is equal to zero and there is only one characteristic length in the problem: $\xi = \sqrt{\kappa/\sigma_0}$ (it is possible to show that the size of the critical hole in the balanced state does not depend on either ε or $\bar{\kappa}$). Therefore:

$$a^*(H_0 = 0) = \rho_1 \sqrt{\frac{\kappa}{\sigma_0}} \qquad (7.36)$$

$$b^*(H_0 = 0) = \rho_2 \sqrt{\frac{\kappa}{\sigma_0}} \qquad (7.37)$$

$$F^*(H_0 = 0) = -4\pi\bar{\kappa} + \rho_3\kappa \qquad (7.38)$$

$$\frac{\mathrm{d}F^*}{\mathrm{d}H_0}(H_0 = 0) = \rho_4 \frac{\kappa^{3/2}}{\sigma_0^{1/2}} \qquad (7.39)$$

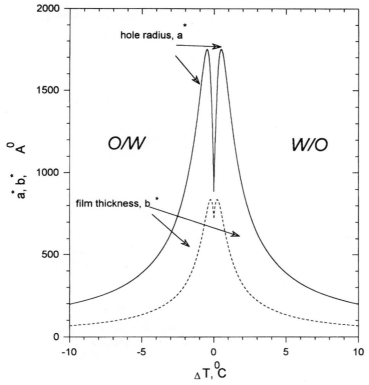

Figure 7.20 *Size of the critical hole for O/W and W/O emulsions*

where ρ_1, ρ_2, ρ_3, and ρ_4 are numerical coefficients. The direct numerical solution of eqns. (7.35) yields the following numerical coefficients: $\rho_1 = 1.15$; $\rho_2 = 0.92$; $\rho_3 = 7.0$; and $\rho_4 = 4\pi^2\rho_1 + 4\pi(\pi - 4)\rho_2 = 35.7$. Therefore, in the vicinity of the balanced state

$$F^* = -4\pi\bar{\kappa} + 7.0\kappa + \frac{35.7\,\kappa^{3/2}}{\sigma_0{}^{1/2}}\Delta H_0 \qquad (7.40)$$

This equation is written for O/W/O films; for W/O/W films, the sign of the third term on the r.h.s. must be reversed.

(ii) Consider now the other limiting case, $\sigma_0 \ll 2\varepsilon H_0{}^2$. In this case the energy barrier cannot explicitly depend on the spontaneous curvature H_0, because there are no other parameters in the problem which have a dimension of length to cancel the H_0 dimension. This explains why the F^* vs. T dependence levels off at large spontaneous curvatures. The energy barrier is, therefore, dependent only on the three parameters having the dimension of energy (it is easy to show that $\bar{\kappa}$ does not enter the second term on the r.h.s.):

$$F^* = -4\pi\bar{\kappa} + \kappa \times f_1(\varepsilon/\kappa) \tag{7.41}$$

and

$$a^* = \frac{1}{H_0} f_2(\varepsilon/\kappa) \tag{7.42}$$

$$b^* = \frac{1}{H_0} f_3(\varepsilon/\kappa) \tag{7.43}$$

Consider only one special case, when $\bar{\kappa} = 0$ and $\varepsilon = \kappa$ (*i.e.* $\sigma = 2\kappa H_0{}^2$). In this particular case, according to numerical simulations, $f_1 = 64.3$ for the normal film, $f_2 = 2.50$, and $f_3 = 0.91$. Note that the numerical coefficient $f_1 = 64.3$ is very big, and the coalescence barrier may become insurmountable even for relatively flexible ($\kappa \approx 1 \ k_B T$) monolayers.

The previous numerical estimates of this section were concerned with the case when the surfactant monolayer is flexible, $\kappa \approx k_B T$. What new features appear if the monolayer is rigid, $\kappa \approx 10-100 \ k_B T$? Essentially the same pattern of emulsion stability is predicted: at positive spontaneous curvatures, O/W emulsions are very stable, whereas at negative values, W/O emulsions are stable (Figure 7.21). The

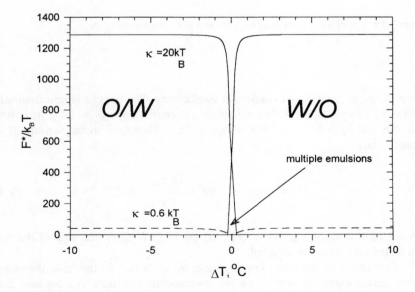

Figure 7.21 *Effect of bending modulus on activation energy. All the parameters of the plots are the same as in Figures 7.17–7.20; $\kappa = 20 \ k_B T$ for the solid-line plot* (Reproduced by permission of the American Chemical Society from ref. 43)

main new feature of the rigid monolayer stabilized emulsions is that, over a narrow region in the vicinity of the balanced state, both O/W and W/O emulsions may be stable. Indeed, the sum of the first two terms on the r.h.s. of eqn. (7.40) may become insurmountably high ($\sim 100-1000\ k_B T$). This, in particular, means that multiple emulsions may be favored close to the balanced state. The region of multiple emulsion stability is, however, limited to a narrow range of spontaneous curvatures in the vicinity of the balanced state. Outside this region, only the 'normal' emulsion type is stable. If an 'inside-out' emulsion is prepared, it will break.

7.4.4   Nucleation Pore in Multilamellar-covered Emulsion Films

Consider now an emulsion film, consisting of N surfactant bilayers. The free energy of the hole, piercing all N bilayers, is equal to (*cf.* Figure 7.15):

$$F = F_1 + N F_2 + N F_3 + N F_4 \qquad (7.44)$$

Here, F_1, F_2, F_3, and F_4 values refer to a single bilayer and are determined by eqns. (7.29), (7.30), (7.32), and (7.33). We discuss below the results of the numerical calculations of the hole nucleation parameters for the case of $N = 2$. This case corresponds to the coalescence of a monolayer-covered droplet with a droplet covered by three monolayers back-to-back. The trend of the F^* *vs.* ΔT dependence is very much the same as for monolayer-covered droplets, but the change in the absolute values of the energy barrier is drastic (Figure 7.22). In particular, in the balanced state the barrier energy is equal to 0.43 $k_B T$ at $N = 1$

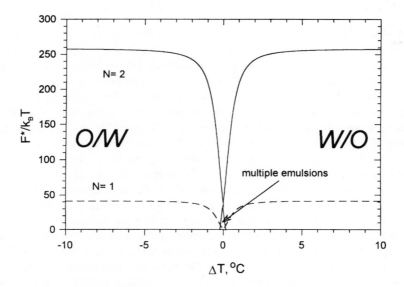

Figure 7.22 *Effect of multilamellar stabilization on emulsion stability. All the parameters are the same as in Figures 7.17–7.20*

and 37 $k_B T$ at $N = 2$! The effect is even larger than one may expect from the simple line tension model, according to which $F^* \approx N^2$. The drastic increase in the absolute value of the energy barrier leads to the appearance of a narrow temperature region, where both O/W and W/O emulsions may be stable. One may expect the formation of multiple emulsions over this region. Note that the temperature stability range of this region is very limited, although it gets broader at higher N.

7.5   The Kabalnov–Wennerström Theory *versus* Experiment

The Kabalnov–Wennerström theory provides a qualitative explanation for the oriented wedge trend in macroemulsion stability, *i.e.* it explains the PIT and 'Optimal Surfactant Formulation' behavior of macroemulsions, as well as the multilamellar stabilization. For a quantitative comparison, the data for only one system is currently available.[114] A detailed comparison will be made below.

7.5.1   Experimental Measurements of the Activation Energy and the Pre-exponent

This system is the same as the one used in computer simulations, shown in Figures 7.18–7.20. The spontaneous curvature can be controlled by temperature and salinity; in both cases, essentially a linear relationship is observed:[24,114]

$$H_0 = -\alpha(T - \overline{T}) - \beta C_{NaCl} \qquad (7.45)$$

where $\alpha = -1.525 \times 10^7$ [m^{-1} K^{-1}] and $\beta = -2.26 \times 10^7$ [m^{-1} (wt%)$^{-1}$] are empirical coefficients, C_{NaCl} is the concentration of NaCl (wt% per water only), and $\overline{T} = 32.65$ °C is the balanced temperature without the salt present. Two different kinetic experiments were conducted. (i) Winsor III experiment in which the kinetics of O/W and W/O emulsion resolution was studied over the Winsor III region. The experiment was conducted in a very narrow temperature range (several tenths of a degree) or salinity range (several tenths of wt%) near the balanced point. (ii) Winsor II experiment where the resolution of inverse W/O emulsions was studied as a function of temperature over the Winsor II region. This experiment was conducted in the temperature range 40–80 °C at zero salt concentration.

   In all the experimental studies discussed below, the macroemulsion stability was characterized by the time for the resolution of one half of the emulsified disperse phase $\tau_{1/2}$, similar to how it had been done before by Salager *et al.*[78] All the macroemulsions contained 50 vol% of disperse phase and were prepared by handshaking in a jacketed beaker under careful temperature control. The emulsification was quite easy to perform because of the very low interfacial tensions in the system. The experimental reproducibility in $\tau_{1/2}$ values was fairly good (within 20%) in the Winsor III experiment, provided that the temperature was controlled to within 0.01 °C. In other words, the way in which the macroemulsions were prepared

(d_0 value) had a very little effect on $\tau_{1/2}$. This was not the case in the Winsor II region, where the reproducibility was much worse, with the deviations up to 100%. The experiments were run at least three times and with three different samples at each temperature to reduce the error.

Finally, only rough estimates for the initial droplet size are available: $d_0 \approx 1-30$ μm, which makes the accuracy of the pre-exponent, at best, within three orders of magnitude.

7.5.1.1 Analysis of Winsor III Experimental Data

According to eqns. (7.21), (7.22), and (7.40), disregarding the numerical coefficient in eqn. (7.21), the O/W macroemulsion lifetime close to the balanced point is equal to:

$$k_B T \ln(\tau_{1/2}) = -k_B T \ln(d_0^2 f_0) - 4\pi\bar{\kappa} + 7.0\kappa + \frac{35.7\kappa^{3/2}}{\sigma_0^{1/2}} H_0 \qquad (7.46)$$

If one plots the macroemulsion lifetime *versus* spontaneous curvature, one can evaluate the spontaneous curvature coefficient of activation energy, $\zeta = 35.7\kappa^{3/2}/\sigma_0^{1/2}$ from the slope of the line and the pre-exponent f_0 from the intercept. This evaluation resides on the approximations that: (i) the pre-exponent itself does not depend on the spontaneous curvature; and (ii) the initial droplet size d_0 is close in all the experiments; and (iii) changes in $k_B T$ are minor. Of these factors, only (iii) is completely negligible. Fortunately, the contributions from (i) and (ii) are under a logarithm and cannot influence the value of ζ significantly; however, they do affect the pre-exponent.

7.5.1.2 Analysis of Winsor II Experimental Data

Over the Winsor II region, the W/O emulsion breakage was analyzed by the conventional Arrhenius plot:

$$\ln(\tau_{1/2}) = -\ln(d_0^2 f_0) + \frac{F^*}{k_B T} = -\ln(d_0^2 f_0) - \frac{S^*}{k_B} + \frac{E^*}{k_B T} \qquad (7.47)$$

where S^* is the activation entropy and E^* is the activation energy, so that $F^* = E^* - TS^*$. The experiment was conducted over the temperature range 40–80 °C, where the activation free energy is predicted to level off and to be temperature-independent, $F^* = 41.2 \, k_B T$ (*cf.* Figure 7.19). One can evaluate the activation energy from the slope of the $\ln(\tau_{1/2})$ vs. $1/T$ plot, and the pre-exponent from the intercept. Again, this construction relies on the assumption that f_0 and d_0 do not depend on temperature. Although this dependence does exist (for instance, d_0 is expected to be smaller at lower temperatures where the interfacial tension is lower), it is not significant for the evaluation of the activation energy. However, the estimates of the pre-exponent can be quite inaccurate.

7.5.2 $C_{12}E_5 - n$-Octane–Water–NaCl System: Overall Trend

Figure 7.23a shows the overall pattern of macroemulsion stability as a function of salinity and temperature. At low salinities, the macroemulsions have an O/W type, and at high salinity, W/O, with the inversion at the balanced point (*ca.* 9 wt% NaCl). Both O/W and W/O emulsions are very stable far away from the balanced point. The behavior is not completely identical when the spontaneous curvature is controlled by temperature: although, at low temperatures, O/W emulsions are very stable, at high temperatures the W/O emulsion stability passes through the maximum and then decreases. This secondary decrease in stability has been also observed by others.[64] It can be, in part, explained by the classical thermal activation effect (see below). Figure 7.23b shows the appearance of the samples at room temperature, as a function of salinity. As the salinity increases, the systems change from very stable O/W to very stable W/O type with the inversion at the three-phase equilibrium range. O/W emulsions can be distinguished from W/O by the fact that the former form a cream layer at the top of the vessel, while the latter form the milky sediment at the bottom.

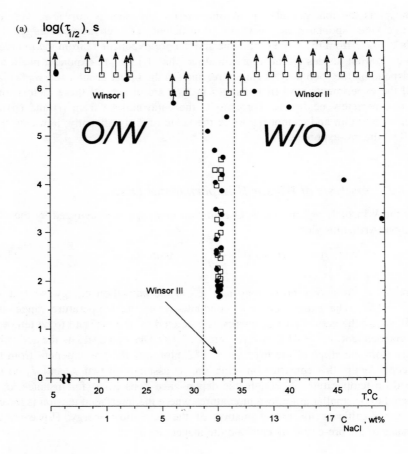

(b)

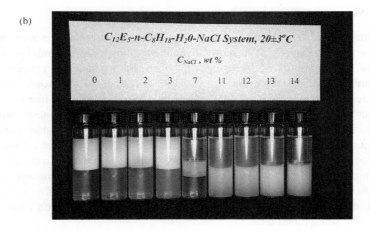

(c) log($\tau_{1/2}$), s

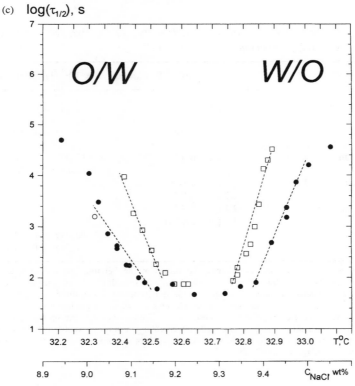

Figure 7.23 *Logarithm of macroemulsion lifetime,* log($\tau_{1/2}$), *s, vs. temperature* T, °C *(upper x-axis) and salinity (lower x-axis) for the $C_{12}E_5$-n-octane–water–NaCl system. Salinity scan data at* T = 19.91 °C *(squares), and the data of the temperature scan experiment for no salt (circles) are shown. (a) Overall shape of the curves. (b) Appearance of the samples in the salinity scan experiment. (c) Detailed behavior in the vicinity of the balanced point*
(Reproduced by permission of the American Chemical Society from ref. 114)

7.5.3   Winsor III Region

In the Winsor III region, the macroemulsion stability is extremely temperature- and salinity-sensitive. Figure 7.23c shows that changes by only several tenths of a degree or several tenths of the weight % of NaCl produce the change in the macroemulsion stability from minutes to days. According to the theory presented in the previous section, the emulsion coalescence barrier is a linear function of the monolayer spontaneous curvature in the vicinity of the balanced state, with the slope determined by eqn. (7.40). The onset of the increase in emulsion stability is somewhat offset from the balanced state because of the pre-exponent contribution. As can be seen from eqn. (7.46), the slope can be evaluated from the semi-logarithmic plots of the macroemulsion lifetime *versus* temperature and salinity. Table 7.5 compares the experimental values of the slopes $\zeta = dF^*/dH_0$ with theory. The agreement is reasonable in view of the uncertainty in the value of the monolayer bending modulus κ.

Table 7.5   *Spontaneous   curvature   coefficients   of   activation   energy* $\zeta = dF^*/dH_0$: experiment versus theory[114]

	$\zeta \times 10^{27}$/J m, O/W systems	$\zeta \times 10^{27}$/J m, W/O systems
Theory: [a] $\zeta = \pm 35.7 \dfrac{\kappa^{3/2}}{\sigma_0^{1/2}}$	7.24	−7.24
Experiment: temperature scan, $C_{NaCl} = 0$	5.6 ($r = 0.99$)	−8.7 ($r = 0.99$)
salinity scan, $T = 19.91$ °C	8.5 ($r = 0.98$)	−13.2 ($r = 0.98$)
		−13.1 ($r = 0.95$)
salinity scan, $T = 19.61$ °C	8.4 ($r = 0.97$)	−
	8.7 ($r = 0.97$)	
salinity scan, $T = 31.71$ °C	9.4 ($r = 0.998$)	−
	9.0 ($r = 0.990$)	
salinity scan, $T = 32.24$ °C	−	−12.0 ($r = 0.998$)
		−11.2 ($r = 0.996$)

[a] Recently, Leitao and co-authors re-evaluated the values of the bending and saddle splay moduli, suggesting instead that $\kappa = 1.00 \, k_B T$ and $\bar{\kappa} = -0.36 \, k_B T$.[19] As result of this, H_0 must also be changed, because the spontaneous curvature is related to the experimentally measurable micellar size r as $1/r = H_0 \kappa/\kappa + \bar{\kappa}/2$, neglecting the entropy term. The changes affect the predictions of the model only slightly due to the mutual compensation of these effects. The agreement in the spontaneous curvature coefficients of the activation energy ζ is equally acceptable. The limiting value of the activation free energy over the Winsor II region is predicted to be $F^* = 61 \, k_B T$, in comparison with the experimental value of activation energy E^* of 42–49 $k_B T$.

The Kabalnov–Wennerström theory predicts that the sharpness of the transition from unstable to stable emulsions is controlled by the value of ζ. When the surfactants are compared in a homologous series, κ increases and σ_0 decreases with the surfactant chain length. Accordingly, the sharpness of transition from unstable to stable systems is expected to increase with the surfactant chain length.

7.5.4   Winsor II Region

The theory predicts that in the Winsor II region the activation energy levels off. However, as the temperature increases, the ratio of the activation energy to $k_B T$ decreases and the emulsions become less stable. Figure 7.24 shows the W/O macroemulsion stability over the region of elevated temperatures (40–80 °C) in Arrhenius coordinates at several surfactant concentrations. If the surfactant concentration is above 4 wt%, the plots yield the activation energy of $E^* = 42-49\ k_B T$, in a good agreement with the theoretical value $F^* = 41.2\ k_B T$. Note that the activation energy as measured by the Arrhenius plot E^* is not identical to the free energy value F^* because the first is the total energy while the second is the free energy, *i.e.* has an entropy contribution. In other words, only a part of the F^* value contributes to the activation energy; the other part contributes to the pre-exponent. Still, the values of F^* and E^* are expected to be close.

A faster emulsion breakage at the lowest surfactant concentration (4%) is probably due to the fact that the surfactant monolayer is no longer saturated. The

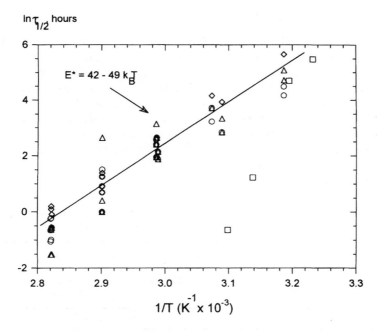

Figure 7.24 *Breakage of W/O emulsions in Arrhenius coordinates in the Winsor II region. The $C_{12}E_5$ concentration in oil is equal to 4% (squares), 15% (circles), 20% (triangles), and 30% (diamonds). Once the surfactant concentration exceeds the critical micelle concentration (ca. 4% at 80 °C), the emulsion breakage is consistent with the Arrhenius effect with an activation energy 42–49 $k_B T$, which is fairly independent of the surfactant concentration. At 4% surfactant concentration the monolayer is not saturated and its bending modulus is very low, which produces a steeper stability decay*

CMC of $C_{12}E_5$ in heptane is reported to increase with temperature from ~ 1 to 4 wt% between 30 and 80 °C.[115] At temperatures above 40 °C the monolayer is no longer saturated, the bending modulus decreases, and the macroemulsion resolves faster, as has been discussed by others.[116]

7.5.5 Pre-exponent

The frequency of the film rupture is the product of the exponent and pre-exponent. So far we have concerned ourselves with the exponent term. The pre-exponent of nucleation phenomena is dependent on the mechanism of the growth of the nuclei. Thus, for the growth of a nucleus of a new phase in a super-saturated solution, the pre-exponent is controlled by the rate of diffusion of the molecules to the growing nucleus.[117,118] For the growth of a neck in an emulsion film, the pre-exponent can be expected to be determined by the viscous flow of the liquid under the action of surface tension. From scaling arguments, one concludes:

$$f_0 \approx \frac{\sigma}{\eta(\xi^*)^3} \tag{7.48}$$

where σ is the interfacial tension, η is the viscosity, and ξ^* is the critical hole size (hole radius or film thickness, which are the same for order-of-magnitude estimates). As one can judge from this equation, not only the exponential factor but also the pre-exponent depends on the spontaneous curvature and, therefore, on temperature. Close to the balanced point, where the interfacial tension is low and the critical size of the hole is large, the pre-exponent is small. Further away from the balanced point, where the interfacial tension is large and the critical size of the hole is small, the pre-exponent is larger. Table 7.6 provides a comparison of theory and experiment for the orders of magnitude. The agreement is fair, despite a considerable uncertainty.

Table 7.6 *Pre-exponent for the $C_{12}E_5$–octane–water system*

	Experiment	Theory[a]
Balanced point	10^8–10^{13} cm^{-2} s^{-1}	5×10^{13} cm^{-2} s^{-1}
Winsor II range, 40–80 °C	10^{17}–10^{21} cm^{-2} s^{-1}	2×10^{20} cm^{-2} s^{-1}

[a] Eqn. (7.48). For balanced point, $\sigma = 4 \times 10^{-4}$ mN m^{-1}, $\eta = 0.01$ Pa s, $\xi^* \approx 1 \times 10^{-5}$ cm; for Winsor II at 60 °C, $\sigma = 0.7$ mN m^{-1}, $\eta = 0.01$ Pa s, $\xi^* \approx 7 \times 10^{-7}$ cm.

In spite of the considerable dependence of the pre-exponent on spontaneous curvature and therefore on temperature, this effect is still minor with respect to the temperature dependence of the exponent, which justifies the evaluation of the activation energy from the Arrhenius plot, with the balanced point region and the region of elevated temperatures treated separately.

7.5.6 Surfactant Concentration

A characteristic example of the macroemulsion stability dependence on the surfactant concentration is shown in Fig. 7.25a and involves almost the same system as the one discussed above, with the oil changed for *n*-heptane. The experiments were conducted at 20 °C, where the system above the CMC is in the Winsor I state and tends to form O/W emulsions. The transition from very

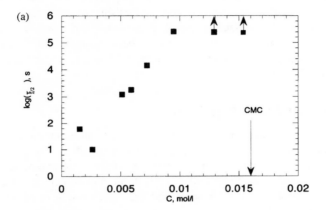

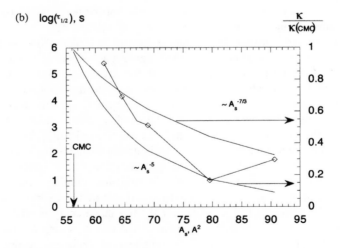

Figure 7.25 *O/W emulsion stability in the* n-*heptane*−$C_{12}E_5$−*water system at 20 °C as a function of the surfactant concentration in the vicinity of the CMC. The concentration C refers to the oil phase (see text). (a) Macroemulsion lifetime vs. surfactant concentration. (b) The data are recast as the macroemulsion lifetime vs. the area per surfactant molecule in the adsorbed layer, calculated from the interfacial tension isotherm.[115] The right y-axis shows the expected relative decrease in the bending modulus, based on the power laws for polymer 'brushes'[119]*

unstable to stable systems occurs in a narrow surfactant concentration range just below the CMC of the surfactant in oil, in agreement with the tight monolayer packing concept. (The CMC in water is several orders of magnitude less and can be neglected. Similarly, the depletion of the surfactant due to adsorption at the oil–water interface has a negligible effect in this concentration range.) Figure 7.25b shows the emulsion lifetime plotted *versus* the area per surfactant molecule in the adsorbed layer. The change from stable to unstable systems occurs in a very narrow range of A_s values, from *ca.* 56 Å^2 at the CMC to *ca.* 80 Å^2, where the emulsion resolves within several seconds. A possible interpretation could be a drastic decrease in the surfactant bending modulus and, therefore, activation energy with A_s. According to the theory of bending elasticity,[119] the bending modulus is a strong power function of A_s, $\kappa \approx A_s^{-\alpha}$, where α can vary from 5 to 7/3 (see Figure 7.25b).

7.6   Summary and Outlook

After many years of emulsion research, several reliable empirical trends in macroemulsion stability to coalescence have been established. The most powerful rules of thumb are (i) the correlation between the surfactant spontaneous curvature and the macroemulsion stability and (ii) the tight monolayer packing concept. Other empirical correlations, HLB scale and Bancroft's rule, can be substantially reduced to these two. In rare cases when the oriented wedge trend and the HLB scale, or Bancroft's rule, disagree, nature usually follows the oriented wedge trend. Thus, Bancroft's rule often fails for nonionic surfactants of the ethylene oxide type below the balanced point, while the oriented wedge trend does not.

On the other hand, mechanistic understanding of macroemulsion stability to coalescence is incomplete. Many researchers agree that viscoelastic properties of the surfactant monolayer are important for the emulsion film rupture. The question of which deformation mode contributes most to the coalescence barrier, and whether the elastic or viscous response is important, remains under discussion. In this review, it has been argued that the bending elastic deformation mode contributes to the coalescence barrier and can be a dominating factor. A very simple model has been presented, which qualitatively explains both the oriented wedge trend and the tight monolayer packing correlation. Other extensions of the model to multilamellar stabilization, multiple emulsion stability, and demulsification by surfactants are also quite straightforward. There are several weak points in the model which require further work. Firstly, the shape analysis of the edge of the nucleation hole is very primitive. It is assumed that it is circular, but in real life it can be different. Also, the hole propagation probably occurs not by adjusting the whole film thickness, but by forming a bulge at the propagating neck. This is quite difficult to tackle theoretically, however. Secondly, it is assumed that the surfactant monolayer covering the neck is in equilibrium with the micellar solution. It is not quite clear if the hole is long-living enough for such an equilibrium to establish. While the lifetime of the flickering hole at the balanced point may be quite long, $\sim 10^{-2}$ s, which normally allows diffusional equilibration at the expected hole sizes (10–100 nm), far away from the balanced point, where the holes are short-lived, this is not the case. Another limiting case of the model can be considered where the rate of the surfactant exchange is negligible. The qualitative predictions of this model could be similar, because the monolayer frustration term will still be present. Thirdly, there must be a disjoining pressure contribution in the nucleation hole formation. This effect is more significant the smaller is the nucleation hole, *i.e.* the further away the system is from the balanced point.

Despite all these shortcomings, the theory seems to be in quantitative agreement with experiment for at least one system studied in detail. Additional experiments with other surfactant systems are quite straightforward to conduct, because the data on the bending moduli, spontaneous curvature and interfacial tensions of several surfactant systems have become available recently.

A major challenge for experimentalists would be to study the microscopic act of the hole rupture, *i.e.* hole flickering and propagation. However, the holes are short-lived and the systems are very temperature sensitive, which makes these experiments difficult.

Acknowledgments

I am grateful to Håkan Wennerström and Reinhard Strey for many discussions on this subject, and to Bernard Binks for a critical reading and careful proof of the manuscript.

7.7 References

1. I.B. Ivanov, ed., 'Thin Liquid Films: Fundamentals and Applications', Surfactant Science Series, vol. 29, Dekker, New York, 1988.
2. D. Langevin and A.A. Sonin, *Adv. Colloid Interface Sci.*, 1994, **51**, 1.
3. A.K. Malhotra and D.T. Wasan, in 'Thin Liquid Films: Fundamentals and Applications', ed. I. B. Ivanov, Surfactant Science Series, vol. 29, Dekker, New York, 1988, p. 829.
4. J. Lucassen, in 'Anionic Surfactants: Physical Chemistry of Surfactant Action', ed. E. H. Lucassen-Reynders, Surfactant Science Series, vol. 29 Dekker, New York, 1981, p. 217.
5. E.H. Lucassen-Reynders, in 'Encyclopedia of Emulsion Technology', ed. P. Becher, Dekker, New York, 1996, vol. 4, p. 63.
6. A. Vrij, J.G.H. Joosten and H.M. Fijnaut, *Adv. Chem. Phys.*, 1981, **48**, 329.
7. H. Sonntag and K. Strenge, 'Coagulation and Stability of Disperse Systems', Halsted Press, New York, 1972.
8. P. Walstra, *Chem. Eng. News*, 1993, **48**, 333.
9. P. Walstra, in 'Encyclopedia of Emulsion Technology', ed. P. Becher, Dekker, New York, 1996, vol. 4, p. 1.
10. J.H. Schulman and J.B. Montagne, *Ann. N. Y. Acad. Sci.*, 1961, **92**, 366.
11. P.G. de Gennes and C. Taupin, *J. Phys. Chem.*, 1982, **86**, 2294.
12. W. Helfrich and M.M. Kozlov, *J. Phys. II*, 1993, **3**, 287.
13. W. Helfrich, *Z. Naturforsch.*, 1973, **28c**, 693.
14. L.A. Turkevich, S.A. Safran and P.A. Pincus, in 'Surfactants in Solution', ed. K. L. Mittal and P. Bothorel, Plenum Press, New York, 1986, p. 1177.
15. J. Meunier, *J. Physique Lett.*, 1985, **46**, L1005.

16. H. Kellay, B.P. Binks, Y. Hendrikx, L.T. Lee and J. Meunier, *Adv. Colloid Interface Sci.*, 1994, **49**, 85.
17. L.T. Lee, D. Langevin, J. Meunier, K. Wong and B. Cabane, *Prog. Colloid Polym. Sci.*, 1990, **81**, 209.
18. R.P. Rand, N.L. Fuller, S.M. Gruner and V.A. Parsegian, *Biochemistry*, 1990, **29**, 76.
19. H. Leitao, A.M. Somoza, M.M. Telo da Gama, T. Sottman and R. Strey, *J. Chem. Phys.*, 1996, **105**, 2875.
20. R.C. Tolman, *J. Chem. Phys.*, 1949, **17**, 333.
21. P.G. de Gennes, *J. Phys. Chem.*, 1990, **94**, 8407.
22. H.N.W. Lekkerkerker, *Physica A*, 1989, **159**, 319.
23. J. Daicic, A. Fogden, I. Carlsson, H. Wennerström and B. Jönsson, *Phys. Rev.*, 1996, **54**, 3984.
24. R. Strey, *Colloid Polym. Sci.*, 1994, **272**, 1005.
25. T. Sottman and R. Strey, *Ber. Bunsenges. Phys. Chem.*, 1996, **100**, 237.
26. D.J. Mitchell and B.W. Ninham, *J. Chem. Soc., Faraday Trans. 2*, 1981, **77**, 601.
27. R. Aveyard, B.P. Binks, P.D.I. Fletcher and J.R. MacNab, *Langmuir*, 1995, **11**, 2515.
28. A. Kabalnov, U. Olsson and H. Wennerström, *J. Phys. Chem*, 1995, **99**, 6220.
29. P.A. Winsor, 'Solvent Properties of Amphiphilic Compounds', Butterworth, London, 1954.
30. T. Sottman and R. Strey, *J. Chem. Phys.*, 1997, **106**, 8606.
31. V.E. Wellman and H.V. Tartar, *J. Phys. Chem.*, 1930, **34**, 379.
32. H. Saito and K. Shinoda, *J. Colloid Interface Sci.*, 1970, **32**, 647.
33. H. Kunieda and K. Shinoda, *Bull. Chem. Soc. Jpn.*, 1982, **55**, 1777.
34. K. Shinoda, *Prog. Colloid Polymer Sci.*, 1983, **68**, 1.
35. R. Aveyard, B.P. Binks and J. Mead, *J. Chem. Soc., Faraday Trans. 1*, 1985, **81**, 2155.
36. R. Aveyard and T.A. Lawless, *J. Chem. Soc., Faraday Trans. 1*, 1986, **82**, 2951.
37. R. Aveyard, B.P. Binks, T.A. Lawless and J. Mead, *J. Chem. Soc., Faraday Trans. 1*, 1986, **82**, 125.
38. R. Aveyard, B.P. Binks, T.A. Lawless and J. Mead, *Can. J. Chem.*, 1988, **66**, 3031.
39. P.D.I. Fletcher and D.I. Horsup, *J. Chem. Soc., Faraday Trans.*, 1992, **88**, 855.
40. D. Langevin, in 'Structure and Dynamics of Strongly Interacting Colloids and Supramolecular Aggregates in Solution', ed. S.-H. Chen, J. S. Huang and P. Tartaglia, Kluwer, Dordrecht, 1992, p. 325.
41. M. Kahlweit, R. Strey and G. Busse, *Phys. Rev. E*, 1993, **47**, 4197.
42. M. Kahlweit, G. Busse and J. Winkler, *J. Chem. Phys.*, 1992, **99**, 5605.
43. A. Kabalnov and H. Wennerström, *Langmuir*, 1996, **12**, 276.
44. M.L. Robbins, in 'Micellization, Solubilization and Microemulsions', ed. K. L. Mittal, Plenum Press, New York, 1976, p. 713.
45. B.P. Binks, J. Meunier, O. Abillon and D. Langevin, *Langmuir*, 1989, **5**, 415.
46. R.C. Tolman, *J. Chem. Phys.*, 1948, **16**, 758.
47. W.D. Harkins, E.C.H. Davies and G.L. Clark, *J. Am. Chem. Soc.*, 1917, **39**, 541.
48. I. Langmuir, *J. Am. Chem. Soc.*, 1917, **39**, 1848.
49. W. D. Bancroft, *J. Phys. Chem.*, 1913, **17**, 501.
50. J.H. Hildebrand, *J. Phys. Chem.*, 1941, **45**, 1303.
51. J.N. Israelachvili, D.J. Mitchell and B.W. Ninham, *J. Chem. Soc., Faraday Trans. I*, 1976, **72**, 1525.
52. R. Aveyard, B.P. Binks, P.D.I. Fletcher and X. Ye, in 'Emulsions—A Fundamental and Practical Approach', ed. J. Sjöblom, Kluwer, Dordrecht, 1992, p. 97.
53. B.P. Binks, *Langmuir*, 1993, **9**, 25.
54. T.T. Traykov and I.B. Ivanov, *Int. J. Multiphase Flow*, 1977, **3**, 471.

55. I.B. Ivanov, *Pure Appl. Chem.*, 1980, **52**, 1241.
56. Z. Zapryanov, A.K. Malhotra, N. Aderangi and D.T. Wasan, *Int. J. Multiphase Flow*, 1983, **9**, 105.
57. W.C. Griffin, *J. Soc. Cosmet. Chem.*, 1949, **1**, 311.
58. W.C. Griffin, *J. Soc. Cosmet. Chem.*, 1954, **5**, 249.
59. J.T. Davies, in 'Proc. 2nd. Int. Congr. Surf. Activity', London, 1957, vol. 1, p. 426.
60. J.T. Davies and E.K. Rideal, 'Interfacial Phenomena', Academic Press, New York, 1963.
61. J.T. Davies, in 'Progress in Surface Science', ed. J. F. Danielli, K. G. A. Parkhurst and A. C. Riddford, Academic Press, New York, 1964, p. 129.
62. P. Becher, 'Emulsions: Theory and Practice', Krieger, Malabar, 1985.
63. P. Becher, in 'Encyclopedia of Emulsion Technology', Dekker, New York, 1983–1996, vol. 2, p. 425; vol. 3, p. 397; vol. 4, p. 337.
64. K. Shinoda and S. Friberg, 'Emulsions and Solubilization', Wiley, New York, 1986.
65. H.T. Davis, *Colloids Surf. A*, 1994, **91**, 9.
66. P. Becher, *J. Disp. Sci. Technol.*, 1984, **5**, 81.
67. J.H. Schulman and E.G. Cockbain, *Trans. Faraday Soc.*, 1939, **36**, 651.
68. J.H. Schulman and E.G. Cockbain, *Trans. Faraday Soc.*, 1940, **36**, 661.
69. A.E. Alexander and J.H. Schulman, *Trans. Faraday Soc.*, 1940, **36**, 960.
70. Th.F. Tadros and B. Vincent, in 'Encyclopedia of Emulsion Technology', ed. P. Becher, Dekker, New York, 1983, vol. 1, p. 129.
71. Th. Forster, W. von Rybinski and A. Wadle, *Adv. Colloid Interface Sci.*, 1995, **58**, 119.
72. K. Shinoda and H. Saito, *J. Colloid Interface Sci.*, 1968, **26**, 70.
73. K. Shinoda and H. Saito, *J. Colloid Interface Sci.*, 1969, **30**, 258.
74. R.P. Enever, *J. Pharm. Sci.*, 1976, **65**, 517.
75. J.L. Salager, J. Morgan, R. Schechter, W. Wade and E. Vasquez, *Soc. Petrol. Eng. J.*, 1979, **19**, 107.
76. M. Bourrel, J.L. Salager, R.S. Schechter and W. H. Wade, *J. Colloid Interface Sci.*, 1980, **75**, 451.
77. J.L. Salager, *Prog. Colloid Polym. Sci.*, 1996, **100**, 137.
78. J.L. Salager, I. Loaiza-Maldonado, M. Minana-Perez and F. Silva, *J. Disp. Sci. Technol.*, 1982, **3**, 279.
79. R.E. Anton and J.L. Salager, *J. Colloid Interface Sci.*, 1986, **111**, 54.
80. L.M. Baldauf, R.S. Schechter, W.H. Wade and A. Graciaa, *J. Colloid Interface. Sci.*, 1982, **85**, 187.
81. M.K. Sharma and D.O. Shah, in 'Macro- and Microemulsions: Theory and Applications', ed. D.O. Shah, American Chemical Society, Washington, 1985, p. 149.
82. A. Beerbower and M. Hill, in 'McCutcheon's Detergents and Emulsifiers', Allured Publishing Company, Carol Stream, 1971.
83. P.M. Kruglyakov and A.F. Koretskii, *Doklady Akad. Nauk SSSR*, 1971, **197**, 1106.
84. P.M. Kruglyakov and Y.G. Rovin, 'Fiziko-Khimiya Chernykh Uglevodorodnykh Plenok', Nauka, Moscow, 1978.
85. J.L. Salager, in 'Encyclopedia of Emulsion Technology', ed. P. Becher, Dekker, New York, 1988, vol. 1, p. 159.
86. L. Cash, J.L. Cayias, G. Fournier, D. McAllister, T. Schares, R.S. Schechter and W.H. Wade, *J. Colloid Interface Sci.*, 1977, **59**, 39.
87. W. Seifritz, *J. Phys. Chem.*, 1925, **29**, 738.
88. A.T. Florence and D. Whitehill, in 'Macro- and Microemulsions: Theory and Applications', ed. D.O. Shah, American Chemical Society, Washington, 1985, p. 359.
89. Y. Sela, S. Magdassi and N. Garti, *Colloids Surf. A*, 1994, **83**, 143.

90. W.D. Harkins, 'The Physical Chemistry of Surface Films', Reinhold, New York, 1952.
91. J.W. McBain, 'Colloid Science', Reinhold, New York, 1950.
92. E.K. Fischer and W.D. Harkins, *J. Phys. Chem.*, 1932, **36**, 98.
93. O. Sonneville, Ph.D. Thesis, L'Université de Paris 6, 1997.
94. B.P. Binks, P.D.I. Fletcher and D.I. Horsup, *Colloids Surf.*, 1991, **61**, 291.
95. M. Nakagaki and K. Shinoda, *Bull. Chem. Soc. Jpn.*, 1954, **27**, 367.
96. D. Exerowa and A. Scheludko, in 'Proc. IV Int. Congr. Surf. Act. Subst.', Gordon and Breach, London, 1964, p. 1097.
97. A.D. Scheludko, *Adv. Colloid Interface Sci.*, 1967, **1**, 391.
98. S. Friberg, L. Mandell and Larsson, *J. Colloid Interface Sci.*, 1969, **29**, 155.
99. S. Friberg and L. Mandell, *J. Pharm. Sci.*, 1970, **59**, 1001.
100. S. Friberg and L. Rydhag, *Colloid Polym. Sci.*, 1971, **244**, 233.
101. N. Krog, N.M. Barfod and R.M. Sanchez, *J. Disp. Sci. Technol.*, 1989, **10**, 483.
102. R. Strey, *Ber. Bunsenges. Phys. Chem.*, 1993, **97**, 742.
103. S. Friberg, P. O. Jansson and E. Cederberg, *J. Colloid Interface Sci.*, 1976, **55**, 614.
104. P.G. de Gennes and J. Prost, 'The Physics of Liquid Crystals', Clarendon Press, Oxford, 1993.
105. B.W. Brooks and H.N. Richmond, *Chem. Eng. Sci.*, 1994, **49**, 1053.
106. B.W. Brooks and H.N. Richmond, *Chem. Eng. Sci.*, 1994, **49**, 1065.
107. E. Dickinson, B.S. Murray and G. Stainsby, *J. Chem. Soc., Faraday Trans. 1*, 1988, **84**, 871.
108. B. Deminiere, Ph.D. Thesis, L'Université Bordeaux 1, Bordeaux, 1997.
109. K.D. Danov, I.B. Ivanov, T.D. Gurkov and R. Borwankar, *J. Colloid Interface Sci.*, 1994, **167**, 8.
110. A.J. de Vries, *Rec. Trav. Chim.*, 1958, **77**, 383.
111. D. Exerowa, D. Kashchiev and D. Platikanov, *Adv. Colloid Interface Sci.*, 1992, **40**, 201.
112. A.V. Prokhorov and B.V. Deryaguin, *J. Colloid Interface Sci.*, 1988, **125**, 111.
113. Y.A. Chizmadzhev and V.F. Pastushenko, in 'Thin Liquid Films: Fundamentals and Applications', ed. I. B. Ivanov, Surfactant Science Series, vol. 29, Dekker, New York, 1988, p. 1059.
114. A. Kabalnov and J. Weers, *Langmuir*, 1996, **12**, 1931.
115. R. Aveyard, B.P. Binks, S. Clark and P.D.I. Fletcher, *J. Chem. Soc., Faraday Trans.*, 1990, **86**, 3111.
116. K.V. Schubert, R. Strey and M. Kahlweit, *Prog. Colloid Polym. Sci.*, 1991, **84**, 103.
117. R. Becker and W. Doring, *Ann. Physik*, 1935, **24**, 719.
118. D.F. Evans and H. Wennerström, 'Colloidal Domain', VCH, New York, 1994.
119. S.T. Milner and T.A. Witten, *J. Phys. II.*, 1988, **49**, 1951.

CHAPTER 8

Lifetime and Destruction of Concentrated Emulsions Undergoing Coalescence

BÉNÉDICTE DEMINIERE, ANNIE COLIN, FERNANDO LEAL
CALDERON AND JÉRÔME BIBETTE

Centre de Recherche Paul Pascal, Av. A. Schweitzer, F-33600 Pessac, France

8.1 Introduction

Emulsions are metastable colloids made out of two immiscible fluids, one being dispersed in the other, in the presence of surface active agents. The droplet volume fraction may vary from zero to almost one; dense emulsions are sometimes called biliquid foams since their structure is very similar to the cellular structure of air–liquid foams for which the continuous phase is very minor. From dilute to highly concentrated, emulsions exhibit very different internal dynamics and mechanical properties. When diluted, droplets are agitated by Brownian motion[1] and behave as viscous Newtonian fluids, whereas when more concentrated, namely above the random close packing volume fraction which is 64% for monodisperse droplets, the internal dynamics are severely restricted and they behave as viscoelastic solids.[2,3] Simple direct emulsions are composed of oil droplets dispersed in water while inverse emulsions are composed of water droplets dispersed in an oil continuous phase. In fact, emulsions are in principle made out of two immiscible phases for which the interfacial tension is therefore non-zero, and may involve other hydrophilic-like or lipophilic-like fluids in the presence of suitable surface active species, each phase being possibly comprised of numerous components. Sometimes, simple emulsions may also contain smaller droplets of the continuous phase dispersed within each droplet of the dispersed phase. Such systems are called double emulsions or multiple emulsions.[4]

Emulsions are artificial and, depending both on the metastability of the freshly formed interfaces and the fragmentation procedure which is employed, various structures may be generated. However, their lifetime may vary considerably: some systems are impossible to prepare whatever the employed procedure, some others disappear within a few seconds or a few hours and some others may stand for many years. Such variety has so far been empirically related to the surfactant solubility within the continuous phase. Bancroft's rule gives a first empirical guide to formulate emulsions: when the surfactant is essentially soluble within one phase, this phase turns out to be the continuous phase, the other being the dispersed phase.[5] As a consequence, a direct emulsion will be preferentially obtained with a water soluble surfactant whereas an inverse emulsion will be more easily obtained with an oil soluble one. On the basis of the hydrophilic–lipophilic balance of each surfactant molecule, as deduced from their preferential solubility with oil and water, HLB numbers from zero (very oil soluble) to 40 (very water soluble) have been attributed to each surfactant.[6] Therefore various surfactants characterized by the same HLB number are supposed to be equally efficient in stabilizing a given oil–water emulsion, though possessing different chemical groups. Hence, a given oil and water emulsion will be preferentially stabilized by a class of surfactant of a given HLB, the so-called optimal HLB. Such a guide is very useful in formulating emulsions and has been extensively accepted throughout the various communities dealing with commercial aspects of these materials.[7] It has also been noticed that mixtures of surfactants belonging to the same HLB class lead to a more stable emulsion compared with one which contains only one type, though characterized by an optimal HLB. This other empirical rule is widely accepted and governs most industrial formulations. Temperature or electrolytes may significantly modify the solubility of nonionic surfactants. Typically the solubility may change by varying the temperature or adding electrolytes, from water soluble to almost oil soluble. For such surfactants, as noticed by Shinoda *et al.*,[8] the stability of emulsions is still in agreement with Bancroft's rule, though the continuous phase may partially solubilize the dispersed phase leading to a microemulsion continuous phase. These nonionic surfactants, by changing the temperature for instance, may alternatively stabilize an oil-in-water or a water-in-oil emulsion, both having the same composition and in both cases being in agreement with Bancroft's rule, reinforcing the range of this empirical link.

The destruction of emulsions may proceed through two distinct mechanisms; one is due to diffusion of the dispersed phase through the continuous phase, the other is due to breaking of the thin domains separating adjacent droplets. The first mechanism is known as Ostwald ripening and is better understood.[9] It takes place when the dispersed phase is soluble enough within the continuous phase and consists of a gradual coarsening of the emulsions. According to the mean field description of Lifshitz and Slyozov,[10] the average droplet diameter varies as t^α, where α is equal to $1/3$, and the size distribution becomes asymptotically self similar in time, in agreement with experiments.[11] Because such coarsening is driven by the Laplace pressure mismatch between different droplets and requires diffusion over distances of about the droplet sizes, the destruction tends to slow down with time as described by the exponent $1/3$. Such an exponent may become even smaller ($\alpha = 1/2$) when instead of the bulk solubility, the surface permeation limits the growth process.[12] When long range surface forces are involved, large differences are expected and the growth may even be arrested,[13] as is the case when the dispersed phase is composed of a binary mixture, one component being almost insoluble outside, therefore retaining the soluble one due to the gradual loss of mixing entropy.[14]

The other mechanism is known as coalescence and consists of the rupturing of the thin film that forms between adjacent droplets, which allows two droplets to transform into one. The thin film that forms when two droplets are in contact is a metastable molecular assembly and its lifetime will be a key factor in determining the lifetime of the bulk emulsion. Breaking of the metastable thin film is an activated process which proceeds through the opening of a tiny molecular-sized hole that further grows under the action of the interfacial tension. Note that the previously exposed Bancroft's rule[5] relates surfactant solubility and emulsion metastability regarding coalescence processes only, and does not concern itself whatsoever with diffusion processes; it is more related to the mutual solubility of dispersed and continuous phases. Understanding thin film metastability and breaking at a more microscopic level should significantly consolidate the fundamental knowledge of emulsions and further allow us to both explain and understand the empirical Bancroft's rule.

8.2   Thin Film Metastability: Background

Metastable thin films or surfactant bilayers are present in various systems. The most famous is certainly the soap film, with its remarkable colors, or the soap bubble. In this case, surfactants adsorb at the two air–water interfaces which face each other with some water in between. These films are easily obtained by, for instance, drawing a frame out of an aqueous surfactant solution. The film rapidly drains under the action of capillary forces and reaches an equilibrium thickness set by repulsive surface forces. These surface forces have been extensively studied. Typical force profiles may vary from long range to very short range and from attractive to repulsive, depending on the nature of the surfactants, temperature and various added components.[15] When short range adhesion exists or when the film is forced to drain under the action of an external osmotic pressure or evaporation, the two surfactant layers may still persist and become almost a bilayer (Newton black films).[16] The same type of thin films is found in dense emulsions and foams, which are cellular materials made of discrete cells separated by thin flat films at each contact. The same variety of attractive or repulsive, short or long range forces have been measured between the droplet interfaces, depending on the nature of the surface active agents, temperature or added components within the continuous phase.[17] Bilayers may also spontaneously form at the contact between adjacent droplets when spontaneous wetting occurs[18,19] or simply by concentrating the system to a high internal phase ratio.[20] However, the interfacial tension within the thin film or the bilayer is reduced when adhesion occurs, whereas it increases when external pressure is required. Surfactant vesicles are one other example for which monolayer adhesion drives the bilayer formation, and are still sometimes a metastable molecular assembly.[21] The lifetime of these films and therefore the origin of the breaking have been viewed so far as possibly due to two distinct mechanisms.

One mechanism is a mechanical instability like that occurring in spinodal decomposition:[22,23] the surfaces are submitted to thermal fluctuations and one capillary mode (peristaltic) may become unstable by compensation of the excess surface energy (induced by undulations) and the short range attractive potential energy (see Figure 8.1). The film becomes unstable with respect to fluctuations with wavelengths larger than a critical wavelength λ_c:

$$\lambda_c = (-2\pi^2\gamma/(\mathrm{d}^2 V/\mathrm{d}h^2))^{1/2} \tag{8.1}$$

where γ is the interfacial tension and $V(h)$ is the free energy of interaction as a function of the film thickness. Let us consider a thin film of radius R_f, thickness x_i, separated by two interfaces of interfacial tension γ, with a fluid of viscosity η in between, and characterized by a Hamaker constant A. When the film is drained to a critical thickness x_c at which the instability takes place, it takes time τ for the two interfaces to come into contact as depicted in Figure 8.1. Now x_c is given by:

$$x_c = 0.22(AR_f^2/f\gamma)^{1/4} \tag{8.2}$$

where f is a numerical factor of about 7 and τ is given by:

$$\tau = 96\pi^2\gamma\eta x_c^2 A^{-2} \tag{8.3}$$

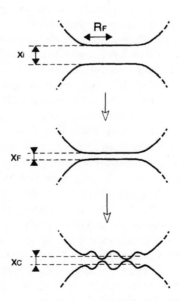

Figure 8.1 *Consider a thin film of radius R_f, initial thickness x_i, separated by two interfaces of interfacial tension γ, with a fluid of viscosity η in between, and characterized by a Hamaker constant A. When the film is drained to a critical thickness x_c at which the instability takes place, it takes a time τ for the two interfaces to come into contact as depicted here*

As an example, for a system characterized by the parameters $\eta = 10^{-3}$ Pa s, $\gamma = 10^{-2}$ mN m^{-1}, $R_f = 1$ μm, $A = 0.5 \times 10^{-20}$ J, $x_c = 3.6$ nm and $\tau = 2.3 \times 10^{-4}$ s; the values of these parameters are roughly those of an aqueous thin air–liquid film. This timescale omits the time it takes for the film to drain down to the critical thickness x_c. This drainage time τ_d may be evaluated[24] by assuming no slipping at boundaries and is given by:

$$\tau_d = 3\eta r R_f^2 / 4\gamma x_c^2 \tag{8.4}$$

where r is the radius of curvature of the curved part of the interface, γ/r being the Laplace pressure which is responsible for the drainage until this pressure becomes equal to the disjoining pressure; τ_d is equal to 5.8×10^{-3} s when x_c is 3.6 nm and $r = 1$ μm. This first mechanism assumes that once interfaces come into contact, breaking spontaneously takes place. Therefore, such a mechanism considers that the limiting step for rupture is only controlled by the thinning of the film up to molecular contact. The occurrence of a mechanical instability in the presence of an attractive contact potential allows us to predict a finite time for the interfaces to touch, while in the absence of this undulation instability it would take an infinite time.[24] Note that this mechanism is relatively fast and would never account for the existence of very metastable emulsions or bilayers, which may stand for years. Such a mechanism is certainly more suitable to account for the faster growth of spinodally decomposing mixtures where no surfactant is added.[25]

A second distinct mechanism consists of the nucleation of a thermally activated hole which reaches a critical size, above which it becomes unstable and grows, leading to the coalescence of two adjacent droplets, as depicted in Figure 8.2. The

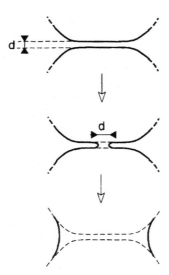

Figure 8.2 *Nucleation of a thermally activated hole which reaches a critical size above which it becomes unstable and grows, leading to the coalescence of two adjacent droplets*

energy cost for a hole of size d is $E(d)$ and is given in a first approximation by $\pi\Gamma - \pi\gamma d^2/4$, where Γ is a line tension that accounts for all kinds of microscopic consequences due to the high curvature that is inherent in the formation of a channel perpendicular to both interfaces.[26] Γ is positive and therefore a maximum in E occurs at a critical distance r^*, $E(r^*)$ being the hole nucleation activation energy E_a. Following the mean field description of Arrhenius, we then define ω, the frequency of opening an 'efficient hole' per unit surface, as $\omega_0 P(E_a)$, where ω_0 is the natural frequency of the process and $P(E_a)$ is the probability for a local energy fluctuation to be greater or at least equal to E_a at a temperature T. The Arrhenius mean field description applied to hole nucleation assumes that there are no spatial correlations between nucleated holes, so that the probability of forming a hole greater than r^* is $\exp(-E_a/kT)$, kT being the thermal energy at a temperature T. The natural frequency ω_0 can be viewed as the product of two terms: ν and ϕ. One (ν) is related to hydrodynamics and we assume a local energy fluctuation greater than E_a, thus allowing a finite time (which defines ν) to reach the critical size r^*. The other (ϕ) is related to entropy and describes the number of possibilities within a unit surface of making distinct holes of size r^*. By contrast with the first mechanism for which all parameters are known, the nucleation frequency ω is difficult to estimate since E_a, r^* and ν are very microscopic quantities.

8.3   Destruction of Concentrated Emulsions Undergoing Coalescence

The lifetime of an emulsion can vary considerably from one system to another; it can change from minutes to many years, depending on the nature of the surfactants, the nature of both phases and their volume ratio. Moreover, the scenario of growth is also very dependent upon the system. For instance, when a silicone oil-in-water emulsion stabilized by SDS above its CMC is stressed by an homogeneous osmotic pressure, coarsening occurs through the growth of a few randomly distributed large droplets that eat up the small ones.[27] In Figure 8.3 we show a microscopic picture of such growth. The initial emulsion is comprised of monodisperse droplets having a diameter σ of about 1.5 µm and has been submitted to an osmotic stress of 0.6 atm during 15 days at room temperature. This observation allows us to conclude that one type of growth is naturally preferred, leading to a very specific scenario of destruction. Another scenario is even more frequently encountered and familiar to everyone who has been working in the field of emulsion formulation. Very often, a concentrated O/W emulsion, which is unstable, rapidly exhibits the formation of a macroscopic domain made of the dispersed phase, which sits on the top of the remaining emulsion. In fact, by measuring the height of this phase as a function of time, the efficiency of various surfactants can be compared.[28] Note that this procedure is currently used in many routine tests of formulation. Such a phenomenon suggests that the growth scenario is dominated by the growth of macroscopic domains.

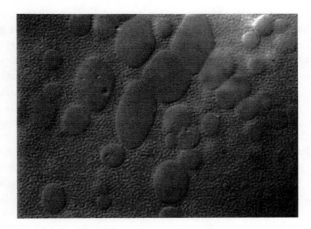

Figure 8.3 *Microscopic picture of an emulsion, initially composed of monodisperse droplets having a diameter σ of about 1.5 μm, which has been submitted to an osmotic stress of 0.6 atm during 15 days at room temperature*

Temperature variation applied to crude polydisperse nonionic dense emulsions sometimes leads to the same kind of growth and the same rapid formation of a macroscopic domain. More recently, Kabalnov and Wennerström[29] have shown that for a direct octane–$C_{12}E_5$–water emulsion with the surfactant concentration being set around a few percent, and probably a quite large initial polydispersity (unknown) at $\phi = 50\%$, the half-life time $\tau_{1/2}$ varies with temperature (at least a few degrees from the PIT) as $\ln \tau_{1/2} \propto 1/kT$. $\tau_{1/2}$ is defined as the time it takes for half the emulsion to transform into an oil layer. By *a priori* assuming that τ is simply proportional to $\exp(-E_a/kT)$ (as defined previously) the authors have identified the slope as an activation energy and deduced the value of $45kT_r$, T_r being room temperature.

In principle, understanding the metastability of dense emulsions, and hence their lifetime, requires two types of information which account for two distinct phenomena. One concerns the thin film and its lifetime, associated with the mechanism of a coalescence event. The other concerns the distribution of these events throughout the whole material, and therefore the growth scenario associated with the time and space size distribution evolution. More important is the fact that the spatial and temporal evolutions of the size distribution are the only accessible functions that may be tracked to reveal the microscopic parameters controlling the thin film metastability. The rather frequent occurrence of heterogeneous types of growth (coexistence between large growing domains and almost colloidal scale droplets) complicates the link to be established between the microscopic properties, such as the thin film lifetime, and the more macroscopic ones, such as the bulk emulsion lifetime. In particular, there is some ambiguity in characterizing the thin film metastability by a unique lifetime or equally by a unique frequency of rupture per unit surface, while the scenario of destruction involves the growth of very specific and relatively rare nuclei. Indeed, such types of growth suggest that some sites are

growing faster and therefore very probably possess different microscopic characteristics from the average emulsion. Besides the difficulty of setting a growth model
which would account for such types of behaviour, by including the required film
characteristics the experimental determination of the size distributions in time and
eventually in space (spatial correlations) is also very complicated. We note,
however, that a crude determination of $\tau_{1/2}$ for nonionic temperature sensitive
polydisperse emulsions, as a function of temperature, has led to a reasonable
Arrhenius plot, according to Kabalnov and Wennerström,[29] suggesting that, in some
cases, all of the previously reported complications may be ignored.

In order to probe the microscopic characteristics of the emulsion thin film
metastability and to conclude about the mechanism of rupture, we believe that it is
essential to measure the evolution of the droplet size distribution and to compare
these data to a realistic growth model. Such a comparison should allow a reliable
determination of the various film rupture characteristics. We also believe that
concentrated emulsions may be helpful in determining these characteristics for
several reasons. Firstly, owing to their high internal phase ratio, the coalescence
time is certainly not controlled by drainage which is already completed. Secondly,
because of the initial colloidal size of droplets and the rather high internal phase
ratio, there is a significant amount of thin film per unit volume and therefore quite
a significant number of coalescence events required for destruction. Indeed, for a
micron sized emulsion at $\phi = 80\%$, the initial surface area per unit volume $6\phi/\sigma$
is about 5 m^2 per cm^3, which corresponds to about 10^{12} coalescence events per cm^3
of destroyed emulsion. This number is still as large as 10^9 for an emulsion which
changes its droplet diameter from 1 to 10 μm. We then conclude that over a
relatively small variation of the average diameter, statistical averaging of coalescence events should be sufficient to further consider averaged quantities only, when
interpreting the kinetics of growth.

One difficulty to overcome is of course to find a system which is stable enough
at room temperature, can be safely prepared and is such that, by changing some
parameters, will exhibit coalescence on a reasonable timescale. One other difficulty
to overcome consists of designing a concentrated emulsion which, at least in the
early course of the destruction, will exhibit a scenario which might be easily
accounted for by a simple growth model.

8.4 Droplet Growth in Dense Emulsions Undergoing Coalescence

If we *a priori* assume that a thermally activated hole nucleation mechanism is
governing the thin film and emulsion metastability, we expect that some systems
may be stable at room temperature and coarsen at a higher temperature. However,
since we aim to probe coalescence only, we have to consider oils that are poorly
soluble in water. We have designed oil-in-water emulsions that fit these criteria.
Figure 8.4 shows a microscopic picture of a silicone oil-in-water emulsion
stabilized by a mixture of two nonionic surfactants $C_{10}E_5$ and $C_{12}E_5$ (Lauropal
205), with $\phi = 85\%$ and $\sigma = 0.85$ μm; the surfactant concentration C is set in the

Figure 8.4 *Microscopic picture of a silicone oil-in-water emulsion stabilized by a mixture of two nonionic surfactants $C_{10}E_5$ and $C_{12}E_5$ (Lauropal 205) at C = 0.06%, with $\phi = 85\%$ and $\sigma = 0.85\ \mu m$*

continuous phase to 0.06%. The interfacial tension between the aqueous surfactant phase and this silicone oil is 8.5 mN m^{-1} (polydimethylsiloxane, 5000 cP). As shown in this picture, the emulsion is fairly monodisperse. It has been obtained from a crude polydisperse one which has been purified by the fractionated crystallization procedure.[30] The monodisperse droplets are densely packed owing to their volume fraction ($\phi = 85\%$), and do not exhibit (as we were expecting) any coarsening at room temperature for months. However, since we chose the surfactant for its intermediate HLB value (HLB = 10), which reflects, as suggested by the empirical HLB scale, a limited ability to stabilize direct emulsions, we also expect that by raising the temperature coalescence will occur on a reasonable timescale.

This emulsion (about 2 cm^3) is loaded into a sealed tube that is further thermostated at various selected temperatures. By taking off, time after time, a minor amount of the sample and cooling it to room temperature, we observe the time evolution of the emulsion. We use direct microscopy and static light scattering (Malvern granulometer) to detect the droplet diameter and discuss the evolution of polydispersity. Figure 8.5 shows the employed surfactant–water phase diagram as a function of temperature. When temperature increases, the surfactant–water phase diagram (when considering, for instance, a 1% mixture) changes as follows: L_1 phase (micellar phase)–$L_1' + L_1''$ (coexistence of two micellar phases)–$L_1 + L_3$ (coexistence of a micellar and a sponge phase)–$L_1 + L_2$ (coexistence of a micellar and a reverse micellar phase).

We did not find any noticeable change in our emulsion within a day at temperatures below 70 °C. However, at precisely 70 °C and above, the emulsion undergoes a very surprising change, which occurs more and more rapidly as the temperature is increased above 70 °C. The emulsion rapidly jumps (within a few seconds when T is above 75 °C) from the initial droplet size to a new larger size,

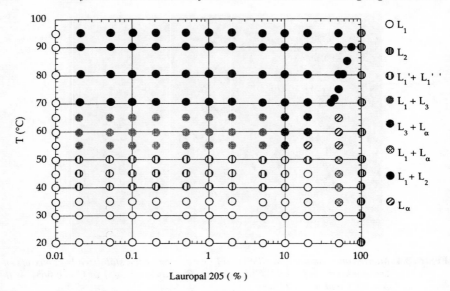

Figure 8.5 *Surfactant–water phase diagram as a function of temperature*

the emulsion remaining surprisingly monodisperse if the initial one is also mono-disperse. Figure 8.6 shows the evolution of the size distribution as a function of time (from $t = 0$ to $t = 5$ minutes) at 90 °C and Figure 8.7 shows a microscopic picture of the emulsion after 1 minute, at which time the transformation is already complete. The transformation proceeds in the following way: after a period of time

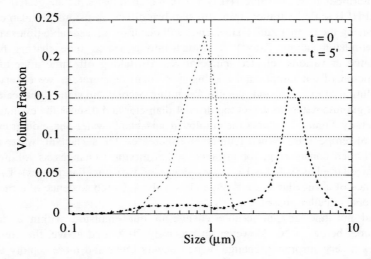

Figure 8.6 *Evolution of the size distribution as a function of time (from* t = 0 *to* t = 5 *minutes) at 90 °C for* C = 0.06% *and* $\phi = 0.85$

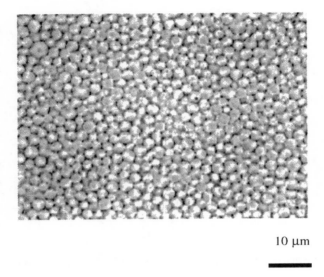

10 μm

Figure 8.7 *Microscopic picture of the emulsion in Figure 8.6 after 1 minute at 90 °C*

of about a few seconds, during which all sizes from the initial to the final coexist, the system reaches a new uniform size which is apparently stabilized and independent of temperature.

Such a transformation is discussed in detail elsewhere.[31] However, we aim here to mention its origin briefly, at least to show that some coalescence phenomena may not be governed by a thermally activated process. By contrast, this coarsening is induced by the nucleation of the L_2 phase within the continuous phase, which takes place above 70 °C. The nucleation of tiny nodules of the L_2 phase very probably induces a dewetting process, by building unstable capillary bridges between droplets, as already found in air–liquid foams.[32] The L_2 phase is essentially a surfactant phase with some water (30%) swelling the reverse micelles. This instability rapidly disappears probably when all of the L_2 phase nodules are expelled within plateau borders. This type of instability is not within the scope of this chapter. We focus here on a more general aspect of thin film metastability: the case of homogeneous thin films in the absence of nucleating heterogeneities. Nevertheless, it is again worth noticing that rupture of metastable thin films may have various origins and mechanisms. One of those is thought be a capillary instability when bridging by one phase is mechanically unstable.

When this rather short time transformation is accomplished the emulsion is comprised of bigger, still fairly monodisperse, droplets at the same volume fraction. With the L_1/L_2 spinodal decomposition being completed, the thin films in between droplets are logically comprised of less water and certainly consist of a surfactant bilayer only. Figure 8.8 shows the long time (up to several hours) evolution of this emulsion at 80 °C. The average droplet radius is again plotted as a function of time. After the very rapid change, due to dewetting, which is almost not noticeable on

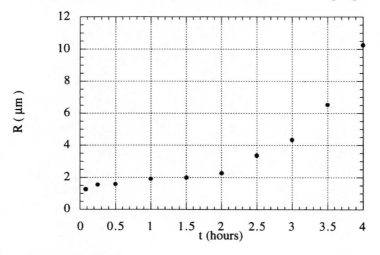

Figure 8.8 *Long time evolution of the radius of the emulsion in Figure 8.6 at 80 °C*

the scale of this plot, the emulsion does not change significantly for say an hour, and at longer times a severe augmentation of the radius occurs. This long time coarsening is apparently speeded up with time and clearly differs from the previously described one which was rapidly saturating. Figures 8.9 and 8.10 both show the droplet size evolution; one is a direct microscopic observation, the other is the small angle light scattering pattern from the same samples. These microscopic pictures and their corresponding scattering patterns are taken at various times at 80 °C. The droplet growth is surprisingly homogeneous and monodisperse, as seen on both microscopic pictures and scattering patterns. Two or three diffraction rings are visible, which reflects the relatively high monodispersity. The presence of Bragg spots, at least on one of these pictures, also suggests that the droplets are so monodisperse that they may also arrange with some long range ordering. It is quite surprising that a destruction process (coalescence) conserves a unique droplet size throughout the system. Intuitively one would think that although the system is initially monodisperse, polydispersity would be enhanced by the growth. In fact this does not happen (at least in some cases), and we will have to come back to this point later. The diverging tendency of $R(t)$ is also of great interest. It definitely tells us that the dominant coarsening mechanism is not diffusion controlled, since if it was then $R(t)$ would saturate as $t^{1/3}$ or $t^{1/2}$.

8.5   Growth Model for Dense Emulsions

We aim to derive the simplest model that would account for the droplet growth within a concentrated emulsion undergoing coalescence when the thin films are characterized by a unique parameter ω, the frequency of rupture per unit surface of film. We assume that this timescale is the only one that limits the growth. We consider the concentrated emulsion to be analogous to a biliquid foam, for which ϕ

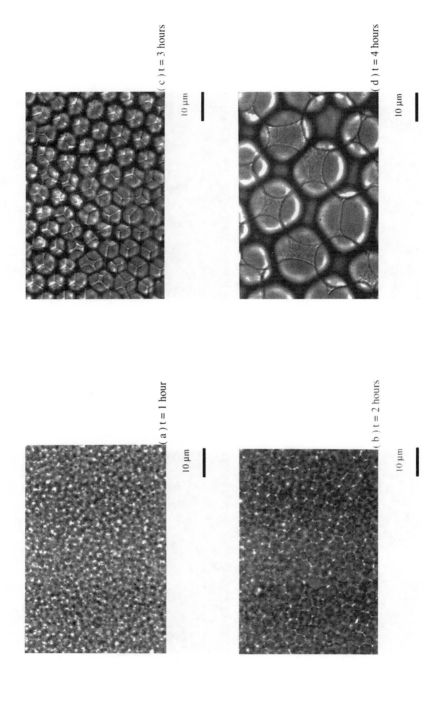

Figure 8.9 *Microscopic observation of the droplet size evolution taken at various times at 80 °C. The initial emulsion was the one presented in Figure 8.4*

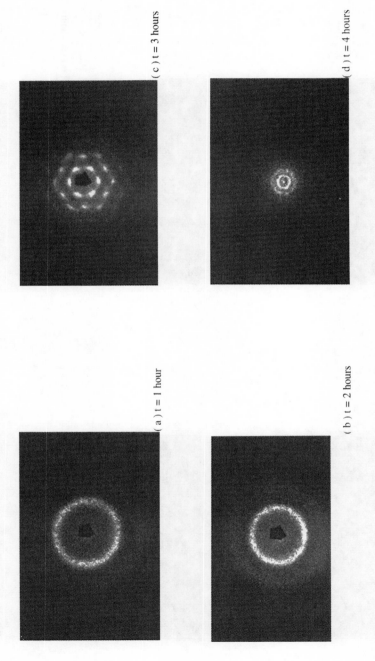

(a) t = 1 hour

(b) t = 2 hours

(c) t = 3 hours

(d) t = 4 hours

Figure 8.10 *Small angle light scattering pattern from the same samples as in Figure 8.9, realized at various times at 80 °C*

is almost one, and for which droplet surfaces are entirely faceted. We consider only one radius R at a time, that may be taken as a mean radius if polydispersity is also to be considered. However, polydispersity is not supposed to change the obtained scaling of R *versus* t. We define n, the number of droplets of size R per unit volume and, therefore the variation of n due to coalescence, dn, during the time, dt, is given by the differential equation:

$$dn = -\omega 4\pi R^2 n\, dt \qquad (8.5)$$

Indeed, the variation of n, which corresponds also to the number of coalescence events, will be proportional (with the previous assumptions) to the droplet surface area times the frequency of rupture ω. The droplet volume fraction ϕ is given by $\phi = n4\pi/3\, R^3$, allowing us to rewrite the previous equation as:

$$dR/R^3 = 4\pi/3\, \omega\, dt \qquad (8.6)$$

By integrating, we obtain:

$$1/R^2(t) = 1/R_0^2 - 8\pi/3\, \omega t \qquad (8.7)$$

where R_0 is the initial droplet radius. From this equation we then define τ_e as the time it takes to get R infinite $(1/R^2 = 0)$:

$$\tau_e = 3/8\pi\, (\omega R_0^2)^{-1} \qquad (8.8)$$

i.e. we identify the time it takes for the emulsion to be destroyed (emulsion lifetime). Such a description predicts a divergence of $R(t)$ at a time τ_e, hence predicting a finite time for the total destruction. The divergence, or equally the dramatic acceleration of the droplet growth with time, is a direct consequence of coalescence. Indeed, the larger the droplets, the more probable is the rupture and the larger the increase in radius, due to droplet volume conservation. As mentioned before, coarsening processes induced by diffusion have a radically different scaling and have a tendency to slow down with time $(t^{1/3})$, by contrast with coalescence induced coarsening which, according to this model, is completed within a finite amount of time. Such a model omits a number of other effects that are inherent to the destruction of emulsions, or more generally, to the destruction of any realistic cellular material undergoing successive rupture of the thin domains separating each cell. Our mean field description neglects any consequences of polydispersity and of spatial correlations between droplets of different sizes. Moreover, the inherent rearrangements that have to proceed throughout the dense cellular system in order to relax local stresses may set another timescale, which is ignored.

8.6 Determination of $\omega(T)$, ω_0 and E_a

We expect that the previously described long time growth may be accounted for by this model and hence allows us to deduce the parameters $\omega(T)$, ω_0 and E_a. We compare the experimental data (shown in Figure 8.8) to the previous mean field equation (8.7). The same data are plotted as $1/R^2$ *versus* time in Figure 8.11. The full line is the best fit to our data where ω is the only adjustable parameter. R_0 is the radius which is obtained immediately after the very short time process that has been described in Section 8.4. The dependence of $R(t)$ is very well accounted for by the model, which confirms that coalescence is actually responsible for this growth. As mentioned before, a diffusion limited process would have led to a completely different scaling. We note that deviations to the prediction occur when the droplet diameter becomes larger than typically 10 μm. There are many reasons that may explain these deviations and which are obviously not taken into account by our growth equation. In particular, when the droplet diameter becomes large enough, the Laplace pressure as well as the osmotic pressure both relax, which may possibly decrease ω as the coarsening goes on, hence resulting in a somewhat similar deviation. Moreover, rearrangements and drainage of freshly formed thin films, after a coalescence event has occurred, may become long enough to slow down the destruction considerably. Therefore we will further restrict our comparison to a droplet size variation of up to 10 microns, for which there are still about 10^9 coalescence events involved per cm^3.

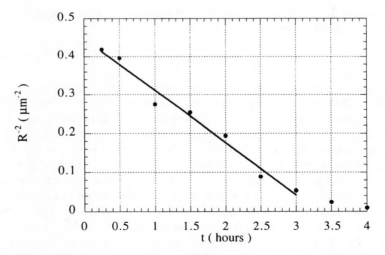

Figure 8.11 *$1/R^2$ versus time at 80 °C. The full line is the best fit to our data where ω is the only adjustable parameter*

The fit is good enough to deduce ω from either the slope or the intercept on the time axis. At 80 °C the intercept gives for the emulsion a lifetime τ_e of about 3.5 hours. From the expression $\tau_e = 3/8\pi(\omega R_0{}^2)^{-1}$ with $R_0 = 3$ μm, we deduce $\omega = 6.4 \times 10^{-2}$ μm^{-2} h^{-1}. We can therefore conclude that a concentrated emulsion with an initial radius of about R_0 will be destroyed within τ_e hours (neglecting the longer time deviations) by simply nucleating $3/8\pi$ holes per $R_0{}^2$ and τ_e. In this case the emulsion will be destroyed within 3.5 hours by opening a hole per μm^2 roughly every 25 hours ($8\pi/3\,\tau_e$). This rather small required frequency is of course one of the consequences of the coalescence process in which the droplet coalescence probability keeps increasing as the droplet size increases. Another equivalent consequence is that the number of coalescence events per unit volume and time keeps decreasing while the droplet size is diverging. In Figure 8.12 we show plots of $R(t)$ obtained at different temperatures T. As expected, the coalescence proceeds faster if the temperature is increased. Nevertheless, the qualitative features of the growth remain unchanged. In Figure 8.13 we plot R^{-2} *versus* t for the three temperatures at which we found some noticeable change in R (75, 80, 90 °C). Again the fits are good enough to deduce for each temperature the values of ω reported in Table 8.1. From 75 to 90 °C, τ_e varies from 3.5 hours to 45 minutes and ω varies from 0.066 to 0.28 $\mu m^{-2}\,h^{-1}$.

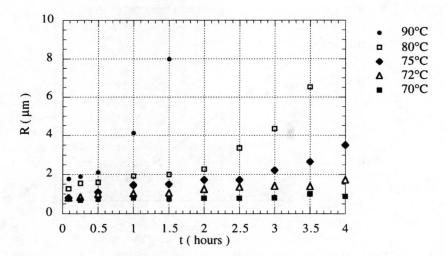

Figure 8.12 *Plots of* R *versus time obtained at different temperatures:* T = 70, 72, 75, 80 *and 90 °C*

We have designed another system for which the same growth characteristics are recovered in order to show the generality of these phenomena. Moreover, silicone oil has a different surface tension from one sample to another due to a possible variation of the percentage of hydroxyl groups or the presence of some very surface active impurities. We have learned that the interfacial tension is very critical in governing the type of growth scenario. Indeed, we have found that silicone oils that have an interfacial tension against the water–surfactant mixture higher than 10 mN m^{-1} lead preferably to a very heterogeneous growth, involving the rapid formation of an oil layer on the top of the sample, or at least to the formation of very large droplets almost coexisting with the initial ones. Silicone oil is replaced by a mixture of hexadecane and oleic acid at 0.5%. Oleic acid is added to the oil

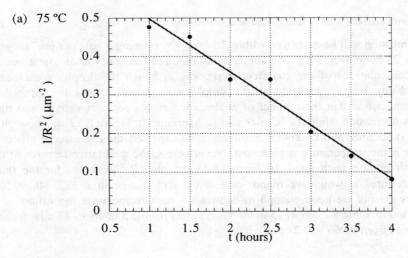

(a) 75 °C

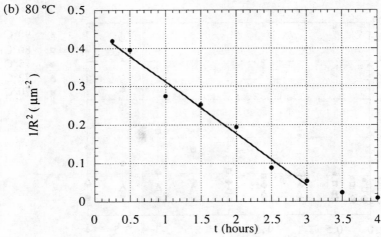

(b) 80 °C

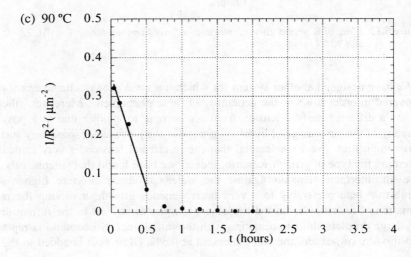

(c) 90 °C

Table 8.1 *ω and τ_e as a function of temperature, for the emulsion of silicone oil–*
Lauropal 205–water

$T/°C$	$\omega/\mu m^{-2}\,h^{-1}$	τ_e
75	6.6×10^{-2}	3.5 h
80	6.4×10^{-2}	3 h
90	2.8×10^{-1}	45 min

phase to lower the interfacial tension, which was found essential (as for silicone oil systems) in order to obtain a fairly monodisperse growth. From 0 to 0.5% oleic acid, the oil–water interfacial tension (in the presence of the same water soluble nonionic surfactant at 0.06%) changes from 11 to 5 mN m^{-1}. In the absence of oleic acid, the growth is very rapidly heterogeneous, exhibiting the formation of a macroscopic oil layer at the top of the sample, precluding any reliable comparison with our model. Such type of growth is, however, as mentioned in Section 8.3, very frequently observed. We have empirically found that adding an oil soluble surfactant, such as oleic acid, changes the growth scenario from a very hetero-geneous to a fairly homogeneous one, for which we can consider that only one size at a time is present. This system is made of the two following phases: hexadecane + 0.5% oleic acid and water + 0.06% Lauropal 205, and exhibits the same transformation when crossing the L$_1$/L$_2$ transition ($T > 70$ °C). The initial emulsion compressed at $\phi = 84\%$ for droplets having a diameter $\sigma = 0.32$ μm is shown in Figure 8.14. When the temperature is increased above 70 °C the emulsion

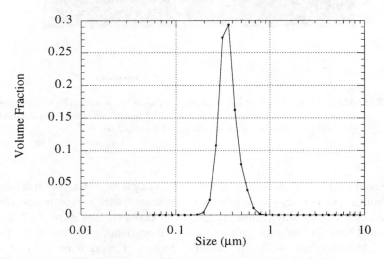

Figure 8.14 *Initial size distribution of emulsion of hexadecane + 0.5% oleic acid in water + 0.06% Lauropal 205 at 70 °C*

Figure 8.13 *1/R² versus time for* (a) 75 °C, (b) 80 °C and (c) 90 °C

transforms into another emulsion at the same oil volume fraction but possessing larger droplets. Again the final size (R_0) is independent of temperature. We show in Figure 8.15 a microscopic picture of this emulsion when the temperature was set at 90 °C for 5 minutes. The timescale of the transformation is only a few seconds once the temperature is not too close (about 5 °C) to the one at which it diverges.[31] Since the temperature, at which a measurable long time growth takes place, is much above this (more than 90 °C for the hexadecane system), we consider the first transformation as instantaneous and again identify $R(t = 0)$ as R_0.

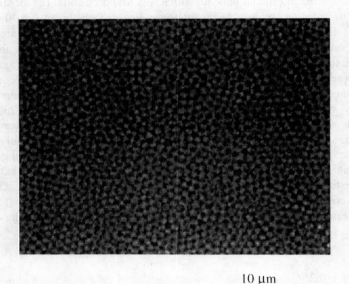

10 μm

Figure 8.15 *Microscopic picture of the emulsion in Figure 8.14 when the temperature was set at 90 °C for 5 minutes: σ = 2.0 μm. The initial emulsion was compressed at φ = 84% for droplets having a diameter σ = 0.32 μm*

In Figure 8.16 we show the curves of R against time for five distinct temperatures. As before, the droplet size increases faster as the temperature is increased, the curves still exhibiting the same features. In Figure 8.17 we plot R^{-2} against time for the same temperatures, the continuous line being the best fit to our data using eqn. (8.7). The deduced values of ω as a function of T are plotted in Figure 8.18 and both $\omega(T)$ and $\tau_e(T)$ are listed in Table 8.2. Again we find ω increases with T, suggesting that the thin film rupture is thermally activated. To confirm this hypothesis we plot, in Figure 8.19, $\ln(\omega)$ as a function of $1/kT$, where k is the Boltzmann constant, and compare it to the Arrhenius equation: $\ln(\omega) = \ln(\omega_0) - E_a/kT$, represented by the continuous line which is the best fit to our data. Ln(ω) as a function of $1/kT$ is clearly well accounted for by a linear function. From this fit we deduce both $\omega_0 = 4 \times 10^{24}$ m^{-2} s^{-1} and $E_a = 2 \times 10^{-19}$ J, which corresponds to $50\,kT_r$, T_r being room temperature (25 °C).

8.7   Discussion

We have so far established a reasonable description of the coalescence induced destruction of dense emulsions, by considering a thermally activated rupture of the thin film (governed by the Arrhenius equation), and by considering a mean field growth equation involving only one frequency so that $R(t)$ is given by:

$$1/R^2(t, T) = 1/R_0{}^2 - 8\pi/3(\omega_0 \exp(-E_a/kT))t \qquad (8.9)$$

and τ_e is given by:

$$\tau_e(T) = 8\pi/3\,(R_0{}^2\omega_0)^{-1} \exp(E_a/kT) \qquad (8.10)$$

These equations have been derived by assuming an homogeneous scenario and concern thin homogeneous films that may be characterized by a constant rupture frequency $\omega(T)$. For the particular system hexadecane/oleic acid/Lauropal 205, we find $\omega_0 = 4 \times 10^{24}$ m^{-2} s^{-1} and $E_a = 2 \times 10^{-19}$ J which corresponds to $50kT_r$. These numbers are of great interest because through the use of eqn. (8.10) we can obtain an estimate of τ_e as a function of E_a at various temperatures. In Figure 8.20 we plot τ_e as a function of E_a for $T = 20$ °C and $T = 100$ °C assuming ω_0 constant and equal to 4×10^{24} m^{-2} s^{-1}. For our particular system (Lauropal 205, HLB = 10), for which we have found $E_a = 50kT_r$, we deduce $\tau_e(T = 20\ °C) = 5$ years. The deduced lifetime at room temperature is in good agreement with experimental observations since samples that have been prepared three years ago and left at room temperature are still to be destroyed. This plot also shows that once the energy of activation is above $50kT_r$, the emulsion lifetime at room temperature is almost infinite; this is also in fairly good agreement with our observations since some samples that have been prepared 10 years ago with surfactants possessing a higher HLB (sodium dodecyl sulfate, nonylphenol (10) ethoxylate with silicone oil) are essentially unchanged. According to this graph, the variety of systems for which the kinetics of growth are measurable, *i.e.* possessing a lifetime ranging from a few minutes to a few days, therefore conveniently accessible to experiments, is quite restricted. Indeed, activation energies slightly above $55kT_r$ lead to systems that are so metastable that their lifetime is inaccessible, while those slightly below $45kT_r$ lead to systems almost impossible to prepare. Nevertheless, depending on how the activation energy varies with chemical changes or concentration variations, a sufficiently large variety of systems may be probed in order to set some links between chemical series (or HLB) and the corresponding activation energy.

The microscopic origins that set the scale of both ω_0 and E_a are certainly very instructive but remain quite elusive so far. The natural frequency ω_0, as mentioned before, may be written as the product of two terms: ν and ϕ. One is related to

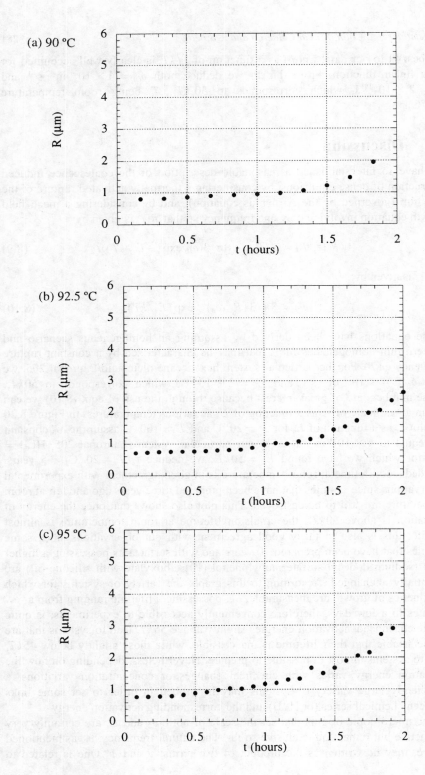

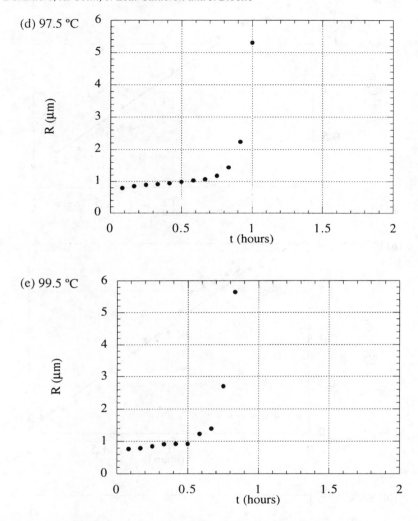

Figure 8.16 R versus *time for five distinct temperatures for the emulsion in Figure 8.14*
        (a) *90 °C,* (b) *92.5 °C,* (c) *95 °C,* (d) *97.5 °C,* (e) *99.5 °C*

hydrodynamics (ν): let us assume a local energy fluctuation which allows a critical hole of size r^* to open. Such a process will take a finite time which defines the frequency ν (s^{-1}). The other, ϕ (m^{-2}), accounts for the number of possibilities per unit surface to make distinct holes of size r^*. The physics which sets the frequency ν is quite unclear. However, one hypothesis consists in writing ν as $\gamma/(\eta r^*)$, where γ is the interfacial tension between the dispersed and continuous phase and η is the continuous phase viscosity. Such an expression assumes that the opening process is driven by interfacial tension but limited by the film viscosity and the critical hole size over which the film has to open. ϕ is simply given by $1/r^{*2}$, being the number

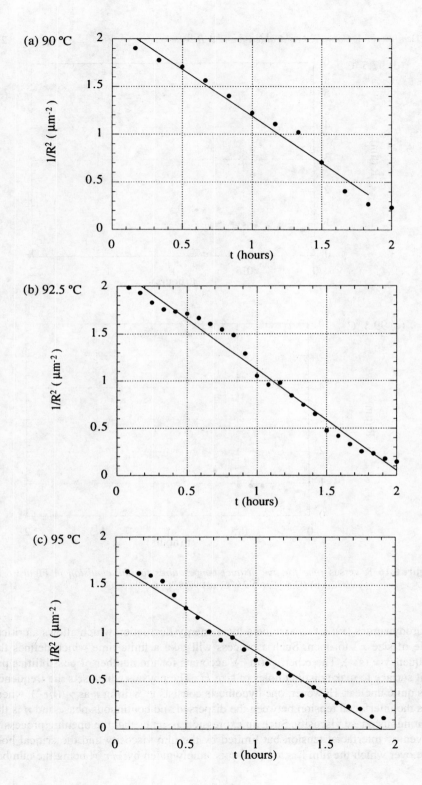

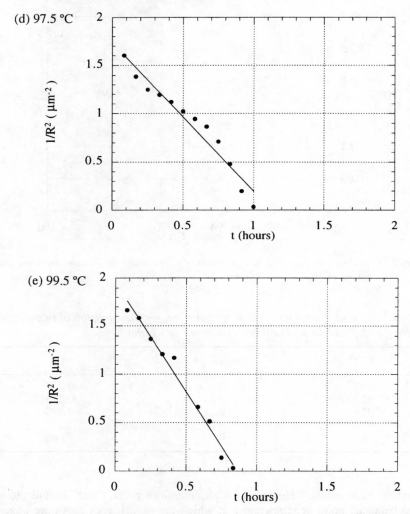

Figure 8.17 *1/R² versus time for the same temperatures given in Figure 8.16. The continuous line is the best fit to our data*

of distinct holes of size r^* that it is possible to pierce within a unit surface. Therefore ω_0 is expressed as $\gamma/(\eta r^{*3})$, to be compared to the experimental value and allowing some estimation of r^*. If η is taken to be 10^{-3} Pa s and $\gamma = 10^{-3}$ mN m^{-1}, we obtain $r^* = 10$ nm, which is about a micelle radius and quite reasonable. The energy of activation E_a certainly involves many distinct properties of the film such as the curvature energy, the interfacial tension and probably also the Gibbs elasticity. Calculations of the energy barrier for nucleating an unstable hole, based on curvature and interfacial tension contributions, but neglecting Gibbs elasticity, have led to a comparable number of about $45kT_r$, for a similar system (octane–$C_{12}E_5$–water). Elastic constants were obtained from experimental and/or

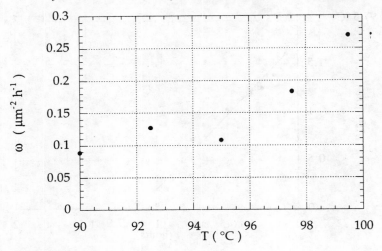

Figure 8.18 *Deduced values of ω as a function of temperature for hexadecane/oleic acid/ Lauropal 205/water emulsion*

Table 8.2 *ω and τ_e as a function of temperature, for the emulsion of hexadecane/ 0.5% oleic acid in 0.06% Lauropal 205/water*

$T/\,°C$	$\omega/\mu m^{-2}\,h^{-1}$	τ_e
90	0.088	3.25 h
92.5	0.13	2.21 h
95	0.11	1.85 h
97.5	0.18	1.2 h
99.5	0.27	53 min

theoretical estimations.[33] However, such calculations predict that, around the so-called balanced point or temperature at which the spontaneous curvature is zero, the activation energy decreases almost to zero. By contrast, we find that the energy of activation is non-zero much above this balanced point ($T = 50°C$ for our system[34]). Such discrepancy at high temperature suggests that some other contributions set the scale of the energy barrier, at least in the case of low surfactant content as we are probing here. One other possibility might be that once the surfactant content is high enough, allowing the formation of a bicontinuous middle phase at the balanced point, the destruction process is entirely different.

We now come back to the scenarios of coalescence which, as already mentioned, may be very different. All of them have been experimentally encountered and are quickly summarised. We propose to distinguish two distinct classes: heterogeneous or homogeneous scenarios. The so-called heterogeneous scenario may be of two types. The first in Figure 8.21 involves the formation of an oil phase at the top of the sample, called 'demixtion growth'. The second in Figure 8.22 (and Figure 8.3)

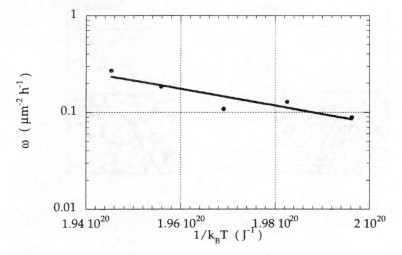

Figure 8.19 *Ln(ω) as a function of $1/kT$ for the emulsion in Figure 8.18. The Arrhenius equation: $ln(\omega) = ln(\omega_0) - E_a/kT$ is represented by the continuous line which is the best fit to our data. From this fit we deduce both $\omega_0 = 4 \times 10^{24}\ m^{-2}\ s^{-1}$ and $E_a = 2 \times 10^{-19}\ J$ which corresponds to $50\ kT_r$, T_r being room temperature (25 °C)*

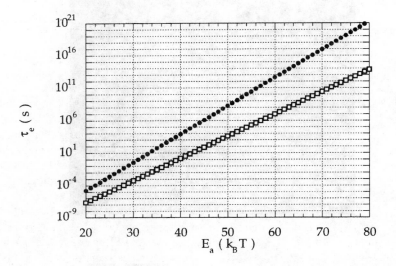

Figure 8.20 *τ_e as a function of E_a for $T = 20\,°C$ (full circles) and $T = 100\,°C$ (empty squares) assuming ω_0 constant and equal to $4 \times 10^{24}\ m^{-2}\ s^{-1}$*

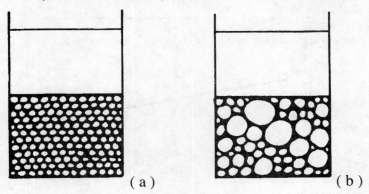

Figure 8.21 *Representation of an heterogeneous scenario of the type 'demixtion growth'.* (a) *and* (b) *involve the formation of an oil phase at the top of the sample*

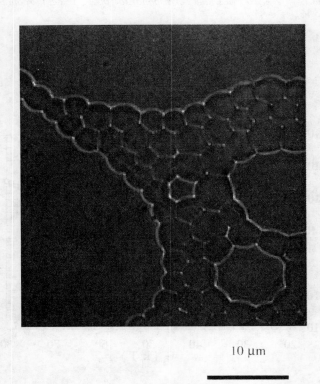

10 μm

Figure 8.22 *Bimodal growth*

corresponds to a bimodal growth where only a few nuclei eat up the initial droplets. The so-called homogeneous scenario may also be of two distinct types: one corresponds to a polydisperse growth involving only one mode as illustrated in Figure 8.23. The other corresponds to a monodisperse growth as previously described (see Figure 8.9). In these latter cases, the destruction proceeds homogeneously throughout the whole sample. We have described throughout this chapter an homogeneous type of growth that occurs when dense emulsions are initially monodisperse in the presence of a relatively low amount of surfactant (about 0.06% in the continuous phase). By extending our experiments, we have found some empirical guide that sets the occurrence of one scenario or the other. Concerning the amount of surfactant C, we come up with the following conclusion: increasing C changes the scenario from heterogeneous (bimodal, Figure 8.22), to homogeneous (monodisperse, Figure 8.9), and polydisperse (Figure 8.23) and finally heterogeneous again (demixtion type, Figure 8.21). Concerning the role of the interfacial tension, we arrive at the following conclusion: increasing γ keeps the scenario homogeneous but changes it from monodisperse (assuming C is initially below 0.06%) to very polydisperse. Finally, we found that traces of ionic surfactant at concentrations as low as 1 ionic molecule for 100 nonionic ones change the scenario from an homogeneous one to an heterogeneous, demixtion type, similar to the one sketched in Figure 8.21(a).

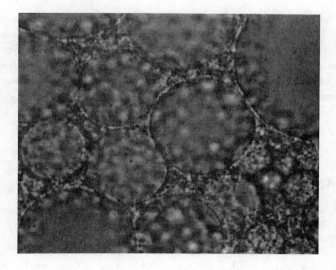

10 μm

Figure 8.23 *Polydisperse growth*

8.8   Conclusion

The destruction of dense emulsions is a rich domain. A large variety of behavior is observed. At the macroscopic scale, the system may exhibit a demixtion, as represented by Figure 8.21, or an homogeneous destruction (the emulsion remains macroscopically homogeneous at any time). At the colloidal scale, the droplet size distribution may change from almost bimodal to polydisperse or very monodisperse. In addition, when the continuous phase spinodally decomposes, one phase may cause severe destruction, as described in Section 8.4.

At this stage of the investigation we aim to relate these various scenarios to microscopic film properties. We believe that bimodal growth (Figure 8.22) or demixtion (Figure 8.21) reflects heterogeneities of the thin film metastability within the dense system. Such differences are certainly caused by some impurities which speed up the coalescence where they sit. By contrast, homogeneous destruction (both polydisperse and monodisperse) reflects the spatial uniformity (homogeneity) of the thin film metastability, which can be therefore characterized by a unique frequency. Monodispersity which persists during the growth certainly originates from a subtle effect whereas polydisperse growth directly originates from the coalescence rate itself and its droplet surface dependence. Monodispersity is obviously due to a feedback effect: to recoalesce again, one bigger droplet has to wait until neighbors have reached the same size. One possible reason for this may be that a freshly formed droplet, because of the local surface reduction, is not instantaneously touching its neighbors, hence the probability of coalescence being temporarily reduced until neighbors rearrange and rebuild a more compact cage around it.

On this basis, one question is unanswered. How much can the lifetime be modified by the presence, number and efficiency of these impurities, when compared to the mean field prediction? Answering this question will allow us to model heterogeneous growths (bimodal or demixtion) reliably and very much extend the knowledge of emulsion stability since these types of destruction are the most frequently encountered. However, we believe the mean field description which is considered here, and its ability to descibe homogeneous growths, gives a firm basis to the theory of emulsion stability.

8.9   References

1. J. Perrin, *C. R. Acad. Sci.*, 1908, **147**, 475; J. Perrin, 'Les Atomes', Gallimard, Paris, 1948, p. 165.
2. H.M. Princen, *J. Colloid Interface Sci.*, 1983, **91**, 160.
3. T.G. Mason, J. Bibette and D.A. Weitz, *Phys. Rev. Lett.*, 1995, **75**, 10.
4. S.S. Davis, J. Hadgraft and K.J. Palin, in 'Encyclopedia of Emulsion Technology', ed. P. Becher, Dekker, New York, 1985, vol. 2, p. 159.
5. W.D. Bancroft, *J. Phys. Chem.*, 1913, **17**, 501; W.D. Bancroft, *J. Phys. Chem.*, 1915, **19**, 275.
6. W.C. Griffin, *J. Soc. Cosmet. Chem.*, 1949, **1**, 311; W.C. Griffin, *J. Soc. Cosmet. Chem.*, 1954, **5**, 249.

7. A. Drecchioni, F. Puisieux and M. Seiller, in 'Galenica, Agents de Surface et Emulsions', eds. F. Puisieux and M. Seiller, Lavoisier, Paris, 1983, vol. 5, p. 155.
8. K. Shinoda and H. Saito, *J. Colloid Interface Sci.*, 1968, **26**, 70; K. Shinoda and H. Saito, *J. Colloid Interface Sci.*, 1969, **30**, 258; H. Saito and K. Shinoda, *J. Colloid Interface Sci.*, 1970, **32**, 647; K. Shinoda and H. Sagitani, *J. Colloid Interface Sci.*, 1978, **64**, 68.
9. W. Ostwald, *Z. Phys. Chem.*, 1901, **37**, 385.
10. I.M. Lifshitz, and V.V. Slyozov, *Sov. Physics JETP.*, 1959, **35**, 331; I.M. Lifshitz and V.V. Slyozov, *J. Phys. Chem. Solids*, 1961, **19**, 35.
11. A.S. Kabalnov, A.V. Pertzov and E.D. Shchukin, *J. Colloid Interface Sci.*, 1987, **118**, 590; A.S. Kabalnov, K.N. Makarov, A.V. Pertzov and E.D. Shchukin, *J. Colloid Interface Sci.*, 1990, **138**, 98.
12. D.J. Durian, D.A. Weitz and D.J. Pine, *Science*, 1991, **252**, 686; D.J. Durian, D.A. Weitz and D.J. Pine, *Phys. Rev. A*, 1991, **44**, 7902.
13. C. Sagui and R.C. Desai, *Phys. Rev. Lett.*, 1995, **74**, 1119.
14. A.S. Kabalnov and E.D. Shchukin, *Adv. Colloid Interface Sci.*, 1992, **38**, 69.
15. D. Exerowa, T. Kolarov and Khr. Khristov, *Colloids Surf.*, 1987, **22**, 171.
16. O. Bélorgey and J.J. Benattar, *Phys. Rev. Lett.*, 1991, **66**, 313.
17. O. Mondain-Monval, F. Leal-Calderon and J. Bibette, *J. Phys. II*, 1996, **6**, 1313.
18. M.P. Aronson and H.M. Princen, *Nature*, 1980, **286**, 370.
19. P. Poulin and J. Bibette, *Phys. Rev. Lett.*, 1997, **79**, 3290.
20. O. Sonneville, Thèse, Université Paris VI, 1997.
21. J.N. Israeachvili, 'Intermolecular and Surface Forces', 2nd edn., Academic Press, London, 1992.
22. A. Scheludko, *Proc. K. Ned. Akad. Wetensch. B*, 1962, **65**, 76.
23. A. Vrij, *Discuss. Faraday Soc.*, 1966, **42**, 23.
24. O. Reynolds, *Philos. Trans. R. Soc. Ser. A*, 1886, **177**, 57.
25. F. Perrot, P. Guenoun, T. Baumberger, D. Beysens, Y. Garrabos and B. Le Neindre, *Phys. Rev. Lett.*, 1994, **73**, 688.
26. L.V. Chernomordik, M.M. Kozlov, G.B. Melikyan, I.G. Abidor, V.S. Markin and Y.A. Chizmadzhev, *Biochim. Biophys. Acta*, 1985, **812**, 643.
27. J. Bibette, D.C. Morse, T.A. Witten and D.A. Weitz, *Phys. Rev. Lett.*, 1992, **69**, 981.
28. P. Becher, 'Emulsions: Theory and Practice', 2nd edn., Krieger, Malabar, 1985.
29. A. Kabalnov and H. Wennerström, *Langmuir*, 1996, **12**, 1931.
30. J. Bibette, *J. Colloid Interface Sci.*, 1991, **147**, 474.
31. B. Deminière, T. Stora, A. Colin, F. Leal Calderon and J. Bibette, 1998, submitted to *Phys. Rev. Lett.*
32. P.R. Garrett, *J. Colloid Interface Sci.*, 1979, **69**, 107; P.R. Garrett, *J. Colloid Interface Sci.*, 1980, **76**, 587; A. Bonfillon-Colin and D. Langevin, *Langmuir*, 1997, **13**, 599.
33. H.N.W. Lekkerkerker, *Physica A*, 1989, **159**, 319; W. Helfrich, *J. Phys. Condens. Matter*, 1994, **6**, A79.
34. A. Kabalnov, personal communication, February 1998.

CHAPTER 9

Molecular Diffusion in Emulsions and Emulsion Mixtures

JEFFRY G. WEERS

Alliance Pharmaceutical Corp., 3040 Science Park Road, San Diego, California 92121, USA

9.1 Ostwald Ripening

Irreversible coarsening in emulsions can occur by two distinct mechanisms: coalescence and Ostwald ripening. Coalescence is the process by which two or more droplets fuse to form a single larger droplet, the fusion being controlled by hole nucleation within the thin films separating droplets. Conversely, Ostwald ripening[1] (also referred to as molecular diffusion or isothermal distillation) is the process by which large droplets grow at the expense of smaller ones due to differences in their chemical potential. In Ostwald ripening the growth occurs by diffusion of the dispersed phase through the continuous phase one molecule at a time.

9.1.1 The Kelvin Equation

The difference in chemical potential or dispersed phase solubility between different sized droplets was first elucidated by Lord Kelvin[2] in 1871, viz:

$$c(r) = c(\infty)\exp\frac{2\gamma^i V_m}{rRT} \approx c(\infty)\left(1 + \frac{2\gamma^i V_m}{rRT}\right) = c(\infty)\left(1 + \frac{\alpha}{r}\right) \quad (9.1)$$

where $c(r)$ is the solubility surrounding a particle of radius r, $c(\infty)$ is the bulk phase solubility, γ^i is the interfacial tension between the dispersed and continuous phases, V_m is the molar volume of the dispersed phase, R is the molar gas constant, and T is the absolute temperature. The quantity $\alpha = 2\gamma^i V_m/RT$ is termed the characteristic length scale. It has an order of ≈ 1 nm or less, indicating that the difference in solubility for a 1 μm droplet is of the order of only 0.1%. Theoretically, Ostwald ripening should lead to the condensation of all droplets into a single droplet (*i.e.* phase separation). This is not observed in practice, however, due to a significant decrease in the rate of growth with increasing droplet size.

9.1.2 Lifshitz–Slezov–Wagner (LSW) Theory

The kinetics of Ostwald ripening is most often described in terms of the theory developed by Lifshitz and Slezov[3] and independently by Wagner[4] (*i.e.* LSW theory). For a detailed derivation of LSW theory the reader is referred to their original manuscripts, and the excellent review by Kabalnov and Shchukin.[5] The LSW theory assumes that: (i) the mass transport is due to molecular diffusion through the continuous phase; (ii) the dispersed phase particles are spherical and fixed in space; (iii) there are no interactions between neighboring particles, *i.e.* the particles are separated by distances much larger than the diameter of the droplets; and (iv) the concentration of the molecularly dissolved species is constant except adjacent to the particle boundaries. It follows then that the rate of Ostwald ripening, ω, is given by:

$$\omega = \frac{d}{dt}(r_c^3) = \frac{8\gamma^i Dc(\infty)V_m}{9RT} f(\phi) = \frac{4Dc(\infty)\alpha}{9} f(\phi) \qquad (9.2)$$

where r_c is the critical radius of a droplet which is neither growing nor decreasing in size, and D is the diffusion coefficient for the dispersed phase in the continuous phase. The factor $f(\phi)$ was not part of the original derivation, and reflects the dependence of ω on the dispersed phase volume fraction, ϕ. Droplets with $r > r_c$ grow at the expense of smaller ones, while droplets with $r < r_c$ tend to disappear.

The unique dependence of droplet growth and shrinkage surrounding the critical radius is best illustrated in the ingenious experiment conducted by Kabalnov *et al.* for 1,2-dichloroethane-in-water emulsions (Figure 9.1).[6] In this experiment, emulsion droplets were fixed to the surface of a glass microscope slide (*i.e.* not allowed to coalesce), and the evolution of the particle size distribution was followed over time. In Figure 9.1a, a double exposure was made on the same frame of film (300 s apart), with the later time appearing in the image to the right. It is apparent that the larger particles grew in size (reflected by the I symbol on the photomicrograph), the smaller ones decreased in size or disappeared (symbol II), while those at the mean radius neither grew nor shrank (symbol III). The time evolution was further reflected in Figures 9.1b–d. Here the three images were taken at 0, 900, and 3600 s, respectively. In these micrographs a quadrangle joining four of the larger droplets is shown to aid in visual comparison. In the early stages, growth of the larger droplets at the expense of the smaller droplets was apparent. At the later time point, some of the droplets, which initially had grown in size, themselves begin to decrease in size and disappear. A schematic representation of the trajectories of individual droplets relative to the critical radius as derived from LSW theory is shown in Figure 9.2.[5] The predicted trajectories are in excellent agreement with the experimental observations.

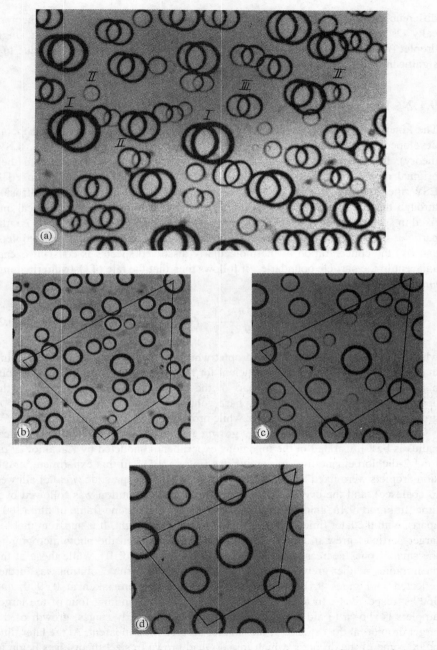

Figure 9.1 *Photomicrographs of 1,2-dichloroethane droplets as a function of time.* (a)
*Double exposure on the same frame, with the image to the left at time zero and
the image to the right at 300 s later.* (b–d) *Time evolution of droplet sizes with
image* (b) *taken at time zero,* (c) *900 s later,* (d) *3600 s later*
(Reproduced by permission of Academic Press from ref. 7)

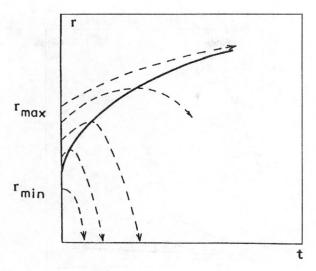

Figure 9.2 *Schematic representation of the trajectories of individual droplets during the course of Ostwald ripening (dashed lines), and changes with time of the critical radius (solid line)*
(Reproduced by permission of Elsevier Science from ref. 5)

One important aspect of LSW theory is the prediction that droplet growth over time will be proportional to $r_c{}^3$. This cubic scaling has been verified repeatedly in laboratories throughout the world.[7-19] An example, taken from the 1,2-dichloroethane study, is shown in Figure 9.3.[7]

Another consequence of LSW theory is the prediction that the size distribution function $g(u)$ for the normalized droplet radius $u = r/r_c$ adopts a time-independent form given by:

$$g(u) = \frac{81 e u^2 \exp[1/(2u/3 - 1)]}{\sqrt[3]{32}(u + 3)^{7/3}(1.5 - u)^{11/3}} \text{ for } 0 < u \leqslant 1.5 \qquad (9.3)$$

and

$$g(u) = 0 \text{ for } u > 1.5 \qquad (9.4)$$

A characteristic feature of the size distribution function is the cut-off at $u > 1.5$. The temporal changes in the particle size distribution of 1,2-dichloroethane-in-water emulsions were obtained by statistical analysis of the photomicrographs shown in Figure 9.1. The results are plotted as the size distribution function, and compared with the theoretical prediction based on LSW theory (Figure 9.4).[6] Although small deviations from LSW theory were noted (*i.e.* the size distribution function broadens slightly), it is clear that the time-independent nature of the size

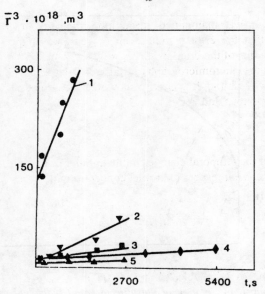

Figure 9.3 *Increase in the cube of the mean radius as a function of time during Ostwald ripening in emulsions of:* (1) *1,2-dichloroethane*; (2) *benzene*; (3) *nitrobenzene*; (4) *toluene; and* (5) p-*xylene*
(Reproduced by permission of Academic Press from ref. 7)

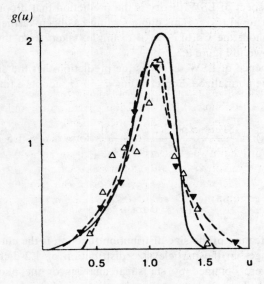

Figure 9.4 *Correspondence between the theoretical function* g(u) *(full line) and the experimentally determined functions obtained for 1,2-dichloroethane droplets at time 0 (open triangles) and 300 s (inverted solid triangles)*
(Reproduced by permission of Academic Press from ref. 6)

distribution function was maintained. In addition, the peak in $g(u)$ occurred near unity, indicating that the critical radius can be reasonably approximated by the number average radius of the droplets, *i.e.* $r_c \approx \bar{r}_n$.

It is clear from the photomicrographs in Figures 9.1 and 9.2 that the number of droplets decreases over time. It is easy to show that the number of droplets per unit volume, n, is related to ϕ and r_c by:

$$n = 3\phi/4\pi r_c^3 \tag{9.5}$$

It follows then that the temporal changes in the number of particles per unit volume will be second order with respect to droplet concentration,[20] *viz*:

$$-\frac{dn}{dt} = \frac{4\pi\omega}{3\phi}n^2 \tag{9.6}$$

9.2 Deviations from LSW Theory

To best illustrate the deviations between LSW theory and experiment, the data derived for hydrocarbon-in-water emulsions will be employed. The significance of Ostwald ripening in the destabilization of hydrocarbon-in-water emulsions was first elucidated by Davis and co-workers.[21,22] They found that emulsion stability depended critically on the chain length of the dispersed hydrocarbon phase. Further, they showed that the emulsions of short-chain hydrocarbons could be stabilized by the addition of a small percentage of long-chain hydrocarbons, in accordance with earlier predictions of Higuchi and Misra[23] for stabilizing emulsions against Ostwald ripening (see Section 9.3).

Shortly thereafter, quantitative comparisons between the experimentally observed growth rates and LSW theory were made by Kabalnov *et al.*[7] Their results for a series of hydrocarbons ranging from *n*-nonane to *n*-hexadecane are shown in Table 9.1. Kabalnov *et al.* showed that emulsion growth rates could be accurately straightened by plots of the cube of the mean number radius with time, in accordance with LSW theory.[7] A dramatic dependence of the observed growth rates on hydrocarbon chain length was observed. Excellent agreement between experiment and LSW theory was observed, although a systematic underestimation of the experimental rates by a factor of *ca.* 2–3 was found.[7] Subsequent experiments in other laboratories have confirmed the cubic scaling and theoretical underestimation of the ripening rates.[9,10,17–19]

9.2.1 Effect of Brownian Motion

The deviation between theory and experiment has been ascribed to the effects of Brownian motion.[7] Recall that LSW theory assumes that the droplets are fixed in space and that molecular diffusion is the only mechanism of mass transfer. In the case of moving particles, the contributions of molecular and convective diffusion are related by the Peclet number, *Pe*, *viz*:

Table 9.1   *Comparison of experimental, ω_e, and theoretical ω_t, Ostwald ripening rates at 25 °C for hydrocarbon-in-water emulsions stabilized by 0.1 M SDS* [a]

Hydrocarbon	$c(\infty)$/ml ml^{-1} [b]	ω_e/cm^3 s^{-1}	ω_t/cm^3 s^{-1} [c]	$\omega_r = \omega_e/\omega_t$
C_9H_{20}	3.1×10^{-7}	6.8×10^{-19}	2.9×10^{-19}	2.3
$C_{10}H_{22}$	7.1×10^{-8}	2.3×10^{-19}	0.7×10^{-19}	3.3
$C_{11}H_{24}$	2.0×10^{-8}	5.6×10^{-20}	2.2×10^{-20}	2.5
$C_{12}H_{26}$	5.2×10^{-9}	1.7×10^{-20}	0.5×10^{-20}	3.4
$C_{13}H_{28}$	1.4×10^{-9}	4.1×10^{-21}	1.6×10^{-21}	2.6
$C_{14}H_{30}$	3.7×10^{-10}	1.0×10^{-21}	0.4×10^{-21}	2.5
$C_{15}H_{32}$	9.8×10^{-11}	2.3×10^{-22}	1.4×10^{-22}	1.6
$C_{16}H_{34}$	2.7×10^{-11}	8.7×10^{-23}	2.2×10^{-23}	4.0

[a] Taken from ref. 7.
[b] Molecular solubilities of hydrocarbons in water taken from: C. McAuliffe, *J. Phys. Chem.*, 1966, **70**, 1267.
[c] For theoretical calculations, the diffusion coefficients were estimated according to the Hayduk–Laudie equation (W. Hayduk and H. Laudie, *AIChE J.*, 1974, **20**, 611) and the correction coefficient $f(\phi)$ was assumed to be equal to 1.75 for $\phi = 0.1$ (P.W. Voorhees, *J. Stat. Phys.*, 1985, **38**, 231).

$$Pe = \frac{rv}{D} \tag{9.7}$$

where v is the velocity of the droplets and can be approximated by $v = (3k_BT/M)^{1/2}$, where k_B is the Boltzmann constant, and M is the mass of the droplets. For $r = 100$ nm, the Peclet number is equal to eight, indicating that the mass transfer will in fact be accelerated with respect to that predicted by LSW theory.[7] Also, recent simulation studies indicate that thermal and convective contributions can significantly enhance the growth rate without affecting the fundamental nature of the Ostwald ripening process.[24] Further, the broadening of the time-independent particle size distribution observed experimentally (*e.g.* for 1,2-dichloroethane droplets) can reasonably be ascribed to stochastic effects on the coarsening due to thermal noise.[25]

9.2.2   Effect of Disperse Phase Volume Fraction

The LSW theory assumes that there are no interactions between droplets and, therefore, is limited to low disperse phase volume fractions. At higher volume fractions the rate of ripening is dependent on the interaction between diffusion spheres of neighboring particles. In general it is expected that emulsions with higher volume fractions of disperse phase will have broader particle size distributions and faster absolute growth rates than those predicted by LSW theory. This has in fact been verified experimentally for cobalt grains.[26] The volume fraction dependence is included in LSW theory simply as a multiplication factor, $f(\phi)$. A number of theoretical studies have examined $f(\phi)$.[27-34] Analytical solutions are not available; hence only numerical solutions have been proposed. The most discussed of these has been the theory of Enomoto *et al.*[27] Their predicted volume fraction dependence is shown in Figure 9.5. The rate is predicted to increase by a factor of 2.5 in the range of $\phi = 0.01$ to 0.3.

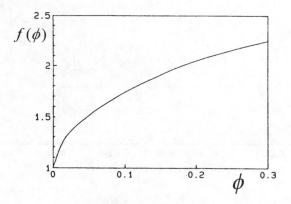

Figure 9.5 *Plot of the correction factor,* f(ϕ), *to LSW theory, which takes into account the volume fraction dependence of the growth rate. The dependence shown is given by Enomoto* et al.[27]
(Reproduced by permission of Elsevier Science from ref. 9)

The predicted volume fraction dependence has not been observed for hydrocarbon-in-water emulsions.[9] Instead, the experimentally determined ripening rates are found to be independent of volume fraction for $0.01 \leqslant \phi \leqslant 0.3$, in clear disagreement with Enomoto's theory (Figure 9.6). Taylor's studies[9] were conducted at a fairly high surfactant concentration (5%). It has been suggested that emulsion droplets may have been effectively screened from one another by micelles present in the solution.[9]

A strong dependence on volume fraction has been observed for fluorocarbon-in-water emulsions. For example, Ni *et al.* found a *ca.* three-fold increase in ω in going from $\phi = 0.08$ to 0.52 for concentrated emulsions of perfluorooctyl bromide.[35] Trevino *et al.* also observed the dependence of ω on ϕ, and in addition verified the cubic scaling and time independence of the distribution function for similar concentrated perfluorooctyl bromide emulsions.[12]

9.2.3 Role of Micelles in Facilitating Molecular Diffusion

Several recent studies have focused on the role that micelles play in mediating the mass transfer between emulsion droplets, the hypothesis being that micelles may facilitate mass transfer by acting as carriers of oil molecules. Karaborni and co-

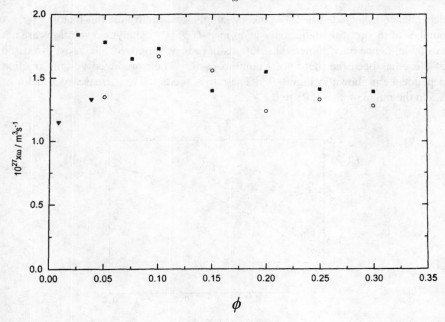

Figure 9.6 *Variation in the rate of Ostwald ripening for decane-in-water emulsions as a function of decane volume fraction. The SDS concentration is constant at 0.174 M. The three sets of points refer to three independent sets of data* (Adapted by permission of Elsevier Science from ref. 9)

workers used molecular dynamics simulations to examine how oil molecules are transferred from a bulk oil phase (*e.g.* emulsion droplets) to micelles.[36,37] Their results indicate that there are at least three mechanisms for this process: (i) oil molecules are transferred *via* direct droplet/micelle collisions; (ii) oil molecules exit the oil droplet and are trapped by micelles in the immediate vicinity of the droplet; (iii) oil molecules exit the oil droplet collectively with a large number of surfactant molecules to form a micelle. Mechanism (iii) will not be considered here since it can account only for the transfer of oil to micelles and not the eventual transfer back to larger emulsion droplets.

In mechanism (i), the micellar contribution to the rate of mass transfer is directly proportional to the number of droplet/micelle collisions, hence to the volume fraction of micelles in the solution. In this case, the molecular solubility term found in the LSW theory (eqn. 9.2) is replaced by the micellar solubility. Since micelles can solubilize thousands of times more oil than the aqueous phase, large increases in the rate of mass transfer would be expected with increases in micelle concentration.[8-10,17,38] Numerous studies indicate, however, that the presence of micelles affects the mass transfer to only a small extent (at least for the case of ionic surfactants).[8-10,17-19,38] This is illustrated for Ostwald ripening in decane-in-water emulsions stabilized by SDS from a recent study of Soma and Papadopoulos (Figure 9.7),[17] who found only a two-fold increase in ω above the CMC. These

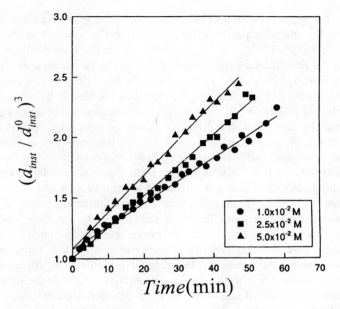

Figure 9.7 *Plot of the normalized cube of the z-average diameter (where d_{inst} = diameter at time t and d^0_{inst} = diameter at time 0) for decane-in-water emulsions as a function of time for different concentrations of SDS above the critical micelle concentration*
(Reproduced by permission of Academic Press from ref. 17)

results are consistent with the earlier studies of Kabalnov[8] and of Taylor,[9] who found increases in the mass transfer rate of only 2–5 times with increasing micelle concentration.

The lack of a strong dependence of the mass transfer on micelle concentration for ionic surfactants may result from electrostatic repulsions between the emulsion droplets and micelles, which would provide a high energy barrier preventing droplet/micelle collisions.[8] Taisne *et al.* estimated that the disjoining pressure between a micelle and emulsion droplet would be of the order of 10^4 Pa, indicating little probability for a direct collisional mechanism.[39] Karaborni *et al.* also noted in their molecular dynamics simulations that the collision mechanism is more likely with nonionic surfactants and less likely with ionic surfactants.[36]

In mechanism (ii), a micelle in the vicinity of an emulsion droplet rapidly takes up dissolving oil from the continuous phase. The 'swollen' micelle then diffuses away to another droplet, where the oil is redeposited. Such a mechanism would be expected to result in an increase in the rate of mass transfer over and above LSW theory by a factor Φ given below:[8]

$$\Phi = 1 + \frac{\phi_s \Gamma D_m}{D} = 1 + \frac{\chi^{eq} D_m}{c^{eq} D} \tag{9.8}$$

where ϕ_s is the volume fraction of micelles in solution, $\chi^{eq} = \phi_s c_m^{eq}$ is the net oil solubility in the micelles per unit volume of micellar solution reduced by the density of the solute, $\Gamma = c_m^{eq}/c^{eq} \approx 10^6 - 10^{11}$ is the partition coefficient for the oil between the micelle and bulk aqueous phase at the saturation point, and D_m is the micellar diffusivity (*ca.* 10^{-6} to 10^{-7} cm^2 s^{-1}). For a decane-in-water emulsion in the presence of 0.1 M SDS, eqn. (9.8) predicts an increase in the rate of ripening by three orders of magnitude, in sharp contrast to experimental observations.

A variation in the molecular diffusion theme was proposed by Miller.[40] He suggested that a 'prohibited zone' exists around the emulsion droplets due to electrostatic repulsions with the micelles. Mass transport within this zone can be accomplished only by molecular diffusion. Such a process would be expected to have a ripening rate lower than that predicted by eqn. (9.8); however, it would be expected that the average particle size would show a quadratic scaling with time.[8] As discussed, experimental studies have shown a cubic scaling, and Miller's mechanism, although important, is not likely to be rate determining.

To reconcile the apparent mismatch between theory and experiment it is important to consider the kinetics of the micellar solubilization process. Kabalnov thought it likely that the micelles are not in equilibrium with the droplets.[8] He proposed that the rate of oil monomer exchange between oil droplets and micelles is slow, and rate-determining. Thus, at low micelle concentrations, only a small proportion of the micelles are able to rapidly solubilize the oil. This leads to a small but measurable increase in the ripening rate with micellar concentration.

Taylor and Ottewill proposed that micellar dynamics may also be important.[10] According to the model of Aniansson *et al.*,[41] micellar growth occurs in a step-wise fashion and is characterized by two relaxation times, the short relaxation time, τ_1, which is related to the transfer of monomers in and out of the micelle, and τ_2, the long relaxation time, which is the time required for break-up and reformation of the micelle. The long relaxation time has been examined for SDS micelles and is highly dependent on total concentration. At low concentrations (*ca.* 0.05 M), $\tau_2 \approx 0.01$ s, while at 0.2 M, $\tau_2 \approx 6$ s.[42] Taylor and Ottewill[10] suggest that, at low concentrations, τ_2 may be fast enough to have an effect on the rate of ripening, but that at 5% SDS in swollen micelles the average lifetime may be as long as 1000 s (taking into account the effect of the solubilizate on τ_2), too long to have a significant effect on the oil transfer rate.

Recently, Kabalnov and Weers have suggested that micelles may simply compress the diffusion path in a manner analogous to the effect of electrolytes in compressing the electrical double layer.[43] This will be discussed in more detail in the section on composition ripening.

Of further interest was the observation by Taylor[9] that the ripening rate reached a plateau at *ca.* 0.174 M SDS, and actually decreased at still higher concentrations. Taylor explained the maximum in rate in terms of variations in the micellar diffusion coefficient. Stigter *et al.*[44] measured the self-diffusion coefficient of SDS micelles as a function of concentration using a dye-tracer technique, and found that the diffusion coefficient fell nearly two-fold in going from 0.34% to 5% w/v SDS. Thus, increasing micellar concentrations and/or micellar shape changes progressively hinder micelle diffusion in solution. Taylor[9] found that in making corrections for the changing diffusion coefficient a roughly linear plot of the corrected ripening rate *vs.* SDS concentration could be attained at low SDS concentrations. The plots still tended to plateau at higher concentrations, indicating that changes in the diffusion coefficient alone were not able to account for the surprising dependence of the ripening rate on micellar concentration.

Although numerous studies have been published on the effect of micellar SDS, no systematic studies have been published to date examining variations in the rate of Ostwald ripening in the presence of nonionic surfactant micelles. Much larger increases in the rate might be expected due to the larger solubilization capacities and absence of electrostatic repulsions between emulsion droplets and micelles. A preliminary study by Weiss *et al.*[45] suggests that significant increases in median diameter can be achieved for tetradecane-in-water emulsions from *ca.* 0.3 μm to 1.7 μm over a period of hours. Little growth is noted for the same tetradecane-in-water emulsions when diluted in water. The experiment involves placing emulsion into an aqueous solution containing 2% of the nonionic surfactant Tween 20. In an experiment of this sort, one has to contend not only with the Ostwald ripening process, but also with the solubilization of oil into the Tween 20 micelles. The driving force for the solubilization process is the tendency for surfactant mono-layers to adopt their optimum curvature, that is for the curvature of the surfactant monolayer to be equivalent to the spontaneous curvature. This is determined principally by the geometry of the surfactant molecule and the need to minimize the contact between polar and nonpolar regions. In the presence of emulsion droplets, the surfactant micelles will incorporate oil molecules until they achieve their optimum curvature.

Both the maximum solubilization capacity and the rate of solubilization depend critically on the hydrocarbon chain length.[46,47] Dramatic differences in micelle facilitated emulsion growth rates were noted as a function of hydrocarbon chain length.[45] The rather astounding increase in droplet diameter noted for the tetra-decane-in-water emulsions suggests that the rate of Ostwald ripening has been dramatically increased.[45] More careful kinetic studies are required, however, to ascertain the mechanism of the mass transfer.

Recent unpublished studies by Cabane and co-workers have found that the rate of Ostwald ripening may be increased dramatically for alkane-in-water emulsions in the presence of nonionic surfactant micelles as the 'balanced state' is approached.[48]

9.3 Stabilization of Emulsions with Respect to Ostwald Ripening

9.3.1 Ostwald Ripening in Emulsions Containing Two Disperse Phase Components

To counteract emulsion growth *via* Ostwald ripening, Higuchi and Misra[23] proposed the addition of a second disperse phase component which is insoluble in the continuous phase. In this case, significant partitioning between different droplets is predicted, with the component having low solubility in the continuous phase being expected to be concentrated in the smaller droplets. During Ostwald ripening in two-component disperse phase systems, equilibrium is established when the difference in chemical potential between different sized droplets, which results from curvature effects, is balanced by the difference in chemical potential resulting from partitioning of the two components (similar to Raoult's law for vapor/liquid equilibria). Higuchi and Misra[23] derived the following expression for the equilibrium condition, wherein the chemical potential of the medium soluble component, $\Delta\mu_1$, is equal for all of the droplets in a polydisperse medium:

$$\Delta\mu_1/RT = (\alpha_1/r_{eq}) + \ln(1 - X_{eq2}) = (\alpha_1/r_{eq}) - X_{02}(r_0/r_{eq})^3 = constant \quad (9.9)$$

where $\Delta\mu_1 = \mu_1 - \mu_1^*$ is the excess of the chemical potential of the first component with respect to the state μ_1^* when radius r is infinite and $X_{02} = 0$, r_0 and r_{eq} are the radii of an arbitrary droplet under initial and equilibrium conditions, respectively, X_{02} and X_{eq2} are the initial and equilibrium mole fractions of the medium insoluble component 2, and α_1 is the characteristic length scale of the medium soluble component, 1.

The equilibrium determined by eqn. (9.9) is stable if the derivative of $\partial\Delta\mu_1/\partial r_{eq}$ is greater than zero for all droplets in a polydisperse system. Based on this the following stability criterion was derived by Kabalnov *et al.*[49]

$$X_{02} > 2\alpha_1/3d_0 \quad (9.10)$$

where d_0 is the initial droplet diameter. If the stability criterion is met for all droplets, two patterns of growth will result, depending upon the solubility characteristics of the secondary component. If the secondary component has zero solubility in the continuous phase, then the size distribution will not deviate significantly from the initial one, and the growth rate will be equal to zero. In the case of limited solubility of the secondary component, the distribution is governed by rules similar to LSW theory (*i.e.* the distribution function is time-invariant), and the mixture growth rate, ω_{mix}, can be approximated by:[49]

$$\omega_{mix} = (\phi_1/\omega_1 + \phi_2/\omega_2)^{-1} \quad (9.11)$$

where ϕ is the volume fraction, and the subscripts 1 and 2 denote the medium soluble component and medium insoluble component, respectively. If the stability criterion is not met, a bimodal size distribution is predicted to emerge from the initially unimodal one. Since the chemical potential of the soluble component is predicted to be constant for all droplets, it is also possible to derive the following equation for the quasi-equilibrium of component 1:

$$X_{02} + 2\alpha_1/d = constant \quad (9.12)$$

where d is the diameter at time t. The theory for two-component dispersed phases has recently been examined experimentally.[16,50]

Weers and Arlauskas[16] utilized sedimentation field-flow fractionation (SdFFF) coupled with gas chromatography (GC) to study the predicted component partitioning. SdFFF utilizes an applied external field to order droplets according to their size, with the larger, slower diffusing droplets accumulating near the wall of a thin ribbon-like channel.[51] The ordered array is then subjected to a laminar flow of a mobile phase so that the particles near the wall move slowly compared to those in mid-stream. This combination of field and flow leads to separation, with the smaller droplets eluting the channel first. The elution volume of the sample is directly proportional to the effective mass of the constituent particles by well established equations, provided that the densities of the particles and mobile phase are known.[51] SdFFF is a unique particle sizing technique in that it is possible to collect monosized fractions of droplets across the entire particle size distribution, and subsequently use another technique to analyze them. In their study examining Ostwald ripening in two-component disperse phases, Weers and Arlauskas[16] used GC to analyze the fluorocarbon composition of monodisperse droplet fractions. The SdFFF fractogram measured for a 90% w/v fluorocarbon-in-water emulsion containing a mixture of perfluorooctyl bromide, PFOB, and perfluorodecyl bromide, PFDB ($\phi_{PFDB} = 0.1$) is shown in Figure 9.8a. Also plotted is the mole fraction of PFDB, X_{PFDB}, present in monosized droplet fractions as determined by GC. Qualitatively, it is clear that the smaller droplets are indeed enriched in the slower diffusing PFDB component, while the larger droplets are enriched in the faster diffusing PFOB component. Kabalnov *et al.*[49] predicted, in their theory for two-component disperse phases, that a plot of X_{PFDB} *vs.* $1/d$ will yield a straight line with a slope given by $2\alpha_{PFOB}$ (Figure 9.8b). The agreement between the theoretical (2α) and the experimentally determined slope (m) is quite good.

In two other studies the predicted evolution of a bimodal particle size distribution was examined.[16,50] Kabalnov *et al.*[50] examined hexane-in-water emulsions stabilized by varying levels of hexadecane ($X_{02} = 0.1, 0.01, 0.001$). For the higher mole fractions, the emulsion growth rates were reliably predicted by eqn. (9.11). Also, these emulsions had a physical appearance identical to that of an emulsion containing only hexadecane, *i.e.* they did not cream. The $X_{02} = 0.001$ sample behaved quite differently, however. Upon storage, this emulsion quickly separated into two layers: a sedimented layer with a droplet size of *ca.* 5 μm and a dispersed population of submicron droplets (*i.e.* a bimodal distribution). The 5 μm droplets were equivalent to the droplet size observed for a fresh perfluorohexane emulsion prepared under similar conditions. In light of the fact that the stability criterion was not met for this mole fraction of hexadecane, the observed bimodal distribution of droplets is predictable. A bimodal size distribution was also observed by the SdFFF technique by Weers and Arlauskas[16] in PFOB-in-water emulsions stabilized by low levels of PFDB.

9.3.2 Effect of the Interfacial Layer in Slowing Ostwald Ripening

According to LSW theory, the presence of surfactants at the oil–water interface has a direct effect on the rate of Ostwald ripening, *via* reductions in the interfacial tension.[3–5] This has been illustrated quite nicely for hydrocarbon-in-water emul-

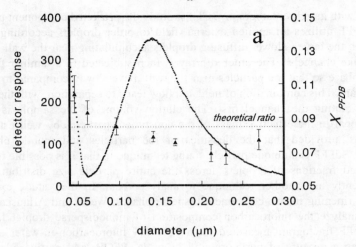

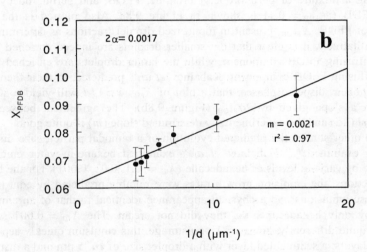

Figure 9.8 (a) *Fractogram obtained for a 90% w/v PFOB/PFDB-in-water emulsion*
($\phi_{PFDB} = 0.1$) by sedimentation field-flow fractionation. The fractogram was
measured immediately after preparation and prior to terminal sterilization. The
right ordinate (given by solid triangles in the plot) represents the mole fraction
of PFDB, X_{PFDB}, in various monosized droplet fractions as determined by gas
chromatography. (b) *Plot of X_{PFDB} vs. $1/d$ for the emulsion in (a). The measured*
slope, m, is in close agreement with the theoretically predicted 2α value
(Reproduced by permission of the American Chemical Society from ref. 16)

sions stabilized by varying levels of SDS below the critical micelle concentration
(CMC), where γ^i decreases proportionally with increasing surfactant concentration
(Table 9.2).[8,9,17] Excellent linear correlations have also been observed for a number
of emulsifiers in a plot of ω *vs.* γ^i.[52]

Table 9.2 *Comparison of the experimentally observed Ostwald ripening rates for decane-in-water emulsions with the theoretical rates at different surfactant (SDS) concentrations*[a]

$[SDS]^b$/M	ω_e/cm^3 s^{-1}	ω_t/cm^3 s^{-1}	$\omega_r = \omega_e/\omega_t$
0.0	2.50×10^{-18}	1.62×10^{-19}	15.4
1.0×10^{-4}	4.62×10^{-19}	1.59×10^{-19}	2.9
5.0×10^{-4}	4.17×10^{-19}	1.35×10^{-19}	3.1
1.0×10^{-3}	3.68×10^{-19}	1.17×10^{-19}	3.2
5.0×10^{-3}	2.13×10^{-19}	5.41×10^{-20}	3.9
1.0×10^{-2}	2.06×10^{-19}	2.82×10^{-20}	7.3
2.5×10^{-2}	2.87×10^{-19}	2.82×10^{-20}	10.2
5.0×10^{-2}	3.64×10^{-19}	2.82×10^{-20}	12.9

[a] Taken from ref. 17.
[b] The critical micelle concentration (CMC) of SDS occurs at 8×10^{-3} M. The surface tension varies with SDS concentration below the CMC and is constant above, leading to the changing theoretical rate below the CMC.

Several groups have proposed that the adsorbed interfacial layer may be able to provide additional contributions to slow Ostwald ripening.[53–57] For example, Walstra[53] hypothesized that emulsions could be effectively stabilized with respect to Ostwald ripening by the use of emulsifiers which are (i) strongly adsorbed to the interface and (ii) do not desorb during the ripening process. In this case, an increase in the surface dilational modulus and decrease in the interfacial tension would be observed for the shrinking droplets. Eventually the difference in γ^i between droplets would balance the difference in capillary pressures (*i.e.* curvature effects), leading to a quasi-equilibrium state. In this case, emulsifiers with low solubilities in the continuous phase (*e.g.* solid particles) would be preferred. Although no systematic studies of this type have been made on emulsions, Prins and co-workers have shown the importance of the concept in studies at the gas/liquid interface.[58,59] Decreases in the disproportionation of gas bubbles by an order of magnitude were achieved by the 'irreversible' adsorption of proteins.[59]

Long-chain phospholipids are natural surfactants with a very low solubility in water (CMC $\approx 10^{-10}$ M). Excellent stabilization of fluorocarbon-in-water emulsions can be achieved with phospholipids, perhaps better than might be expected from interfacial tension effects alone. Kumacheva *et al.*[55] argued that Ostwald ripening in phospholipid-stabilized emulsions may be controlled by molecular diffusion of the phospholipid, and not the fluorocarbon. Kabalnov *et al.*[52] were able to show that in order for the molecular diffusion of phospholipid to be rate limiting, the phospholipid would have to have a solubility in water which is three orders of magnitude lower than that of the oil. Thus, in most instances, the kinetics are oil-diffusion controlled. Moreover, an interfacially controlled ripening process would be accompanied by a quadratic scaling with time of the average radius, in contrast to the cubic scaling observed experimentally.[52] A recent study by Kabalnov *et al.* has shown that the decreased ripening rates observed for phospholipid stabilized emulsions can be accounted for simply by reductions in γ^i.[52]

9.3.3   Other Ways of Slowing Ostwald Ripening

Walstra[53] also proposed that Ostwald ripening could be stopped if a suitable gelling agent was introduced into the continuous phase. In this context 'suitable' is an agent which introduces a yield stress which is larger than the Laplace pressure of the droplets. This stabilization method is not considered practical, however, since even for a droplet of radius 100 nm and $\gamma^i = 10$ mN m^{-1}, a yield stress of 2×10^5 Pa would be required, a value not readily attainable with most gelling agents.

Perhaps the easiest way to slow molecular diffusion in emulsions, provided it does not cause irreversible damage, is to freeze them. Freezing will also stop sedimentation, aggregation, and coalescence of the droplets as well.

9.4   Effect of Ostwald Ripening on Initial Droplet Size

The importance of Ostwald ripening in determining initial droplet size in sub-micron miniemulsions can be realized by examining the particle size dependence of the characteristic time, τ_{OR}, for Ostwald ripening, *viz:*

$$\tau_{OR} \approx r^3/\alpha c(\infty)D \approx r^3/\omega \tag{9.13}$$

Values of τ_{OR} when $r = 100$ nm are tabulated for various hydrocarbon oils in Table 9.3. The dramatic dependence of τ_{OR} on hydrocarbon chain length is readily apparent. What is not as apparent from Table 9.3 is the logarithmic dependence of τ_{OR} on r_c. Figure 9.9 shows that Ostwald ripening can be extremely rapid for small droplet sizes, thereby providing a key component in determining initial droplet size. For example, it is not likely that droplet sizes less than 100 nm will be observed for decane-in-water emulsions since the droplets would ripen to this size on the timescale of a few minutes. This was confirmed by Kabalnov *et al.*,[7] who noted large differences in initial droplet size for their hydrocarbon-in-water emulsions as the chain length of the hydrocarbon was decreased. For instance,

Table 9.3   *Characteristic time for Ostwald ripening in hydrocarbon-in-water emulsions stabilized by 0.1 M SDS*[a]

Hydrocarbon	ω_e/cm^3 s^{-1}	$\tau_{OR} \approx (r^3/\omega_e)$[b]
C_9H_{20}	6.8×10^{-19}	25 min
$C_{10}H_{22}$	2.3×10^{-19}	73 min
$C_{11}H_{24}$	5.6×10^{-20}	5 h
$C_{12}H_{26}$	1.7×10^{-20}	16 h
$C_{13}H_{28}$	4.1×10^{-21}	3 d
$C_{14}H_{30}$	1.0×10^{-21}	12 d
$C_{15}H_{32}$	2.3×10^{-22}	50 d
$C_{16}H_{34}$	8.7×10^{-23}	133 d

[a] Taken from ref. 7.
[b] Characteristic lifetimes measured for $r = 100$ nm.

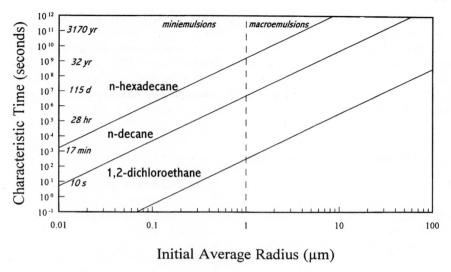

Figure 9.9 *Plot of the characteristic time for Ostwald ripening* vs. *droplet size*

nonane-in-water emulsions had an initial droplet size of 178 nm, decane-in-water emulsions had a size of 124 nm, and undecane-in-water emulsions a size of 88 nm. It is also apparent from Figure 9.9 that the driving force for Ostwald ripening decreases dramatically with increasing droplet size. This is why phase separation is not observed for Ostwald ripening controlled droplet growth. For the most part, Ostwald ripening is an important destabilization mechanism only for submicron droplet populations.

The importance of Ostwald ripening in determining the initial particle size was examined in two other experimental studies. Weers *et al.*[15] examined the role of ripening in determining the initial droplet size in fluorocarbon-in-water emulsions stabilized by egg yolk phospholipids. Whereas droplet sizes of approximately 0.2 μm were achievable with the eight-carbon perfluorooctyl bromide molecule as the dispersed phase, less than 0.1 μm droplets were obtained for mixtures of perfluorooctyl bromide with its less water soluble ten-carbon homolog, perfluoro-decyl bromide. Perfluorodecyl bromide, being virtually insoluble in the continuous phase, served to inhibit molecular diffusion by reducing the water solubility of the dispersed phase in small droplets of the polydisperse dispersion. This in effect was equivalent to shifting the curves in Figure 9.9 between decane and hexadecane.

Taylor and Ottewill[38] examined miniemulsions formed by dilution of oil-in-water microemulsions composed of SDS/dodecane/pentanol/water. The microemulsion had a droplet size of less than 20 nm. Following dilution from the balanced state, the microemulsion droplets were no longer thermodynamically stable and tended to grow rapidly to a new quasi-equilibrium. The observed growth can occur by either a coalescence or Ostwald ripening mediated process. In the case of the dodecane microemulsion described above, growth occurred by ripening. Growth to *ca.* 70 nm occurred quickly, followed by a slow growth with cubic scaling of the average radius over time.

Ostwald ripening also has an effect on the initial particle size distribution. As the steady state equilibrium is approached, the particle size distribution will begin to order into the time-independent distribution given by eqns. (9.3) and (9.4). Thus, an emulsion which was initially nearly monodisperse will broaden over time, while a broad distribution will actually narrow over time.

9.5   Composition Ripening

Mass transfer in emulsions is driven not only by differences in droplet curvature, but also by differences in their composition. This is observed in most instances when two oils, A and B, are emulsified separately and the resulting emulsions are mixed. Mass transfer from emulsion A to emulsion B and *vice versa* will occur until the compositions of the droplets are identical.

Perhaps the best evidence of composition ripening in emulsions comes from the so-called 'reverse recondensation' experiments performed by Pertsov *et al.*[60] These experiments involved the preparation of two emulsions which differed significantly both in their initial size and in their rate of molecular diffusion. Pertsov *et al.*[60] were able to show that if the larger sized emulsion was composed of the faster diffusing oil, then molecular diffusion would occur in the 'reverse' direction (*i.e.* from large to small drops). Such a disappearance of the large droplet fraction cannot occur by a coalescence mediated process. The driving force for the 'composition ripening' is the free energy gain associated with oil mixing.

Arlauskas and Weers[61] used SdFFF/GC to study the reverse recondensation process. The results of a typical experiment are shown in Figure 9.10. In this study, separate emulsions of 47 vol% bis(perfluorohexyl)ethene (F66E, $c(\infty) \approx 1 \times 10^{-12}$ ml/ml) and perfluorohexyl bromide (PFHB, $c(\infty) \approx 8 \times 10^{-8}$ ml/ml) were prepared. The initial average drop sizes of the PFHB and F66E emulsions were 0.49 and 0.18 μm, respectively. Within 10 minutes of mixing equivolumes of the two emulsions, the drop size was again determined by SdFFF. None of the *ca.* 0.5 μm droplets remained, and the mean particle size was shifted to 0.22 μm (*i.e.* reverse recondensation had occurred). The increase in particle size to 0.22 μm reflects that all of the volume of the PFHB has been incorporated into the F66E droplets, *i.e.* $d = 0.18(\sqrt[3]{2})$. The composition of the mixed emulsion as a function of particle size was determined by SdFFF/GC, and the results are also plotted in Figure 9.10 (inset). Equal amounts of F66E and PFHB were measured across the entire particle size distribution, further proof that composition ripening had in fact occurred.

In the same study, Arlauskas and Weers[61] examined composition ripening for several pairs of oils. The characteristic time, τ_{CR}, for molecular diffusion between emulsions can be estimated by:

$$\tau_{CR} \approx r^2/c(\infty) \qquad (9.14)$$

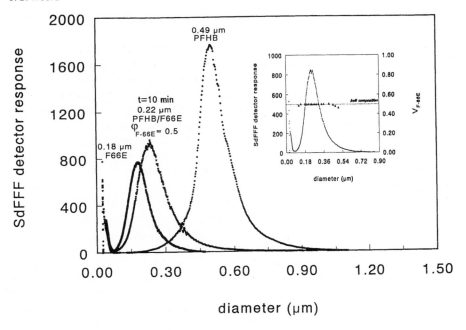

Figure 9.10 *SdFFF fractograms obtained for a 47% v/v F66E emulsion, a 47% v/v PFHB emulsion, and an equivolume mixture of the two, 10 minutes after mixing. Diffusional mass transfer occurs from the PFHB droplets, which have moderate water solubility, to the F66E droplets, which have low water solubility. The inset shows that equivolume ratios of the two fluorocarbons (as determined by gas chromatography) are found across the entire particle size distribution (triangles, right ordinate where V_{F66E} is the volume fraction of F66E in the droplets) for the mixed emulsion sample. The dashed line represents the predicted composition for the case where all of the PFHB has diffused into the F66E droplets (i.e. a 50/50 bulk ratio)*
(Reproduced by permission of the American Chemical Society from ref. 61)

For the F-66E/PFHB emulsion, τ_{CR} was estimated to be about 12 min, in close agreement with the rapid mass transfer observed experimentally. Excellent agreement with experiment was also observed for the molecular diffusion of perfluorooctyl bromide ($\tau_{CR} \approx 3$ h), hexadecane ($\tau_{CR} \approx 520$ h), and the absence of ripening observed for triolein ($\tau_{CR} \approx 4$ years) emulsions.

The initial rate of composition ripening, Λ, was derived by Taisne *et al.*,[39] *viz:*

$$\Lambda = \frac{3Dc(\infty)}{2r^2} \exp \frac{\alpha}{r} \qquad (9.15)$$

The composition of emulsion droplets during composition ripening processes has also been determined by changes in the crystallization temperatures of the oils using differential scanning calorimetry (DSC),[62–64] and by differences in turbidity for mixtures of oils with vastly different refractive indices.[39]

A comparison of τ_{CR} and τ_{OR} for various hydrocarbon oils is shown in Table 9.4. It is clear that the characteristic time for Ostwald ripening is two to three orders of magnitude longer than that of composition ripening. The differences in characteristic times are governed by the driving force for the mass transfer. Whereas composition ripening is controlled by the entropy of mixing, Ostwald ripening is driven by small differences in capillary pressures (Kelvin effect), the key difference between the two being the characteristic length scale present in the Ostwald ripening equation. It has an order of $\alpha \approx 10^{-7}$ cm, effectively decreasing the rate of the Ostwald ripening process relative to that of composition ripening.

Table 9.4 *Characteristic times for composition ripening, τ_{CR}, and Ostwald ripening, τ_{OR}, in hydrocarbon-in-water emulsions stabilized by 0.1 M SDS* [a]

Hydrocarbon	ω_e/cm^3 s^{-1}	$\tau_{OR} \approx (r^3/\omega_e)$ [b]	$\tau_{CR} \approx r^2/c(\infty)D$
C_9H_{20}	6.8×10^{-19}	25 min	1 min
$C_{10}H_{22}$	2.3×10^{-19}	73 min	4 min
$C_{11}H_{24}$	5.6×10^{-20}	5 h	15 min
$C_{12}H_{26}$	1.7×10^{-20}	16 h	1 h
$C_{13}H_{28}$	4.1×10^{-21}	3 d	4 h
$C_{14}H_{30}$	1.0×10^{-21}	12 d	16 h
$C_{15}H_{32}$	2.3×10^{-22}	50 d	61 h
$C_{16}H_{34}$	8.7×10^{-23}	133 d	230 h

[a] Taken from ref. 7.
[b] Characteristic lifetimes measured for $r = 100$ nm.

Further evidence for the rather minor effect of micelles on molecular diffusion in emulsions comes from composition ripening experiments. McClements and Dungan[62] examined the role of Tween 20 micelles on composition ripening in hexadecane/octadecane emulsion mixtures by monitoring changes in crystallization temperatures of the oils using differential scanning calorimetry (DSC). This is illustrated in Figure 9.11. Increases in the rate of oil exchange by a factor of five were observed with increasing micelle concentration. It should be noted that Tween 20 is a nonionic surfactant. Thus, the minor effect of micelles on molecular diffusion processes cannot be ascribed solely to repulsions between emulsion droplets and micelles.

The mass flux of oil transferring between droplets as a function of Tween 20 concentration was examined theoretically by Kabalnov and Weers.[43] These authors derived the following equation to describe the mass flux:

$$J = D_{oil}\tilde{n}(1/\delta + \kappa) \qquad (9.16)$$

where J is the mass flux of oil in units of molecules cm^{-2} s^{-1}, D_{oil} is the molecular diffusion coefficient of the oil in water, $\tilde{n}_{oil}$ is the concentration of oil molecules in

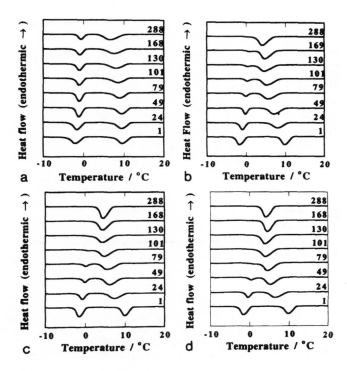

Figure 9.11 *DSC measurements of the dependence of the crystallization behavior of emulsion mixtures over time (1–288 h). A 10% w/w emulsion of hexadecane is mixed in equal volumes with a 10% w/w octadecane emulsion. The different traces reflect the additional amount of Tween 20 which is added after emulsification:* (a) 0%; (b) 1%; (c) 2%; and (d) 5%
(Reproduced by permission of the American Chemical Society from ref. 62)

units of molecules cm^{-3}, δ is the diffusion path, *i.e.* the radius of the spherical droplet, and κ is the reciprocal of the decay length for the steady state diffusion profile.

Figure 9.12 shows a plot of the initial rate of mass transfer *vs.* the square root of the surfactant volume fraction in micelle form.[43] According to the theory, the intercept of the line is equal to: $B = D_{oil} \tilde{n}_{oil}/\delta$, and the ratio of the slope to the intercept $A/B = \delta\sqrt{3}/\lambda$, where λ is the micellar radius. The agreement between the experimentally determined values of B and A/B and theory are good, and support the notion that composition ripening is controlled by molecular diffusion and not by micelle-droplet fusion/scission processes. In short, micelles tend to compress the diffusion path in a manner analogous to how electrolytes compress the electrical double layer. The same arguments apply to Ostwald ripening. Note that if the micellar mass transfer is dominant, $\kappa \gg 1/\delta$, there must be a crossover from cubic to quadratic scaling of the average particle size with time, because the diffusion path is no longer equal to the particle radius but to $1/\kappa$. The experimental

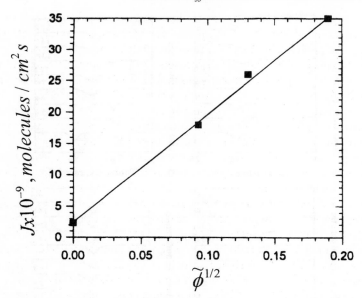

Figure 9.12 *Mass flux of hexadecane transferred into octadecane droplets as a function of the square root of the volume fraction of Tween 20 micelles in solution. Theoretical fit is from the study of Kabalnov and Weers.*[43]
(Reproduced by permission of the American Chemical Society from ref. 62)

data available so far indicate that the micellar effects on Ostwald ripening are rather limited, presumably because the particle size of the systems studied was rather small, so that $\kappa \approx \delta$.[8–10,17–19,38]

9.6 Practical Importance of Molecular Diffusion

9.6.1 Ostwald Ripening in Fluorocarbon Emulsions

Owing to their extraordinary capacity for dissolving gases and their remarkable chemical and biochemical stability, fluorocarbons (FC) are currently being explored as antihypoxic agents in a number of applications.[65] Of note to the current analysis is the use of these materials for intravascular applications ('blood substitutes').[65] In order to avoid producing fatal emboli in the blood, the FC oil must be emulsified to form a FC-in-water emulsion. The formulation of FC emulsions for biomedical applications is not a trivial task because of the large number of constraints imposed by the intended usage. For a detailed discussion of the formulation of FC emulsions for biomedical applications the comprehensive review by Krafft *et al.* should be consulted.[65] One important constraint is that the mean droplet size must be in the range of 50–200 nm, in order to decrease uptake by cells of the reticuloendothelial system (RES) and reduce the particulate related side effects (*e.g.* febrile response, flu-like syndrome) often found for larger particle sized emulsions.[66,67] For submicron FC-in-water emulsions, Ostwald ripening is the predominant mechanism of destabilization.

It was recognized very early by researchers in the blood substitute field that FC emulsion stability depended critically on the nature of the dispersed FC phase.[68] Whereas optically transparent emulsions with a small droplet size (*ca.* 100 nm) could be prepared for both perfluorotributylamine (FTBA) and perfluorodecalin (FDC) oils, the FTBA emulsion remained transparent for at least 1 year, while the FDC emulsion clouded within one day (the droplet size increasing to *ca.* 200 nm over that period). Unfortunately, FTBA was not a suitable candidate for the intended use, owing to a prolonged retention in the organs of the RES ($t_{1/2} = 900$ d).[69] Although it had poor emulsion stability, FDC had an acceptable retention time ($t_{1/2} = 7$ d). Researchers at the Green Cross Corp. (Osaka, Japan) found empirically that the addition of perfluorotripropylamine (FTPA) to perfluorodecalin emulsions improved emulsion stability significantly, to the point where a product containing a 70/30 ratio of FDC/FTPA was commercialized as a synthetic oxygen carrier (Fluosol®) designed to deliver oxygen to hypoxic tissues during invasive percutaneous transluminal coronary angioplasty (PTCA) procedures.[70] Unfortunately, Fluosol did not meet with commercial success. The main reason for its failure was user friendliness: the emulsion was so unstable that it had to be shipped and stored in the frozen state. Prior to injection, Fluosol then had to be thawed and reconstituted by admixing two annex solutions; after which it had to be utilized within eight hours or discarded. In addition to going through this cumbersome procedure, a test-dose had to be given in order to detect those patients who were sensitive to Pluronic F-68, one of the components of the emulsion that can provoke complement activation. Little wonder that surgeons were not inclined to use the product. Also, in spite of the fact that FTPA did not improve emulsion stability to the point where room temperature storage was possible, it still had a fairly long half-life ($t_{1/2} = 65$ d) in the RES.[69] The big impetus for the second generation products was to improve their user-friendliness, which necessitated improving the physical stability of the emulsion, while maintaining fast elimination.

The rate-determining step in elimination of FCs from the body is the dissolution of the FC into circulating lipid carriers.[71-73] As you might imagine, this process depends critically on the lipophilic nature of the FC, with more lipophilic molecules being eliminated more quickly.[74] In order to improve elimination, the solubility in lipids must be maximized. Thus, decreasing the chain length of the FC oil to improve solubility is beneficial. On the other hand, emulsion stability depends critically on decreasing the water solubility of the FC molecule,[75,76] which can be best accomplished by increasing the chain length and/or molecular weight of the FC (Table 9.5). Therein lies the dilemma facing the formulator. It was proposed by Kabalnov *et al.*[77] and independently by Weers *et al.*[14] that the best way to overcome the dilemma was to use a mixture of highly lipophilic FCs. Lipophilic character can be introduced into FCs in two ways: by introducing a highly polarizable halogen atom (*e.g.* Cl or Br) or by the introduction of a hydrocarbon fragment (*e.g.* CH_2CH_3) into the molecule. Figure 9.13 shows a plot of the RES half-life *vs.* FC

Table 9.5 *Emulsion growth rates and water solubilities of some representative perfluorochemicals*[a]

Fluorochemical	$\omega_e{}^b$/cm^3 s^{-1}	$c(\infty)^c$/M
n-C$_5$F$_{12}$	1.4×10^{-18}	4.0×10^{-6}
n-C$_6$F$_{14}$	1.3×10^{-19}	2.7×10^{-7}
n-C$_7$F$_{16}$	1.7×10^{-20}	3.1×10^{-8}
n-C$_8$F$_{18}$	2.5×10^{-21}	3.8×10^{-9}
(structure)	2.1×10^{-19}	5.0×10^{-7}
(structure)	1.6×10^{-21}	2.5×10^{-9}
(structure)	2.6×10^{-19}	6.7×10^{-7}
(structure)	3.6×10^{-20}	7.7×10^{-8}
(structure)	1.4×10^{-21}	2.2×10^{-9}
(structure)	3.6×10^{-22}	5.1×10^{-10}
(structure)	6.0×10^{-19}	2.2×10^{-6}
(structure)	5.3×10^{-21}	9.9×10^{-9}
N(C$_2$F$_5$)$_3$	5.7×10^{-20}	1.2×10^{-7}
N(C$_3$F$_7$)$_3$	1.9×10^{-22}	2.8×10^{-10}
C$_8$F$_{17}$Br	1.9×10^{-21}	5.1×10^{-9}

[a] Taken from ref. 5.
[b] 10% v/v fluorochemical-in-water emulsions stabilized by 0.1 M SDS at 25 °C.
[c] Water solubilities estimated from emulsion growth rates using LSW theory.

molecular weight showing how lipophilic FCs exhibit a shorter RES half-life than would be predicted for their molecular weight.[65]

Weers *et al.*[14] proposed the use of the highly lipophilic PFDB molecule as the stabilizer for PFOB emulsions. Indeed, a small percentage of PFDB is being used to stabilize PFOB emulsions in Oxygent™ (AF0144; Alliance Pharmaceutical Corp, San Diego, CA; Johnson and Johnson, New Brunswick, NJ), a formulation expected to be in Phase III clinical trials shortly. Oxygent is designed to provide oxygen temporarily to tissues during general surgical and cardipulmonary bypass procedures.[78]

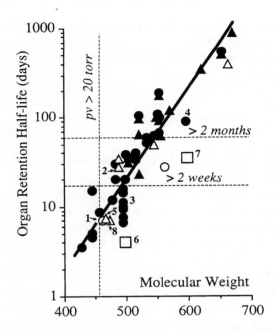

Figure 9.13 *Half-life of fluorochemicals in the organs of the reticuloendothelial system* vs. *fluorochemical molecular weight. Linear fluorochemicals are represented by open triangles, cyclic by open circles, linear with heteroatom by closed triangles, and cyclic with heteroatom by closed circles. Highly lipophilic fluorocarbons exhibit a much shorter half-life than would be predicted for their molecular weight. Fluorocarbons with a vapor pressure,* pv > 20 *torr, are not suitable due to severe pulmonary embolism. 1, perfluorodecalin; 2, perfluoro-tripropylamine; 3, perfluoro-N-methyldecahydroisoquinoline; 4, perfluoro-N-methylcyclohexylpiperidine; 5, bis(perfluorobutylethene); 6, perfluorooctyl bromide; 7, perfluorodecyl bromide; 8, perfluorodichlorooctane*
(Adapted by permission of Raven Press from refs. 74 and 65)

Another formulation based upon the lipophilic perfluorodichlorooctane (PFDCO) molecule has been developed by Hemagen PFC (St. Louis, MO).[79,80] This formulation contains no secondary FC as a stabilizer. The PFDCO is formulated into a 'three-phase emulsion' containing a core of PFDCO surrounded by a layer of triglyceride (*e.g.* safflower oil) in a water continuous phase. Egg yolk phospholipid is present at the triglyceride/water interface. Its improved adhesion at the triglyceride/water interface (relative to the FC/water interface) is thought to provide the driving force for improved physical stability. Unfortunately, this formulation has a fairly large drop size and suffers from a high degree of particle-related side-effects.[79] Nonetheless, it is currently in early Phase II a clinical trials.

Formulations containing small percentages of highly lipophilic fluorocarbon–hydrocarbon diblocks are also being explored.[54,81-83] It has been proposed that the diblock compounds are interfacially active at fluorocarbon/water interfaces and orient in such a way so as to extend their fluorinated portion into the FC phase, and their hydrocarbon portion between the phospholipid acyl chains. In this way they are thought to act as molecular dowels,[54,82,83] anchoring the interfacial phospholipid layer to the FC/water interface. The authors suggest that the 'dowel' molecules impede the molecular diffusion of the FC molecules out of the droplets.[54,82,83] However, if this was indeed the case, quadratic scaling of the average drop size with time would be expected. This is again not observed in practice, suggesting that the kinetics are oil diffusion controlled.

9.6.2 Ostwald Ripening in Triglyceride Emulsions with Added Ethanol

The stability of triglyceride-in-water emulsions with milk proteins has received a considerable amount of attention.[84] Recently, Agboola and Dalgleish[85] examined the stability of such emulsions in the presence of a less polar ethanol solvent. Such emulsions are of practical importance in cream liqueurs. Agboola and Dalgleish[85] showed that, in the presence of ethanol, droplet growth occurred by an Ostwald ripening mechanism. Thus, the solubility of triglycerides is increased dramatically in ethanol/water mixtures.

9.6.3 Miniemulsion Polymerization

In ordinary batch macroemulsion polymerization reactions, monomer macroemulsion droplets ($d = 1-10 \ \mu m$) are in equilibrium with excess surfactant in the form of micelles ($d \approx 0.01 \ \mu m$). The emulsion polymerization reaction can be divided into three intervals (Figure 9.14).[86] In Interval I, nucleation of particles takes place by invasion of radicals from the aqueous phase into micelles or by precipitation of oligomer particles in the aqueous phase outside the particles. Macroemulsion droplets play little role in Interval I owing to the fact that they are of a large size

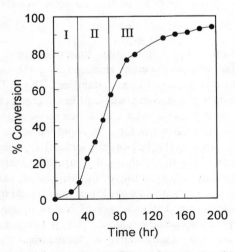

Figure 9.14 *Intervals involved in macroemulsion polymerization (simulated data)*

and low surface area, and cannot effectively compete for the radicals being formed in the aqueous phase. Since most of the monomer is present in the emulsion droplets, conversion in Interval I is low (*ca.* 2–10%). In Interval II, monomer in the macroemulsion droplets diffuses to the nucleated particles. Finally, in Interval III, all of the monomer is now in the polymer particles (*i.e.* there are no remaining macroemulsion droplets), and the polymerization goes to completion.

Because nucleation of particles is often irreproducible, commercial emulsion polymerizations are often seeded with polymer particles of known size and concentration. The seed particles when exposed to monomer in the form of monomer droplets swell to an equilibrium size. Under proper polymerization conditions no new particles form, and the polymerization consists of growing these seed crystals. Thus the polymerization consists of Intervals II and III only.

It is not difficult to imagine potential advantages of using miniemulsions ($d = 0.01-0.5$ µm) in place of macroemulsions for the polymerization.[87–92] Their inherent greater surface area may allow them to compete far more effectively for radicals than macroemulsion droplets. The problem is that typical monomers are of low molecular weight, hence finite water solubility, leading to fast growth of the droplets by molecular diffusion. Ugelstad and co-workers overcame these early difficulties by the addition of a 'cosurfactant' to the dispersed monomer phase.[87] For a detailed discussion of miniemulsion polymerization the reader is referred to the comprehensive review by Ugelstad *et al.*[87] The 'cosurfactant' is a molecule which has limited solubility in the continuous aqueous phase, but which is soluble in the dispersed monomer phase. It is not a cosurfactant in the strict sense, since it generally has no surface activity, but is simply a secondary disperse phase component in the genera of Higuchi and Misra.[23] The secondary disperse phase components used have included hexadecane, hexadecanol, dodecanethiol, and, more recently, monomer soluble polymers. The polymers studied thus far are simply the polymer form of the dispersed monomer phase.[90] Generally, polymers are soluble in their monomers, but insoluble in water, making them an interesting class of stabilizers for this application.

As discussed, miniemulsions are characterized by a rather large surface area, sufficient to alter the principal mechanism of nucleation. In miniemulsion polymerization, nucleation occurs predominantly by radical entry into monomer in the interior of the miniemulsion droplets. In addition, because of the improved physical stability of the miniemulsion droplets, it is possible to adjust the total surfactant concentration so as to limit the total number of micelles in solution, thereby limiting aqueous phase nucleation. For example, Fontenot and Schork[91] studied the batch polymerization of methyl methacrylate with SDS as the surfactant and hexadecane as the secondary disperse phase component. The mechanism of particle nucleation could be altered from droplet control to micellar control as the concentration of SDS was increased. Relative to micelles, droplets can have higher radical numbers (due to their larger size), and can, therefore, significantly enhance the early stages of the polymerization process. The rate of polymerization per particle is faster, and the systems are converted faster. Following nucleation, the reaction proceeds with polymerization of the monomer in the miniemulsion droplets. Thus, there is no distinct Interval II. Droplet nucleations have been found to be more robust and less sensitive to variation in recipe or contaminant level. Also, particles prepared by miniemulsion polymerization tend to be more stable to shear, more monodisperse, and more reproducible than those prepared by macro-emulsion polymerizations.

9.6.4 Use of Composition Ripening to Produce Large Monodisperse Latex Particles

Ugelstad *et al.*[87] have also described a method for producing monodisperse latex particles with very large sizes, including monodisperse, porous particles. The method involves preparation of the droplets by molecular diffusion. Generally, polymer particles are able to be swelled with monomer to only a small extent, *ca.* 0.5 to 5 times their own volume. Ugelstad *et al.*[87] found that water insoluble compounds could be swelled with monomer to volume ratios greater than 1000. In the first step, an aqueous emulsion of the water insoluble compound is prepared. In the second stage, the slightly water-soluble monomer ·is added with additional surfactant under gentle stirring. This results in a population of large droplets containing the more soluble dispersed phase, and smaller droplets with an insoluble dispersed phase. These are the classic conditions for reverse recondensation. Composition ripening quickly begins to swell the small droplets. The key to achieving a monodisperse population of particles is to start with a fixed number of nuclei (droplets in this case) which subsequently grow to a much larger size without any additional nucleation occurring during the growth period. This is especially true if one operates under Interval II conditions where the rate of polymerization is independent of droplet size. Monodisperse latex particles up to 2 μm have been prepared from small particles (0.1 μm) by such techniques (Figure 9.15).

9.6.5 Stabilization of Gas Emulsions

Gas emulsions (*kugelschaum*) are nearly spherical gas bubbles separated by rather thick liquid films in a continuous liquid phase. Air-filled gas emulsions are excellent reflectors of sound waves, and can be used as contrast agents for diagnostic ultrasound procedures. When injected into the bloodstream, gas emulsions can be used to enhance the visualization of blood flow through small vessels and blood-rich organs. Air dissipates rapidly from the gas emulsion bubbles, however, limiting their applicability.[93] Only recently has it become apparent that gas emulsions can be stabilized *in vivo* by the addition of a slow diffusing gas osmotic agent, in a manner analogous to the stabilization of liquid emulsions with a secondary disperse phase component.[23,49]

Under atmospheric conditions the stability of gas emulsions is critically dependent on their size. Surface tension effects lead to an elevated total pressure (*i.e.* Laplace pressure) inside the bubble, quickly leading to dissolution. For a 1 μm bubble with $\sigma = 50$ mN m^{-1}, the Laplace pressure, $p_{\text{Laplace}} = 2\sigma/r$, is nearly twice the atmospheric pressure (p_{atm}). When injected into the bloodstream, efflux of gas from the bubbles is further enhanced by blood pressure, p_{blood}.

In order to overcome the overpressure effects on small gas emulsion bubbles, a

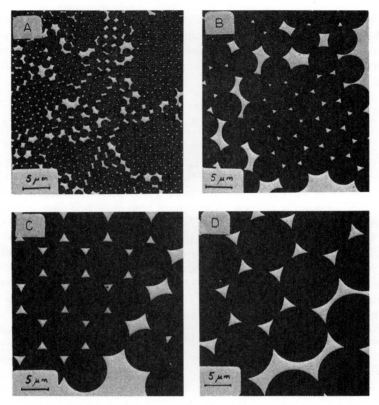

Figure 9.15 *Monodisperse polystyrene latexes prepared by reverse recondensation method:*
(A) 1.5 μm polystyrene seed particles; (B) 5.2 μm; (C) 7.5 μm; (D) 10.2 μm
particles swelled from (A) particles
(Reproduced by permission of Elsevier Science from ref. 87)

gas osmotic agent, which will slowly diffuse from the bubble, may be introduced.
The partial pressure of the gas osmotic agent, p_F, must be sufficient to overcome
the overpressure effects due to Laplace pressure and blood pressure, *viz.*

$$p_{\text{blood}} + p_{\text{Laplace}} = p_F \qquad (9.17)$$

The presence of the gas osmotic agent in the bubble lowers the partial pressures of
blood gases in the bubble, thereby decreasing their tendency to diffuse out of the
bubble.

The kinetics for gas emulsion dissolution in the bloodstream were first elucidated
by Epstein and Plesset[94] in 1950, *viz.*

$$\tau_{\text{EP}} \approx \frac{r_0^2 \rho}{2Dc(\infty)} \approx \frac{r_0^2}{2DL} \qquad (9.18)$$

where τ_{EP} is the Epstein–Plesset lifetime, r_0 is the initial bubble diameter, and ρ is the density of the bulk gas phase. The dimensionless ratio of the solubility of the gas in the liquid to the gas density is known as the Ostwald coefficient, L. It can be rewritten in terms of the ratio of the solubility of the solute between the gas phase and a dilute aqueous solution, *viz.*

$$L = \frac{c_{water}}{c_{gas}} \qquad (9.19)$$

Using eqn. (9.18), the predicted lifetime for an air bubble with a radius of 2.5 μm is only 0.13 s. From the Epstein–Plesset formalism it is clear that gas emulsion persistence can be improved by incorporating gases with small Ostwald coefficients and slow diffusivity. Ostwald coefficients have been published for a large number of gases, some of which are detailed in Table 9.6. It is clear that highly polar gases (*e.g.* CO_2 and NH_3) partition selectively into the aqueous phase, while gases of low polarity (*e.g.* hydrocarbons and fluorocarbons) partition selectively into the vapor phase. The highly biocompatible fluorocarbons have become the gases of choice for this application, with a number of groups independently exploring their potential in diagnostic ultrasound.[95–100] The quadratic dependence of the bubble lifetime with bubble radius has been verified in several other theoretical studies, including those of de Vries,[101] van Liew and Burkhard,[102,103] and Kabalnov *et al.*[104,105]

Table 9.6   *Ostwald coefficients for some representative gases[a]*

Gas	Ostwald coefficient ($L \times 10^6$)
N_2	14480 (35 °C)
O_2	27730 (35 °C)
C_2F_6	1272 (25 °C)
C_4F_{10}	202 (25 °C)
C_5F_{12}	66 (25 °C)
C_6F_{14}	45 (25 °C)

[a]Taken from ref. 104.

The theory of Epstein and Plesset[94] assumes that the bubble contains only a single gas. It fails to take into account the equilibration of the gas osmotic agent with blood gases following administration. van Liew and Burkhard[102,103] examined the case of dissolution of bubbles containing a sparingly soluble gas osmotic agent. Their theory predicts that following injection the bubbles will first swell due to osmosis of blood gases into the bubble. After the bubble is re-equilibrated, the dissolution is determined by the physical properties (diffusivity and Ostwald coefficient) of the gas osmotic agent. van Liew and Burkhard[102,103] also examined the effect of oxygen metabolism on bubble persistence. The inherent unsaturation resulting from oxygen metabolism further promotes the diffusion of gases out of bubbles into the bloodstream, although the effect is minor in comparison with blood and Laplace pressure effects.

The work of van Liew and Burkard[102,103] fails to take into account another key factor in the bubbles' fate, *i.e.* the possibility that the gas osmotic agent may condense into a liquid. Once this occurs, all ultrasound contrast is lost. This effect can be important, especially in light of the fact that in order to achieve low values of the Ostwald coefficient, the saturated vapor pressure must be decreased. A theory taking this into account was recently put forward by Kabalnov *et al.*[104,105] In essence, the partial pressure of the gas osmotic agent in the bubble cannot be higher than the saturated vapor pressure at the given temperature, $p_{F,sat}(T)$. The partial pressure of the gas osmotic agent in the bubble will be intimately affected by the initial re-equilibration, and by oxygen metabolism. With this in mind it is possible to define a collapse radius, $r_{collapse}$, for which condensation occurs, *viz.*

$$r_{collapse} = \frac{2\sigma}{p_{F,sat}(T)} \tag{9.20}$$

The effects of condensation can be dramatically decreased by decreasing the surface tension on the bubble surface, thereby lowering the Laplace pressure.

A curve showing the sequence of events for the dissolution of gas emulsions is shown in Figure 9.16. As discussed, the first stage is characterized by rapid swelling of the osmotic agent bubbles in air (the bubbles may in fact shrink if the partial pressure of air in the microbubbles is greater than atmospheric). After this stage of rapid equilibration, the partial pressure of air, p_A, equalizes to the atmospheric pressure and stays constant during further dissolution. The partial pressure of the gas osmotic agent increases over time, as the bubble shrinks. It is at this stage (stage 2) that the square of the particle radius decreases linearly with time. Ultimately, it is stage 2 that determines the overall lifetime of the bubble. In stage 3, the partial pressure of the osmotic agent becomes high enough that it condenses into a liquid. Figure 9.16 shows the theoretical predictions for a bubble filled with perfluorobutane. The theories predict that the ultrasound signal duration will increase in the order $C_2F_6 < C_4F_{10} < C_5F_{12} < C_6F_{14}$. Although the effects of condensation are not that important for perfluorobutane, they become increasingly important for longer chain lengths. Thus, the partial pressure of perfluorohexane must be limited in the microbubble, so as to reduce the effects of condensation.

Recent experimental studies[105] have confirmed that microbubble persistence is controlled primarily by the dissolution of bubbles and not by removal by the RES. Although the qualitative trends are in agreement with theoretical models, there are some quantitative differences.

The time decay to baseline of the ultrasound intensity increases with decreasing Ostwald coefficient of the filling gas. The efficacy is lost, however, for high boiling

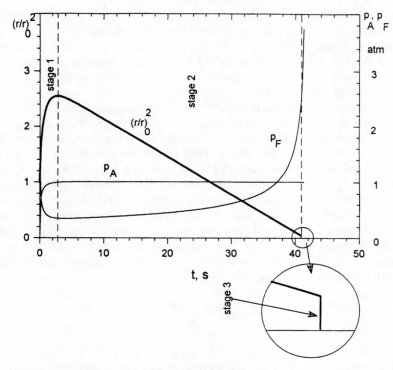

Figure 9.16 *Theoretical dissolution of perfluorobutane filled gas emulsions (σ = 70 mN m⁻¹, r₀ = 2.5 μm, and X_F = 1.0)*
(Reproduced by permission of Pergamon Press from ref. 104)

osmotic agents whose partial pressure is below *ca.* 0.2 atm at 37 °C, even though the Ostwald coefficient is low.[105] The bubble lifetime and the ultrasound scattering intensity increase with the mole fraction of osmotic agent in the bubble. There are quantitative differences in terms of the absolute times of the microbubble persistence (which are underestimated), and in the dependence of the lifetime on the volume fraction of gas osmotic agent, which is weaker than predicted.[105]

Gas emulsions can also be stabilized by altering the surface viscosity of the interfacial layer. Ronteltap *et al.*[58,59] have shown (both theoretically[58] and experimentally[59]) that the surface tension can become very low when the surface dilational viscosity is high. Thus, the incorporation of surface-active agents, which remain firmly bound to the interface as the bubbles shrink, leads to increases in the surface excess, thereby making the interface more rigid and decreasing σ.

Acknowledgement

The author wishes to acknowledge his collaborations with scientists at Alliance Pharmaceutical Corp., in particular with special thanks to Rebecca Arlanskas, Alexey Kabalnov and Tom Tarara.

9.7 References

1. W. Ostwald, *Z. Phys. Chem. (Leipzig)*, 1900, **34**, 295.
2. W. Thomson (Lord Kelvin), *Philos. Mag.*, 1871, **42**, 448.
3. I.M. Lifshitz and V.V. Slezov, *Sov. Phys. JETP*, 1959, **35**, 331.
4. C. Wagner, *Z. Electrochem.*, 1961, **35**, 581.
5. A.S. Kabalnov and E.D. Shchukin, *Adv. Colloid Interface Sci.*, 1992, **38**, 69.
6. A.S. Kabalnov, A.V. Pertsov and E.D. Shchukin, *J. Colloid Interface Sci.*, 1987, **118**, 590.
7. A.S. Kabalnov, K.N. Makarov, A.V. Pertsov and E.D. Shchukin, *J. Colloid Interface Sci.*, 1990, **138**, 98.
8. A.S. Kabalnov, *Langmuir*, 1994, **10**, 680.
9. P. Taylor, *Colloids Surf. A*, 1995, **99**, 175.
10. P. Taylor and R.H. Ottewill, *Colloids Surf. A*, 1994, **88**, 303.
11. C. Varescon, C. Arlen, M. LeBlanc and J.G. Riess, *J. Chim. Phys. Phys.-Chim. Biol.*, 1989, **86**, 2111.
12. L. Trevino, L. Sole-Violan, P. Daumur, B. Devallez, M. Postel and J.G. Riess, *New J. Chem.*, 1993, **17**, 275.
13. V.M. Sadtler, M.P. Krafft and J.G. Riess, *Angew. Chem., Int. Ed. Engl.*, 1996, **35**, 1976.
14. J.G. Weers, J. Liu, T. Fields, P. Resch, J. Cavin and R.A. Arlauskas, *Art. Cells Blood Subs. Immob. Biotech.*, 1994, **22**, 1175.
15. J.G. Weers, Y. Ni, T.E. Tarara, T.J. Pelura and R.A. Arlauskas, *Colloids Surf. A*, 1994, **84**, 81.
16. J.G. Weers and R.A. Arlauskas, *Langmuir*, 1995, **11**, 474.
17. J. Soma and K.D. Papadopoulos, *J. Colloid Interface Sci.*, 1996, **181**, 225.
18. Y. De Smet, J. Malfait, C. De Vos, L. De Riemaaker and R. Finsy, *Bull. Soc. Chim. Belg.*, 1996, **105**, 789.
19. L. Bremer, B. DeNijs, L. Deriemaeker, R. Finsy and E. Geladé, *Part. Part. Syst. Charact.*, 1996, **13**, 350.
20. B.P. Binks, *Annu. Rep. R. Soc. Chem. Sect. C*, 1996, **92**, 97.
21. R. Buscall, S.S. Davis and D.S. Potts, *Colloid Polym. Sci.*, 1979, **257**, 636.
22. S.S. Davis, H.P. Round and T.S. Purewal, *J. Colloid Interface Sci.*, 1981, **80**, 508.
23. W.I. Higuchi and J. Misra, *J. Pharm. Sci.*, 1962, **51**, 459.
24. K. Parbhakar, L. Lewandowski and L.H. Dao, *J. Colloid Interface Sci.*, 1995, **174**, 142.
25. F.P. Ludwig, J. Schmelzer and J. Bartels, *J. Mater. Sci.*, 1994, **29**, 4852.
26. C.H. Kang and D.N. Yoon, *Metall. Trans. A*, 1981, **12A**, 65.
27. Y. Enomoto, K. Kawasaki and M. Tokuyama, *Acta Metall.*, 1987, **35**, 907.
28. P.W. Voorhees, *J. Stat. Phys.*, 1985, **38**, 231.
29. C.W.J. Beenakker, *Phys. Rev. A*, 1986, **33**, 4482.
30. J.A. Marquisee and J. Ross, *J. Chem. Phys.*, 1984, **80**, 536.
31. A.D. Brailsford and P. Wynblatt, *Acta Metall.*, 1979, **27**, 489.
32. H.W. Yarranton and J.H. Masliyah, personal communication.
33. A.J. Ardell, *Acta Metall.*, 1972, **20**, 61.
34. G. Venzl, *Ber. Bunzenges. Phys. Chem.*, 1983, **87**, 318.
35. Y. Ni, T.J. Pelura, T.A. Sklenar, R.A. Kinner and D. Song, *Art. Cells Blood Subs. Immob. Biotech.*, 1994, **22**, 1307.
36. S. Karaborni, N.M. van Os, K. Esselink and P.A.J. Hilbers, *Langmuir*, 1993, **9**, 1175.
37. K. Esselink, P.A.J. Hilbers, N.M. van Os, B. Smit and S. Karaborni, *Colloids Surf. A*, 1994, **91**, 155.
38. P. Taylor and R. Ottewill, *Prog. Colloid Polym. Sci.*, 1994, **97**, 199.

39. L. Taisne, P. Walstra and B. Cabane, *J. Colloid Interface Sci.*, 1996, **184**, 378.
40. D.D. Miller, paper presented at the Symposium on Diffusion and Mass Transfer in Colloidal Systems, American Chemical Society Fall National Meeting, Chicago, 1993.
41. E.A.G. Aniansson, S.N. Wall, M. Almgren, H. Hoffmann, I. Kielmann, W. Ulbricht, R. Zana, J. Lang and C. Tondre, *J. Phys. Chem.*, 1976, **80**, 905.
42. S.G. Oh and D.O. Shah, *J. Disp. Sci. Technol.*, 1994, **15**, 297.
43. A. Kabalnov and J. Weers, *Langmuir*, 1996, **12**, 3442.
44. D. Stigter, R.J. Williams and K.J. Mysels, *J. Phys. Chem.*, 1955, **59**, 330.
45. J. Weiss, J.N. Coupland, D. Brathwaite and D.J. McClements, *Colloids Surf. A*, 1997, **121**, 53.
46. A.J.I. Ward and K. Quigley, *J. Disp. Sci. Technol.*, 1990, **11**, 143.
47. B.J. Carroll, *J. Colloid Interface Sci.*, 1981, **79**, 126.
48. L. Taisne, Doctoral Dissertation, Université Paris VI, 1997.
49. A.S. Kabalnov, A.V. Pertsov and E.D. Shchukin, *Colloid Surf.*, 1987, **24**, 19.
50. A.S. Kabalnov, A.V. Pertsov, Yu. D. Aprosin and E.D. Shchukin, *Kolloidn. Zh.*, 1985, **47**, 1048.
51. J.C. Giddings, *Science*, 1993, **260**, 1456.
52. A. Kabalnov, J. Weers, R. Arlauskas and T. Tarara, *Langmuir*, 1995, **11**, 2966.
53. P. Walstra, in 'Encyclopedia of Emulsion Technology', ed. P. Becher, Dekker, New York, 1996, vol. 4, p. 1.
54. J.G. Riess and M. Postel, *Art. Cells Blood Subs. Immob. Biotech.*, 1992, **20**, 819.
55. E.E. Kumacheva, E.A. Amelina and E.D. Shchukin, *Kolloidn Zh.*, 1989, **51**, 1214.
56. J.G. Weers, *J. Fluorine Chem.*, 1993, **64**, 73.
57. E.A. Amelina, E.E. Kumacheva, A.V. Pertsov and E.D. Shchukin, *Kolloidn Zh.*, 1990, **52**, 216.
58. A.D. Ronteltap, B.R. Damsté, M. DeGee and A. Prins, *Colloids Surf.*, 1990, **47**, 269.
59. A.D. Ronteltap and A. Prins, *Colloids Surf.*, 1990, **47**, 285.
60. A.V. Pertsov, A.S. Kabalnov, E.E. Kumacheva and E.A. Amelina, *Kolloidn Zh.*, 1988, **50**, 616.
61. R.A. Arlauskas and J.G. Weers, *Langmuir*, 1996, **12**, 1923.
62. D.J. McClements and S.R. Dungan, *J. Phys. Chem.*, 1993, **97**, 7304.
63. D.J. McClements, S.R. Dungan, J.B. German and J.E. Kinsella, *J. Colloid Interface Sci.*, 1993, **156**, 425.
64. D.J. McClements, S.R. Dungan, J.B. German and J.E. Kinsella, *Colloid Surf. A*, 1993, **81**, 203.
65. M.P. Krafft, J.G. Riess and J.G. Weers in 'Submicronic Emulsions in Drug Delivery', ed. S. Benita, Horwood, Amsterdam, 1998, in press.
66. S.F. Flaim, *Art. Cells Blood Subs. Immob. Biotech.*, 1994, **22**, 1043.
67. P.E. Keipert, S. Otto, S.F. Flaim, J.G. Weers, E.A. Schutt, T.J. Pelura and D.H. Klein, *Art. Cells Blood Subs. Immob. Biotech.*, 1994, **22**, 1169.
68. R.P. Geyer, in 'Proceedings of the Xth International Congress for Nutrition: Symposium on Perfluorochemical Artificial Blood, Kyoto, Japan', Igakushobo Medical Publisher, Osaka, Japan, 1975, p. 3.
69. K. Yamanouchi, M. Tanaka, Y. Tsuda, K. Yokoyama, S. Awazu and Y. Kobayashi, *Chem. Pharm. Bull.*, 1985, **33**, 1221.
70. R. Naito and K. Yokoyama, 'Technical Information Ser. No. 5', The Green Cross Corp., Osaka, Japan, 1981.
71. V.V. Obraztsov, A.S. Kabalnov, A.N. Sklifas and K.N. Makarov, *Biophysics*, 1992, **37**, 298.

72. Y. Tsuda, K. Yamanouchi, K. Yokoyama, T. Suyama, M. Watanabe, H. Ohyanagi and Y. Saitoh, *Biomat. Art. Cells Immob. Biotech.*, 1988, **16**, 473.

73. Y. Ni, D. Klein and D. Song, *Art. Cells Blood Subs. Immob. Biotech.*, 1996, **24**, 81.

74. J. Riess, *Artif. Organs*, 1984, **8**, 44.

75. A.S. Kabalnov, K.N. Makarov and O.V. Shcherbakova, *J. Fluorine Chem.*, 1990, **50**, 271.

76. S. Rudiger, *J. Fluorine Chem.*, 1989, **42**, 403.

77. A.S. Kabalnov, K.N. Makarov and E.D. Shchukin, *Colloids Surf.*, 1992, **62**, 101.

78. P.E. Keipert, *Art. Cells Blood Subs. Immob. Biotech.*, 1995, **23**, 381.

79. R.J. Kaufman, in 'Blood Substitutes: Physiological Basis of Efficacy', ed. R. M. Winslow, K.D. Vandegriff and M. Intaglietta, Birkhäuser, Boston, 1995, p. 53.

80. R.J. Kaufman, in 'Emulsions: A Fundamental and Practical Approach', ed. J. Sjöblom, Kluwer, Amsterdam, 1992, p. 207.

81. H. Meinert, R. Fackler, A. Knoblich, J. Mader, P. Reuter and W. Rohlke, *Biomat. Art. Cells Immob. Biotech.*, 1992, **20**, 805.

82. C. Cornélus, M.P. Krafft and J.G. Riess, *J. Colloid Interface Sci.*, 1994, **163**, 391.

83. J.G. Riess, L. Sole-Violan and M. Postel, *J. Disp. Sci. Technol.*, 1992, **13**, 349.

84. D.G. Dalgleish, *Curr. Opin. Colloid Interface Sci.*, 1997, **2**, 573.

85. S.O. Agboola and D.G. Dalgleish, *J. Sci. Food Agric.*, 1996, **72**, 448.

86. R.G. Gilbert and D.H. Napper, *JMS-REV. Macromol. Chem. Phys.*, 1983, **C23**, 127.

87. J. Ugelstad, P.C. Mørk, K. Herder Kaggerud, T. Ellingsen and A. Berge, *Adv. Colloid Interface Sci.*, 1980, **13**, 101.

88. D. Mouran, J. Reimers and F.J. Schork, *J. Polym. Sci. A*, 1996, **34**, 1073.

89. J. Reimers and F.J. Schork, *J. Appl. Polym. Sci.*, 1996, **59**, 1833.

90. S. Wang and F.J. Schork, *J. Appl. Polym. Sci.*, 1994, **54**, 2157.

91. K. Fontenot and F. Schork, *J. Appl. Polym. Sci.*, 1993, **49**, 633.

92. Y.T. Choi, M.S. El-Aasser, E.D. Sudol and J.W. Vanderhoff, *J. Appl. Polym. Sci.*, 1985, **23**, 2973.

93. R.S. Meltzer, E.G. Tickner and R.L. Popp, *Ultrasound Med. Biol.*, 1980, **6**, 263.

94. P.S. Epstein and M.S. Plesset, *J. Chem. Phys.*, 1950, **18**, 1505.

95. J.M. Correas and S.D. Quay, *Clin. Radiol.*, 1996, **51**, 11.

96. R. Lohrmann, *U.S. Pat.* 5 562 893, 1996.

97. S.C. Quay, *U.S. Pat.* 5 393 524, 1995.

98. M. Schneider, F. Yan, P. Grenier, J. Puginier and M.-B. Barrau, *U.S. Pat.* 5 413 774, 1995.

99. E.A. Schutt, D.P. Evitts, R.A. Kinner, J.G. Weers and C.D. Anderson, *U.S. Pat.* 5 639 443, 1997.

100. E.C. Unger, T. Matsunaga, V. Ramaswami, D. Yellowhair and G. Wu, *Int. Pat. Appl.* PCT/US 94/05792.

101. A.J. deVries, *Recl. Trav. Chim. Pays-Bas*, 1958, **77**, 283.

102. H.D. van Liew and M.E. Burkhard, *Invest. Radiol.*, 1995, **30**, 315.

103. H.D. van Liew and M.E. Burkhard, *J. Appl. Physiol.*, 1995, **79**, 1379.

104. A. Kabalnov, D. Klein, T. Pelura, E. Schutt and J. Weers, *Ultrasound Med. Biol.*, 1998, in press.

105. A. Kabalnov, J. Bradley, S. Flaim, D. Klein, T. Pelura, B. Peters, S. Otto, J. Reynolds, E. Schutt and J. Weers, *Ultrasound Med. Biol.*, 1998, in press.

CHAPTER 10

Interactions and Macroscopic Properties of Emulsions and Microemulsions

DIMITER N. PETSEV

Department of Chemistry, 1393 Brown Building, Purdue University, West Lafayette, Indiana 47907-1393, USA

10.1 Introduction

Emulsions and microemulsions are dispersions of liquid in liquid. Therefore, an important feature is their interfacial fluidity and deformability, which distinguishes them from suspensions of solid particles. The stability of the latter is usually treated in the framework of the Derjaguin–Landau–Verwey–Overbeek (DLVO) theory,[1-3] which accounts for the electrostatic and van der Waals interactions between solid particles. During recent years it was shown that other types of interparticle forces may often play an important role for the stability of dispersions—hydrodynamic interactions, hydration and hydrophobic forces, oscillatory structure forces, *etc*.[4,5] It was proven both experimentally and theoretically that steric[6,7] and depletion[8-10] interactions may sometimes be a decisive factor for the dispersion stability.

The situation with emulsions and microemulsions is more complex (compared to suspensions of solid particles) due to their fluidity and deformability. It is known that these two features may have a great impact on the hydrodynamic interactions and, hence, on the dynamic properties of such large emulsion drops.[11-16] Therefore, they could be particularly important for the kinetic stability of emulsions against coalescence.[11-13] Along with the hydrodynamic interactions the direct interactions, due to surface forces, can be strongly affected by the deformation as well.[17-19] An approach for calculation of the different contributions (van der Waals, electrostatic, steric, depletion, *etc*.) to the interaction energy, when deformation takes place, was recently developed.[20] The shapes of two deformed drops in contact can be approximated by truncated spheres separated by a planar film. Such geometry allows the derivation of a general explicit expression for the van der Waals interaction energy between two deformed droplets[20,21] following the microscopic method of

Hamaker.[22] The contribution of the surface extension energy and/or bending elasticity to the pair interaction potential is also included. The extension of the drop surface upon the deformation corresponds to a soft interdroplet repulsion. All the remaining possible interactions (electrostatic, steric, depletion, etc.) can usually be treated in the framework of Derjaguin's approximation,[2,3,23] which allows one to account for the two contributions of the total interaction energy: (i) across the flat film and (ii) between the spherical surfaces surrounding the film.[20] Combined with relevant expressions for the hydrodynamic interactions, this approach could be used for studying the coalescence of Brownian emulsion and microemulsion droplets.[24]

The proper account of the droplet deformability is particularly important for systems with low interfacial tension. Microemulsions are a typical example as well as vesicles. Larger emulsion droplets, on the other hand, are prone to deformation even at relatively higher (compared to microemulsions) interfacial tensions because of their lower capillary pressure. Even small deformations are sufficient to introduce a remarkable change in the pair interdroplet energy. The impact of these changes on the macroscopic properties (*e.g.* the osmotic pressure) of emulsions and microemulsions could be great[25] and is by no means to be neglected.

This chapter summarises the recent efforts[19-21,24,26,27] to define and calculate the pair energy of interaction between two droplets, including the effect of their deformation. It also shows how the deformability contributes to the macroscopic properties of the systems using integral equation methods[25] and Brownian dynamics simulations.

10.2 Pair Energy of Interaction

This section outlines the approach for calculation of the pair interaction energy between two equally sized deformable fluid droplets (the generalisation to droplets of different size is often straightforward). The fluid particles are assumed to acquire the shape of truncated spheres when the distance between them is small enough (see Figure 10.1). The accuracy of this idealisation is discussed at the end of this section.

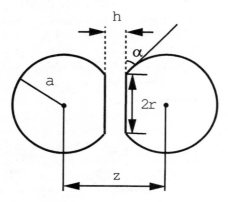

Figure 10.1 *Sketch of two interacting deformable oil droplets*

Let us consider the overall pair interaction energy between the droplets as being a sum of different contributions according to their origin

$$W(h, r) = \sum_i W_i(h, r) \qquad (10.1)$$

The total energy $W(h, r)$ as well as all the terms $W_i(h, r)$ generally depend on the thickness, h, and radius, r, of the plane parallel film, formed between the droplets—see Figure 10.1. The different contributions $W_i(h, r)$ are considered separately below.

10.2.1   van der Waals Interaction Energy

The van der Waals (VW) energy of interaction between two truncated spheres (as depicted in Figure 10.1) could be calculated explicitly for arbitrarily great deformation[20,21] following the general method of Hamaker.[22] Hence, the only simplifications introduced in the final result (besides the idealised geometry) are those due to the method itself: (i) pairwise additivity of the interactions between the molecules of the two droplets and (ii) disregarding the electromagnetic retardation.

The general result for the van der Waals energy reads[20] (*cf.* also ref. 21)

$$W^{\mathrm{VW}}(h, r) = -\frac{A_{\mathrm{H}}}{12} \left\{ \frac{2al}{(l + h)^2} + \frac{2al}{h(2l + h)} + 2\ln\left[\frac{h(2l + h)}{(l + h)^2}\right] + \frac{r^2}{h^2} - \frac{2r^2}{h(l + h)} \right.$$

$$\left. -\frac{2r^2}{(l + h)[2(l - a) + h]} + \frac{(h^2 + 4r^2)(\sqrt{h^2 + 4r^2} - h)}{2h[2(l - a) + h]^2} - \frac{2r^2 a(2l^2 + lh + 2ah)}{h(l + h)^2[2(l - a) + h]^2} \right\}$$

$$(10.2)$$

where $l = a + \sqrt{(a^2 - r^2)}$ and A_{H} is the Hamaker constant. Expression (10.2) holds for droplets of identical radii, a. However, expressions for the case of different sized droplets as well as for a droplet interacting with a solid semi-infinite slab are also available.[20] Eqn. (10.2) could be simplified by assuming small droplet deformation, $r^2/a^2 \ll 1$. In this case the result for the van der Waals interaction energy between two deformed droplets reads

$$W^{\mathrm{VW}}(h, r) =$$

$$-\frac{A_{\mathrm{H}}}{12} \left\{ \frac{4a^2}{(2a + h)^2} + \frac{4a^2}{h(4a + h)} + 2\ln\left[\frac{h(4a + h)}{(2a + h)^2}\right] + \frac{128a^5 r^2}{h^2(2a + h)^3(4a + h)^2} \right\}$$

$$(10.3)$$

The change in the radius of the spherical part of the droplet with the deformation is negligible for small film radii.[20] At moderate interdroplet separations the expression for the van der Waals energy obtains an even simpler form

$$W^1\text{VW}(h,\,r) = -\frac{A_\text{H}}{12}\left[\frac{3}{4} + \frac{a}{h} + 2\ln\left(\frac{h}{a}\right) + \frac{r^2}{h^2} - \frac{2r^2}{ah}\right] \qquad (10.4)$$

The numerical comparison between the exact expression (10.2) and the approximate one (10.4) shows that the latter is rather accurate for $h/a \ll 0.3$ and $r/a \ll 0.5$. In many practically important cases one may use the simplest formula (10.4) without substantial loss of accuracy.

10.2.2 Interactions Calculated by Means of Derjaguin's Approximation

Most of the droplet interactions are amenable to analytical calculation only in the case of relatively weak droplet deformations, $r^2/a^2 \ll 1$, and/or small separations, $h/a \ll 1$. This could be accomplished by employing the well known Derjaguin's approximation[23] which in general form reads

$$W(h,\,r) = 2\pi \int_0^\infty \mathrm{d}\rho\,\rho\,f[H(\rho)] \qquad (10.5)$$

where $W(h,\,r)$ is the interaction energy, and ρ is the running radial co-ordinate in the plane of one of the walls of the plane parallel film formed between the droplets. $H(\rho)$ is the corresponding running local film thickness and in the specific case, shown in Figure 10.1, is defined as[20]

$$H(\rho) = h \text{ for } \rho < r$$

$$H(\rho) = h + \frac{1}{a}(\rho^2 - r^2) \text{ for } \rho > r \qquad (10.6)$$

Introducing eqn. (10.6) into (10.5) we obtain

$$W(h,\,r) = \pi r^2 f(h) + \pi a \int_h^\infty \mathrm{d}H\,f(H) \qquad (10.7)$$

With $f(h)$ we denote the free energy per unit area for an infinite plane parallel film. The above formula means that the energy of droplet interaction consists of two terms: the first is related to the interaction across the film formed between the droplets, and the second is due to the contribution of the spherical surfaces surrounding the film. Eqn. (10.7) could be applied in general to every type of interaction for which the free energy density, $f(h)$, is known. This is used below for calculating different terms contributing to the overall interdroplet energy.

10.2.2.1   Electrostatic Interactions

The electrostatic free energy of a thin liquid film can be approximated with different expressions, depending on the specific conditions.[1-6] For arbitrarily charged droplets and high electrolyte concentrations (*i.e.* weakly overlapped double layers) one may use the non-linear superposition approximation[1,6]

$$f^{\text{EL}}(h) = 64 C_{\text{el}} \kappa^{-1} \tanh^2 \left(\frac{ze\Psi_0}{4kT} \right) \exp(-\kappa h) \qquad (10.8a)$$

For low but constant surface potentials and arbitrary separations the energy density reads[1-3]

$$f^{\Psi}(h) = \varepsilon_0 \varepsilon \kappa \Psi_0^2 \left[1 - \tanh \left(\frac{\kappa h}{2} \right) \right] \qquad (10.8b)$$

while for low and constant surface charge densities, σ, it is[1-3]

$$f^{\sigma}(h) = -\frac{\sigma^2}{\varepsilon_0 \varepsilon \kappa} \left[1 - \coth \left(\frac{\kappa h}{2} \right) \right] \qquad (10.8c)$$

where for a $z{:}z$ electrolyte

$$\kappa^2 = \frac{2e^2 z^2}{\varepsilon_0 \varepsilon k T} C_{\text{el}} \qquad (10.9)$$

where e is the elementary charge, $\varepsilon_0 \varepsilon$ is the dielectric permittivity of the medium, kT is the thermal energy and C_{el} is the electrolyte concentration. With Ψ_0 we denote the droplet surface potential. The weak overlap of the double layers means $\kappa h \gg 1$. At the same time the interdroplet separation must remain small ($h/a \ll 1$). Despite the apparent restrictions, expression (10.8a) is widely used for calculating the electrostatic energy, since the necessary requirements are fulfilled for a wide range of colloidal (emulsion) systems. Substituting (10.8a) into (10.7) one obtains the electrostatic energy for two deformed droplets[20,26,27]

$$W^{\text{EL}}(h, r) = \frac{64\pi C_{\text{el}} kT}{\kappa} \tanh^2 \left(\frac{ze\Psi_0}{4kT} \right) \exp(-\kappa h) \left[r^2 + \frac{a}{\kappa} \right] \qquad (10.10)$$

In the absence of deformation ($r = 0$), the result for charged spheres is recovered.[1-3] A recent calculation, based on numerical solution of the Poisson–Boltzmann equation, was performed[28] and showed identical results to those from eqns. (10.7) and (10.8a) for micrometer sized droplets with high surface potentials ($\Psi_0 \approx 100$ mV). Hence, for many cases of practical importance one may obtain accurate results using the approximate expressions (10.7) and (10.8a).

If the conditions for applying the non-linear superposition approximation are violated, but the particles are weakly charged (or with low surface potential), other simple expressions such as (10.8b) or (10.8c) could be used.[2,3] The respective results for the interaction energy between deformed droplets are given in ref. 20. The case of droplets bearing different surface charge and/or potential is also discussed there.

10.2.2.2 Forces Due to the Presence of Adsorbing Polymers—Steric Interactions

The adsorption of polymers at the surface of emulsion droplets is often an important factor, with a great impact on their stability.[29] The particular interaction energy per unit film area, however, depends on the solvent quality with respect to the polymer.[30] Thus, for theta solvents, the steric interaction energy $f^{St}(h)$ can be calculated using the theory of Dolan and Edwards[31]

$$f^{St}(h) = \Gamma kT \left[\frac{\pi^2}{3} \left(\frac{L_0}{h} \right)^2 - \ln \left(\frac{8\pi}{3} \frac{L_0^2}{h^2} \right) \right] \text{ for } h < \sqrt{3}L_0$$

$$f^{St}(h) = 4\Gamma kT \exp \left[\left(-\frac{3}{2} \frac{h^2}{L_0^2} \right) \right] \text{ for } h > \sqrt{3}L_0$$

$$(10.11)$$

where $L_0 = l\sqrt{N}$ is the mean-square end-to-end distance of the portion of the polymer molecule dissolved in the film, say the polyoxyethylene portion of a nonionic surfactant molecule; l and N are the length of a fragment unit and their number in the polymer chain. Typical ranges of values of l and N are $0.5 \text{ nm} \leqslant l \leqslant 1.5 \text{ nm}$ and $10 \leqslant N \leqslant 10^4$,[6] which correspond to a thickness of the polymer layer from few nanometres to more than 100 nm. The interfacial area per polymer molecule, $A = 1/\Gamma$, is typically several times larger than l^2. By combining eqn. (10.11) with Derjaguin's approximation (10.7) one obtains the energy of steric interaction between two deformed drops[20,27]

$$W^{St}(h, r) = 4\pi\Gamma kT \left[r^2 \exp(-3/2\tilde{h}^2) + \sqrt{\frac{\pi}{6}} aL_0 \operatorname{erfc} \left(\sqrt{\frac{3}{2}}\tilde{h} \right) \right] \text{ for } h > L_0\sqrt{3}$$

$$(10.12)$$

$$W^{St}(h, r) = \pi r^2 \Gamma kT \left[\frac{\pi^2}{3\tilde{h}^2} + \ln \left(\frac{3\tilde{h}^2}{8\pi} \right) \right]$$

$$+ \pi a \Gamma kT L_0 \left[\frac{\pi^2}{3\tilde{h}} \left(1 - \frac{\tilde{h}}{\sqrt{3}} \right) + 2\tilde{h} \left(1 - \ln \sqrt{\frac{3\tilde{h}^2}{8\pi}} \right) - 5.235 \right] \text{ for } h < L_0\sqrt{3}$$

$$(10.13)$$

where $\tilde{h} = h/L_0$.

At theta conditions the steric interaction corresponds to a repulsion and in some aspects resembles the electrostatic force between charged surfaces. The range of steric interaction is proportional to the thickness of the adsorbed layer, $L_0/\sqrt{6}$ (see ref. 4, p. 304). Thus, the variation of the molecular mass of the polymer leads to an effect similar to that due to the variation of the electrolyte concentration in the case of electrostatic repulsion. At sufficiently thick adsorption layers the van der Waals attraction becomes negligible and the drops interact as non-deformable spheres.

In the case of good solvents the interaction energy between two deformed droplets covered with brush layers can be determined by means of Alexander–de Gennes theory[32] as follows:

$$f^{St}(h) = 2kT\Gamma^{3/2}L_g\left[\frac{4}{5}\tilde{h}_g^{-5/4} + \frac{4}{7}\tilde{h}_g^{7/4} - \frac{48}{35}\right] \text{ for } h < 2L_g \qquad (10.14)$$

which leads to the following expression for the interdroplet energy[20,27]

$$W^{St} = \pi r^2 f^{St}(h) + 4\pi akT\Gamma^{3/2}L_g^2[1.37\tilde{h}_g - 0.21\tilde{h}_g^{11/4} + 3.20\tilde{h}_g^{-1/4} - 4.36]$$

$$\text{for } h < 2L_g \quad (10.15)$$

where L_g is the polymer layer thickness in a good solvent defined by

$$L_g = N(\Gamma l^5)^{1/3} \text{ and } \tilde{h}_g = h/2L_g \qquad (10.16)$$

A quantitative experimental verification of the theory of Alexander–de Gennes was performed by Taunton *et al.*,[33] who measured the forces between two polystyrene brush layers in toluene and demonstrated a very good agreement between the theory and experiment. The steric interaction in good solvents corresponds to a monotonic repulsion whose range is characterised by the layer thickness L_g. Therefore, the influence of the polymer molecular mass and surface concentration is qualitatively similar to that in the case of a theta solvent considered above. However, the thickness of the adsorbed layer at theta conditions L_0 (and therefore the range of steric interactions) does not depend on the interfacial polymer concentration, while in good solvents L_g (and the range of steric interaction) increases with the polymer adsorption.

In a poor solvent one can observe a minimum in the steric interaction energy at separations close to the polymer layer thickness.[6,34] This minimum is due to the attraction between the polymer segments in a poor solvent. At smaller separations, strong steric repulsion between the adsorbed layers appears. This interesting case can be treated only numerically.[6,34]

10.2.2.3   Forces Due to the Presence of Non-adsorbing Polymers— Depletion and Structural Interactions

Depletion interactions appear when smaller colloidal species (polymer molecules, micelles, microemulsions) are present in the disperse medium along with the larger emulsion drops. Their presence leads to an attraction between the drops at small separations.[10,20,35] This attraction has an osmotic origin and the corresponding interaction energy can be presented in the form

$$W^D = -P_0 V_E \qquad (10.17)$$

where P_0 is the osmotic pressure created by the micelles (or the polymer molecules) and V_E denotes the volume in the gap between the drops, which is excluded for access by the smaller species.

The simpler case is when nonionic micelles (or polymers), behaving as hard spheres of diameter d, are present. Then P_0 can be expressed accurately enough by means of the Carnahan–Starling[36] formula

$$P_0 = \xi C_M kT \text{ where } \xi = \frac{1 + \Phi + \Phi^2 - \Phi^3}{(1 - \Phi)^3} \qquad (10.18)$$

where C_M is the number density of the micelles (or polymer molecules) and Φ is their volume fraction. The excluded volume V_E in this case is determined mostly by geometrical constraints and for two deformed spheres like those shown in Figure 10.1 can be expressed as[26,27]

$$V_E = \pi \left[r^2(d - h) + \frac{a}{2}(d - h)^2 \right] \text{ for } h \leqslant d \text{ and } d/a \ll 1 \qquad (10.19)$$

In the absence of deformation the first term on the right-hand side of eqn. (10.19) is zero and the excluded volume, V_E, is smaller. This results in weaker depletion attraction.

When both the large droplets and the small species (polymer molecules or micelles) are charged, one may use (for estimations) again expressions (10.17) and (10.18) but introducing an effective volume fraction, Φ_0, defined by

$$\Phi_0 = \frac{4}{3}\pi \left(\frac{d}{2} + \kappa^{-1} \right)^3 \qquad (10.20)$$

In addition, the micelles are electrostatically repelled from the drop surfaces, which leads to an even larger excluded volume, V_E, than that predicted by eqn. (10.19). A reasonable approximation for V_E in this case can be obtained by utilising the idea of Richetti and Kekicheff,[9] that the micelles cannot approach the drop surface closer than a certain distance b (in their experiments, b was determined to be about $2.5\kappa^{-1}$). Therefore, eqn. (10.19) should be replaced by[27]

$$V_E = \pi \left[r^2(d + 2b - h) + \frac{a}{2}(d + 2b - h)^2 \right] \text{ for } h \leqslant d + 2b \qquad (10.21)$$

Hence, the depletion effect due to ionic micelles is greater than that of nonionic micelles at the same micelle number concentration and is strongly dependent on the electrolyte concentration.

At higher volume fractions of the small species (polymers or micelles) the depletion may transform into oscillatory structural interaction. Experimental evidence for such interactions between solid surfaces was demonstrated by Richetti and Kekicheff and Parker *et al.*[9] A similar phenomenon is the stratification (stepwise thinning) of foam films. It was explained as layer by layer expulsion of micelles from the film interior.[37] The semi-empirical approach[38] yields the oscillatory energy density in the form

$$
f^{Os}(h) = \begin{cases} P_0 \dfrac{d \exp\left(1 - \frac{h}{d}\right)}{(4\pi^2 + 1)} \left[\cos\left(\dfrac{2\pi h}{d}\right) - 2\pi \sin\left(\dfrac{2\pi h}{d}\right) \right], & h > d \\[2ex] P_0 \left(h - d + \dfrac{d}{4\pi^2 + 1} \right), & 0 \leqslant h \leqslant d \end{cases}
\tag{10.22}
$$

Introducing the above expression into eqn. (10.7) one may calculate the oscillatory contribution to the respective interdroplet energy,[27] *viz.*

$$
W^{Os}(h, r) = \begin{cases} \pi r^2 P_0 \dfrac{d \exp\left(1 - \frac{h}{d}\right)}{(4\pi^2 + 1)} \left[\cos\left(\dfrac{2\pi h}{d}\right) - 2\pi \sin\left(\dfrac{2\pi h}{d}\right) \right] \\[2ex] \quad - \pi a P_0 \dfrac{d^2 \exp\left(1 - \frac{h}{d}\right)}{(4\pi^2 + 1)^2} \left[(4\pi^2 + 1)\cos\left(\dfrac{2\pi h}{d}\right) + 4\pi \sin\left(\dfrac{2\pi h}{d}\right) \right], \\[1ex] \hspace{7cm} h > d \\[2ex] \pi r^2 P_0 \left(h - d + \dfrac{d}{4\pi^2 + 1} \right) + \pi a P_0 \dfrac{d^2(4\pi^2 - 1)}{(4\pi^2 + 1)^2} \\[2ex] \quad + \pi a P_0 (d - h) \left(\dfrac{d}{4\pi^2 + 1} - \dfrac{d}{2} + \dfrac{h}{2} \right), \quad 0 \leqslant h \leqslant d \end{cases}
$$

$$\tag{10.23}$$

This formula could be extended to ionic micelles or polymers as in the depletion case. For that purpose, the micellar (or polymer molecular) diameter d should be replaced by $d_0 = d + 2/\kappa$ and the film thickness h, by $h - 2b$. The pressure P_0 is calculated as explained above (see eqn. 10.18). Eqn. (10.23) is valid for low volume fractions of the small species (or effective volume fraction, when charge effects are present) not exceeding 0.1, and takes into account both oscillatory and depletion interactions. At higher concentrations one may use more elaborate expressions.[38]

10.2.3 Interaction Energy Contributions Due to the Interfacial Properties of the Individual Droplet—Surface Area Extension and Bending Energies

The formation of a doublet of droplets (as shown in Figure 10.1) is accompanied by a shape change from spherical to that of a truncated sphere. This deviation from the optimal (spherical) shape also contributes to the pair interdroplet energy. The interfacial contribution may be considered as due to (i) an increase of the droplet surface (the droplet volume remains constant) and (ii) an increase of the *local* droplet radius of curvature in the film region. These two components are considered separately below.

10.2.3.1   Surface Area Extension Energy

Assuming that the droplet volume remains constant during the deformation, one may define the free energy increase in the following form[19]

$$W^S = 2\int_{\text{spherical drop}}^{\text{deformed drop}} dS \, \gamma(S) \tag{10.24}$$

where S is the droplet surface area and $\gamma(S)$ is the interfacial tension of the deformed drop which can be expressed by the interfacial tension of a *single spherical droplet*, γ_0, and the effective Gibbs elasticity of the adsorbed surfactant layer, E_G

$$\gamma(S) = \gamma_0 + E_G \ln\left(\frac{S}{S_0}\right) \text{ and } E_G = \left(\frac{d\gamma}{d\ln S}\right)_{S=S_0} \tag{10.25}$$

Eqn. (10.25) is valid for small deformations and E_G is assumed to be constant. Substituting (10.25) into (10.24) we obtain

$$W^S = 2(\gamma_0 - E_G)(S - S_0) + 2E_G S \ln\left(\frac{S}{S_0}\right) \tag{10.26}$$

where $S_0 = 4\pi a^2$ is the surface area of a non-deformed drop.

Our main interest is in small drops, where the capillary pressure is high and the relative surface extension is small, *viz.* $(S - S_0)/S_0 \ll 1$. In this case, eqn. (10.26) becomes

$$W^S = 2\gamma_0(S - S_0) + E_G S_0 \left(\frac{S - S_0}{S_0}\right)^2 \tag{10.27}$$

The first term on the right-hand side of eqn. (10.27) accounts for the surface area extension at constant interfacial tension, while the second is due to the possible increase in the interfacial tension along with the deformation. The surface extension energy could be expressed as a function of the film radius and for small deformations $(r/a \ll 1)$ reads[20]

$$W^S(r) = \pi a^2 \left(\frac{\gamma_0}{2}\frac{r^4}{a^4} + \frac{E_G}{64}\frac{r^8}{a^8}\right) \text{ for } \frac{r^2}{a^2} \ll 1 \tag{10.28}$$

Again, all terms that might be due to the increase of the initial droplet radius with the deformation are neglected since they are of higher order of magnitude.[20] In emulsions, stabilised by surfactants, soluble in the continuous and/or the droplet phase, the adsorption surfactant flux could maintain constant interfacial tension during the deformation. If this is the case, the effective Gibbs elasticity [see eqn. (10.25)] is zero and only the first term in (10.28) remains; hence

$$W^{\mathrm{S}}(r) = \pi a^2 \left(\frac{\gamma_0}{2} \frac{r^4}{a^4} \right) \tag{10.29}$$

Even when the Gibbs elasticity, E_{G}, is of comparable magnitude with the interfacial tension, γ_0, one may still use eqn. (10.29) since the second term in (10.28) decays much faster for small film radii and is being divided also by a factor of 64. There are systems, however, where the Gibbs elasticity could dominate. These are some microemulsions or vesicles, where the interfacial tension itself is extremely low.

10.2.3.2 Surface Bending Energy

The bending energy contribution is usually supposed to be important when studying lipid bilayers and microemulsions.[39,40] In some cases, however, it appears to be significant even for larger droplets. According to Helfrich,[39] the bending energy per unit area of a spherical interface is given by the following expression

$$w^{\mathrm{c}} = 2k_{\mathrm{c}}(H - H_0)^2 = B_0 H + 2k_{\mathrm{c}}H^2 + 2k_{\mathrm{c}}H_0^2 \tag{10.30}$$

where k_{c} is the bending elasticity constant, $H = -1/a$ is the interfacial curvature and $H_0 = -1/R_0$ is the spontaneous curvature (a and R_0 are the respective radii of curvature), and $B_0 = -4k_{\mathrm{c}}H_0$ is the interfacial bending moment of a flat interface (see *e.g.* ref. 41). Then the corresponding contribution to the droplet deformation energy can be found[20] from the difference between the droplet curvature energy before and after the film formation, respectively:

$$W^{\mathrm{c}}(r) = -2\pi r^2 B_0 H \left(1 - \frac{H}{2H_0} \right) \text{ for } (r/a)^2 \ll 1 \tag{10.31}$$

An alternative form of eqn. (10.31) is

$$W^{\mathrm{c}}(r) = 8\pi k_{\mathrm{c}} \left(\frac{r}{a} \right)^2 \frac{a}{R_0} \left(1 - \frac{R_0}{2a} \right) \tag{10.32}$$

The Gaussian term has been neglected in the above consideration since there is no relevant change in the topology of the system (see Figure 10.1). For microemulsion interfaces, k_{c} was measured[42] to be of the order of the thermal energy, kT. The theoretical calculations[43] show that usually $|B_0|$ is of the order of 5×10^{-11} N for emulsion interfaces. If such is the case, one can estimate that for emulsion systems H_0 is of the order of nm^{-1}. Since the curvature of the emulsion droplets, H, is typically several orders of magnitude smaller ($H \approx 10^{-3}$ nm^{-1}), the second term in the parentheses in eqn. (10.31) can be neglected and one obtains

$$W^c(r) = -2\pi r^2 B_0 H \quad (r/a)^2 \ll 1 \text{ and } |R_0/a| \ll 1 \qquad (10.33)$$

Assuming $r \approx a/50$, $a = 3 \times 10^{-5}$ cm and $|B_0| = 5 \times 10^{-11}$ N, one finds $|W^c| \approx 10kT$.[27] In other words, the bending effects can be important for the interaction between relatively large, submicron emulsion droplets as well.

The signs of the curvature H and the bending moment B_0 are a matter of convention. In the present consideration we accept that positive is the curvature of an oil droplet in water. However, the sign of the product $B_0 H$ (and W^c) is independent of the choice of definition (see *e.g.* ref. 27). In the case considered above, $W^c < 0$ for aqueous droplets in oil, whereas $W^c > 0$ for oil droplets in water. The bending energy contribution can vary with variation in the experimental conditions (electrolyte concentration for ionic surfactants or temperature for nonionic surfactants).

The significant contribution of the interfacial bending energy (eqn. 10.33) to the total interaction between emulsion droplets of micrometre size may seem quite surprising. Nevertheless, this is due to the fact that the bent area increases faster ($r^2 \propto a^2$) than the bending energy per unit area decreases ($H \propto 1/a$) when the droplet radius, a, increases, *cf.* eqn. (10.33), and the relatively small bending energy density per unit area is multiplied by the large film area to give the overall contribution. Hence, the bending effects are also important in emulsion systems of low interfacial tension.

10.2.4 Applications of the Expressions for the Pair Interaction Energy

10.2.4.1 Emulsions Stabilised by Ionic Surfactants—van der Waals, Electrostatic and Surface Area Extension Contributions

We start with considering emulsion (or microemulsion) systems stabilised with ionic surfactants, where the van der Waals attraction and electrostatic repulsion are the dominating forces, determining the dispersion stability. This resembles the well known DLVO theoretical model.[1-3] The droplet deformability, however, requires us to account also for the (i) surface area extension contribution and (ii) modification of the expressions for the van der Waals attraction and electrostatic repulsion which are taking place along with the droplet shape change. For clarity the bending contribution will not be taken into account, although it might be important and its consideration is straightforward (see above). The pair interdroplet energy, W, reads

$$W(h, r) = W^{VW}(h, r) + W^{EL}(h, r) + W^S(r) \qquad (10.34)$$

where different terms on the right-hand side of the above expression could be given by eqns. (10.3), (10.10) and (10.29), respectively.[26,27] An illustrative calculation is depicted in Figure 10.2.

The electrolyte concentration for all the cases shown in the figure corresponds to 0.1 M monovalent salt. Figure 10.2a presents a contour diagram of the pair interaction energy $W(h, r)$ for 1 micrometre sized droplets with surface potential $\Psi_0 = 100$ mV, interfacial tension $\gamma = 1$ mN m^{-1} and Hamaker constant $A_H = 1 \times 10^{-20}$ J. Figure 10.2b is for droplets with the same set of parameters except the Hamaker constant, which is $A_H = 2 \times 10^{-20}$ J. It is seen that the stronger attraction leads to a well defined minimum, corresponding to the formation of a plane parallel film between the droplets with finite radius and thickness. We should stress that this minimum corresponds to the far minimum in the classical DLVO theory, where the longer ranged van der Waals attraction exceeds the electrostatic repulsion. Figure 10.2c has the same parameters as Figure 10.2a except for the surface potential which is halved, *i.e.* $\Psi_0 = 50$ mV. The decreased electrostatic repulsion also gives rise to a well defined minimum in the energy, corresponding to two flocculated droplets, separated by a planar film. Note also that this minimum is shifted towards

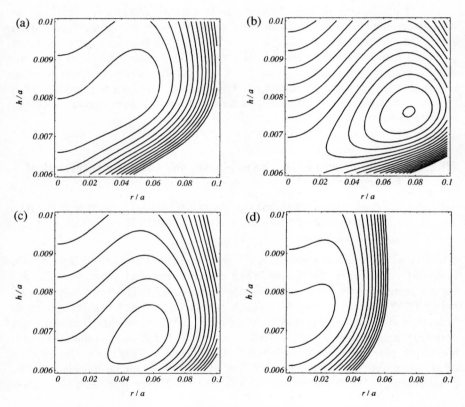

Figure 10.2 *Contour diagram of the interaction energy of two charged deformable droplets (radius, a $= 1$ μm) as a function of the film radius r and film thickness h. Only the negative values of the energy are shown starting from 0. The spacing between the contours corresponds to 2kT. (a) $A_H = 1 \times 10^{-20}$ J, $\gamma = 1$ mN m^{-1}, $\Psi_0 = 100$ mV; (b) $A_H = 2 \times 10^{-20}$ J, $\gamma = 1$ mN m^{-1}, $\Psi_0 = 100$ mV; (c) $A_H = 1 \times 10^{-20}$ J, $\gamma = 1$ mN m^{-1}, $\Psi_0 = 50$ mV; (d) $A_H = 1 \times 10^{-20}$ J, $\gamma = 5$ mN m^{-1}, $\Psi_0 = 100$ mV. The rest of the parameters are given in the text*

smaller film thickness because the electrostatic repulsion decays more rapidly with the separation when the surface potential is lower. Figure 10.2d illustrates the effect of increasing the interfacial tension of the droplet surface. The parameters are again as those in Figure 10.2a, but the interfacial tension is five times greater, *i.e.* $\gamma = 5$ mN m^{-1}. This makes the droplet interface 'harder' and therefore less prone to extension and deformation upon interdroplet interactions. That is why all the negative energy values shown in the contour diagram are shifted towards smaller values of *r*. Larger deformations are rather unfavourable, because of the greater interfacial tension. The film formation could depend also on other parameters such as the electrolyte concentration or the droplet radius, which is shown in Figure 10.3. The parameters used in the calculation are as in Figure 10.2a with $A_H = 1 \times 10^{-20}$ J. The different curves are for different droplet sizes: $a = 0.5$ μm (dashed curve), $a = 1.0$ μm (solid curve) and $a = 2.0$ μm (dotted curve). These results imply that micrometre and even submicron droplets could deform (forming a plane parallel film) under the action of the direct interdroplet interactions at reasonable conditions. The general conclusion from Figures 10.2 and 10.3 is that the stronger the repulsion the weaker the deformation.

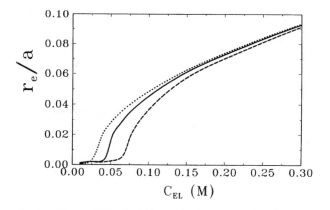

Figure 10.3 *Equilibrium droplet radius* vs. *electrolyte concentration. The full line is for* a $= 1$ μm, *the dashed is for* a $= 0.5$ μm *and the dotted is for* a $= 2$ μm. *The remaining parameters used in the calculations are given in the text* (Reproduced by permission of Academic Press from ref. 27)

In the case discussed above, we may roughly consider the plane (film) parts of the droplet to be in the repulsive region of the curve (the barrier), while the spherical surfaces surrounding the film are in the range of the secondary minimum. The balance between these two contributions provides a mechanical equilibrium, necessary for the stable doublet of droplets to be formed (Figure 10.1). Often the electrostatic repulsion is strong enough and such consideration (flocculation in the secondary minimum) is completely relevant. If the barrier is not sufficiently high, it could be overcome and the droplets may flocculate in the primary minimum, which is due to the van der Waals attraction term (dominating again the electrostatic repulsion at very short distances) and balanced by a short ranged repulsion, W^{SR}, due to the excluded volume of ions and molecules in the film. For the latter we choose the following formal expression

$$W^{SR}(h) = \frac{\pi r^2 B_R a^5}{h^7} \qquad (10.35)$$

where B_R is a constant of short ranged repulsion. The above expression corresponds to the repulsive part in the Lennard-Jones intermolecular potential, after appropriate integrations over the volume of the interacting macroscopic colloidal particles.[44] This term should be added to eqn. (10.34) in order to obtain

$$W(h, r) = W^{SR}(h) + W^{VW}(h, r) + W^{EL}(h, r) + W^S(r) \qquad (10.36)$$

There are experimental observations that emulsion droplets may exhibit such a transition[45] at high electrolyte concentrations. At lower electrolyte concentrations, however, the transition over the electrostatic barrier (see Figure 10.4) may lead to coalescence (see ref. 46).

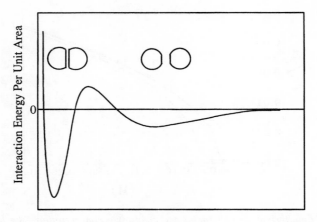

Figure 10.4 *A sketch of DLVO energy* vs. *distance curve*

The high salt concentration leads to a considerable screening of all electrostatic repulsions, including the lateral ones between the charged surfactant headgroups. This results in higher surfactant adsorption at the droplet interface, since the excluded area per adsorbed surfactant molecule effectively decreases, and combined with the counterions in the film may lead to an ordered quasi-crystalline structure.[47] Such a structure may provide repulsion interaction effectively expressed by eqn. (10.35). At lower salt concentrations the adsorption of surfactants should be lower due to the stronger lateral repulsions, and the formation of such ordering and the short range repulsion may not be present. That is why the droplets coalesce once they overtake the electrostatic barrier.[46] All these phenomena are well known in macroscopic foam and emulsion films, where the transition from common (secondary) to Newton black (primary) films was observed and studied in detail.[48]

Equation (10.36) allows us to calculate the contour diagram corresponding to the case where the energy per unit area has the form schematically depicted in Figure 10.4. Such a contour diagram is shown in Figure 10.5. The different terms in eqn. (10.36) are given by eqns. (10.3), (10.5), (10.8b), (10.29) and (10.35). The parameters are: $a = 1$ μm, $A_H = 6 \times 10^{-21}$ J, $B_R = 5$ kT, $C_{el} = 0.6$ м, $\Psi_0 = 25$ mV, $\gamma = 4$ mN m^{-1} and $T = 298$ K. The result obtained (Figure 10.5) should be regarded as a qualitative one. Still, it illustrates the basic features of a doublet of flocculating droplets quite successfully. At the far secondary minimum the droplets are separated by a relatively thick plane parallel film, $h/a = 0.002$, but as the system moves to the primary minimum the film thickness decreases and becomes $h/a = 0.0004$. The latter value equals 0.4 nm, which is of the order of a hydrated ion diameter. It is important to emphasise that this reasonable value is achieved by the choice of the repulsion parameter B_R only, since our approach does not account for the structure and excluded volume of the ions explicitly. Moreover, as mentioned above, the electrostatic energy expression we employed is not strictly correct at such separations. The film radius also shows a rather reasonable (qualitative) behaviour. It is smaller at the secondary minimum ($r/a = 0.081$) and larger at the primary one ($r/a = 0.198$). This is also true for the contact angle, α, calculated from

$$\alpha = \arcsin(r/a) \qquad (10.37)$$

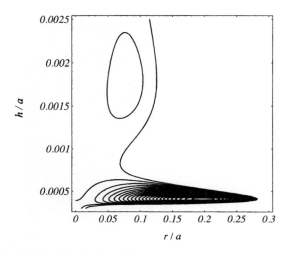

Figure 10.5 *Contour diagram for the interaction energy between two droplets with two minima corresponding to the primary and secondary minima in the DLVO model. The parameters used in the calculation are given in the text. Only the negative values of the energy are shown, starting from 0. The spacing between the contours corresponds to 100 kT*

Thus $\alpha = 4.6°$ at the secondary minimum and $\alpha = 11.0°$ at the primary one. These values also seem to be reasonable and close to those experimentally observed.[45] It is seen from Figure 10.5 that the trajectory of the transition between the two minima is complex compared to the case of solid particles, where no deformation is present. The film radius decreases almost to a zero value in the transition region, which is due to the fact that repulsive interactions do not favour formation of a plane parallel film. The exact trajectory could be important if one is interested in the kinetics of the process. We recall that the whole consideration is based on the idealised, truncated spherical shape of the interacting droplets. A more realistic approach is to solve the Laplace equations for the droplet surfaces. This, however, would lead to much greater computational difficulties without changing qualitatively and in many cases even quantitatively the main results and conclusions[49] (see also below).

10.2.4.2   Depletion Flocculation and Steric Stabilisation

The addition of smaller colloidal species (polymer molecules, micelles, microemulsions) to an emulsion may lead to destabilisation of the system due to depletion attraction or steric stabilisation if these species adsorb at the droplet interface. A model calculation, illustrating the depletion destabilisation, is presented in Figure 10.6. All the parameters are as in Figure 10.2a but in this case nonionic micelles are also present [see eqns. (10.17)–(10.19)]. The micellar diameter is chosen to be 10 nm and the volume fraction equals 0.1.

It is seen that the addition of micelles leads to the formation of a well defined minimum which is about $10kT$ deeper than that in the absence of depletion

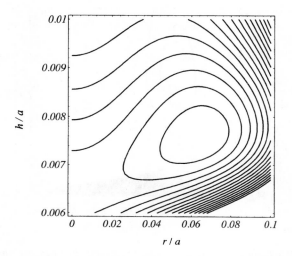

Figure 10.6 *Contour diagram of the interaction energy with depletion interactions, due to the presence of nonionic micelles (see text for the parameters). Only the negative values of the energy are shown, starting from 0. The spacing between the contours corresponds to 2 kT*

interactions. Hence, these interactions can enhance the flocculation in emulsion systems. This fact has been used to develop a procedure for size fractionation of emulsion droplets in order to obtain monodisperse samples.[50] At higher volume fractions of the added colloids the depletion attraction may transform into oscillatory (structural) interactions[27] [see eqn. (10.23) above]. Such is the case presented in Figure 10.7.

The calculations are for a real heptane-in-water emulsion system, stabilised by sodium bis(2-ethylhexyl)sulfosuccinate (AOT) in the presence of microemulsion droplets.[51] The electrostatic and van der Waals energy contributions are also included. The parameters of the system are experimentally determined or taken from the literature: drop radius $a = 1.55$ μm, microemulsion droplet diameter $d = 10$ nm and volume fraction $\Phi = 0.04$, $\Psi_0 = 80$ mV, $\gamma = 0.12$ mN m^{-1} and $A_H = 4 \times 10^{-21}$ J. Figure 10.7a reveals the situation where the most stable is a doublet of droplets with a layer of microemulsion droplets in the separating film. The depletion minimum, where all the microemulsion droplets are expelled from the film, is less deep. This is due to the fact that for that particular electrolyte concentration, 0.025 M, the microemulsion droplet diameter, including the counter-ion atmosphere κ^{-1}, is roughly equal to the film thickness determined from the balance between the electrostatic and van der Waals interactions, and a kind of synergism is observed. If the amount of added electrolyte is decreased by a factor of 2 (0.0125 M) the depletion minimum disappears completely because of the strong electrostatic repulsion dominating at such distances, shown in Figure 10.7b. On the other hand, if the electrolyte concentration is doubled (0.05 M) the depletion attraction becomes dominant and stable flocculation takes place without micro-emulsion droplets in the film (see Figure 10.7c).

The case of steric stabilisation of emulsions is illustrated in Figure 10.8. The energy contributions for this calculation are given by eqns. (10.3), (10.12) and (10.29) with the following parameters: number of monomer units in a polymer chain $N = 25$, length of a monomer unit $l = 0.6$ nm, $\Gamma = 1.15 \times 10^{-14}$ mol m^{-2} ($A_0 = 1.44$ nm^2), $L_0 = 3$ nm. These parameters roughly correspond to an emulsion stabilised by nonionic surfactant, *e.g.* alkylpolyoxyethylene with 25 ethoxy groups. The remaining parameters are the same as in Figure 10.2a. The adsorbed layer acts similarly to an electrostatic repulsion.

10.2.4.3 Temperature Dependent Interactions in Nonionic Micro-emulsions — Bending Energy Contribution

Microemulsions are thermodynamically stable systems of oil and water stabilised by surfactant. Such systems may exhibit different microstructures like droplet (oil in water or water in oil) or bicontinuous, depending on the conditions.[52] In the consideration outlined below, we confine ourselves to single phase microemulsions consisting of oil droplets in water, stabilised by nonionic surfactant.[53] The experimental study[54] of the phase behaviour of such microemulsions shows that they are stable within a certain temperature range which depends also on the ratio of oil to surfactant. Turbidity measurements were performed which showed that the droplets become apparently less repulsive with temperature. The latter fact could be due to

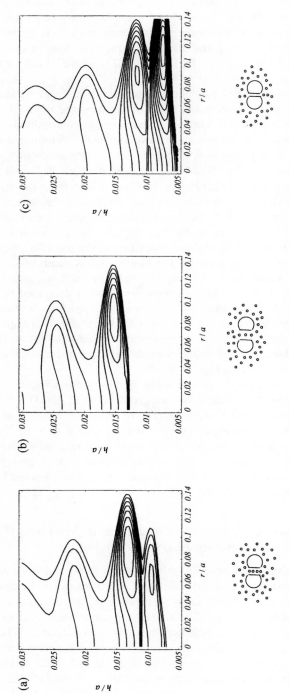

Figure 10.7 *Contour diagram of the interaction energy of two deformable emulsion drops in the presence of microemulsion droplets. Only the negative values of the energy are shown, starting from 0. The spacing between the contours corresponds to 2 kT. The sketches below each contour diagram represent a qualitative illustration of the respective droplet configuration in the deepest minimum. Neither the film dimensions nor the volume fraction of micelles in the sketches are scaled correctly: (a) The electrolyte concentration corresponds to 0.025 M; (b) the electrolyte concentration is 0.05 M; (c) the electrolyte concentration is 0.0125 M. The remaining parameters are given in the text*

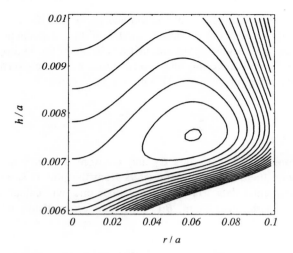

Figure 10.8 *Energy of interaction between two sterically stabilised droplets. The parameters used in the calculation are given in the text*

several reasons: (i) the droplets coalesce and form prolate spheroids,[55] (ii) the droplets become increasingly attractive with temperature and tend to form clusters[56] and (iii) the droplets become less repulsive due to 'softening' of the oil/water interface which manifests itself as an apparent attraction (compared to hard sphere interaction) but in fact is repulsion between soft spheres. Our treatment of the problem, suggested below, is based on the last hypothesis,[53] although the other explanations could be also valid to a certain extent.

We will consider only the bending energy contribution, W^C, to the overall interdroplet energy, since estimates show that other possibilities like the van der Waals, W^{VW}, and/or surface extension, W^S, seem to be negligible for this particular case.[53] It is convenient to write the bending energy contribution as a function of the distance between the droplet mass centres, z [see Figure 10.1 and eqn. (10.32)]

$$W^C(\tilde{z},\,T) = 8\pi\left[1 - \left(\tilde{z} - \frac{\tilde{\Delta}}{2}\right)^2\right]k_C\frac{1}{\tilde{R}_0(T)}\left(1 - \frac{\tilde{R}_0(T)}{2}\right) \qquad (10.38)$$

where $\tilde{z} = z/2a$ and $\tilde{R}_0(T) = R_0(T)/a$ are the dimensionless interdroplet distance and radius of spontaneous curvature, respectively. With $\tilde{\Delta} = \Delta/a$ is denoted the dimensionless film thickness, which is equal to twice the length of the hydrophilic surfactant headgroups and is constant as the droplets move towards each other. This means that the decrease of $\tilde{z}$ is due to the respective increase of the film radius, r, only. The latter quantity, in dimensionless form, is determined from geometrical considerations (see Figure 10.1)

$$\tilde{r} = \frac{r}{a} = \sqrt{1 - \left(\tilde{z} - \frac{\tilde{\Delta}}{2}\right)^2} \qquad (10.39)$$

The temperature dependence of the interaction energy, W^C, is due to the radius of spontaneous curvature, $\tilde{R}_0(T)$.[54] A simple model of this dependence was suggested.[53] It is based on calculating the interactions between the hydrophilic headgroups of the surfactant covering the microemulsion droplet surface, in the framework of Flory–Huggins theory.[57] This model predicts that

$$\frac{1}{R_0(T)} = C_1 + \frac{C_2}{T} \tag{10.40}$$

Explicit expressions for the constants C_1 and C_2 were derived[53] but it is more convenient to obtain them as fitting parameters from the plot $1/R_0$ vs. $1/T$ from independent experimental data. Having these values one may calculate $W^C(\tilde{z}, T)$ and also the second virial coefficient, $B_2(T)$ of the microemulsion, which has the form[57]

$$
\begin{aligned}
B_2(T) &= \frac{3}{2} \int_0^\infty \left[1 - \exp\left(-\frac{W(\tilde{z}, T)}{kT} \right) \right] \tilde{z}^2 \, d\tilde{z} \\
&= \frac{3}{2} \int_0^{(1+\frac{\delta}{2})} \left[1 - \exp\left(-\frac{W(\tilde{z}, T)}{kT} \right) \right] \tilde{z}^2 \, d\tilde{z} \\
&\quad + \frac{3}{2} \int_{(1+\frac{\delta}{2})}^\infty \left[1 - \exp\left(-\frac{W(\tilde{z}, T)}{kT} \right) \right] \tilde{z}^2 \, d\tilde{z}
\end{aligned}
\tag{10.41}
$$

Knowing the second virial coefficient one may also calculate the compressibility and, hence, the turbidity of the microemulsion.[58] For our purpose it seems convenient to scale the turbidity of the system under consideration with that at the lower temperature boundary, assuming the latter to be as for hard spheres, *viz.*

$$\frac{\tau}{\tau^{HS}} = \frac{1 + 2B_2^{HS}\Phi}{1 + 2B_2(T)\Phi} \tag{10.42}$$

Figure 10.9 shows the application of the above expression to some turbidity data of Fletcher and Morris.[54]

The simple model[53] allows a qualitative description of the experimental results on the basis of the concept of 'softening' of the microemulsion droplets with temperature. Furthermore, looking at eqn. (10.38) one observes that the pair energy contribution to the interdroplet interaction becomes zero for $R_0(T) = 2a$. It was shown[53] that the respective temperature where the bending contribution becomes zero corresponds to the transition from an isotropic droplet phase to a bicontinuous structured microemulsion. This implies that the higher order terms in the expansion of the curvature energy are negligible.[39] The accomplishments of the model make us believe that the temperature dependence of the droplet deformability is an important factor governing the nature of the microemulsion droplet interactions and the macroscopic behaviour of such systems (see also ref. 25 and Section 10.3).

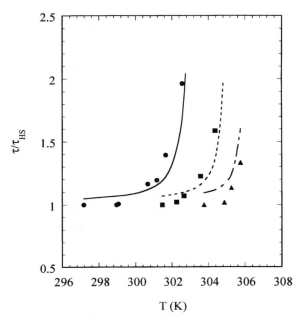

Figure 10.9 *Turbidity* vs. *temperature for nonionic O/W microemulsion droplets. The details are given in the text. The three sets correspond to three different volume fractions and droplet sizes, viz.* a = 6.68 nm *and* ϕ= 0.0485, *solid curve;* a = 10.06 nm, ϕ = 0.073, *dashed curve; and* a = 13.71 nm, ϕ = 0.0995, *dashed-dotted curve. The radius of spontaneous curvature varies with tempera-ture according to eqn. (10.40). The constants are* C_1 = −4.3 nm *and* C_2 = 1320 K nm⁻¹
(Reproduced by permission from ref. 53)

10.2.5 Accuracy of the Model Shape — Actual Geometry of Two Interacting Deformable Droplets

All the results described and used above are based on the assumption that the interacting deformable droplets acquire the shape of truncated spheres when sufficiently strong attraction is present. This is certainly an approximation and needs to be compared with results based on the actual shape of the droplets. The latter could be obtained by solving the augmented (including interactions) Laplace equations[59] for the fluid droplet interfaces

$$\frac{\mathrm{d}[\gamma(x)\sin\phi(x)]}{\mathrm{d}x} + \frac{1}{x}\gamma(x)\sin\phi(x) = P_\mathrm{c} - \Pi(x)$$

$$\frac{\mathrm{d}H(x)}{\mathrm{d}x} = 2\tan\phi(x) \qquad (10.43)$$

$$\frac{\mathrm{d}\gamma(x)}{\mathrm{d}x} = -\Pi(x)\sin\phi(x)$$

where $H(x)$ is the unknown function of the droplet surface shape, $\gamma(x)$ is the interfacial tension, $\phi(x)$ is the running slope angle, $\Pi(x)$ is the disjoining pressure between the surfaces and P_c is the capillary pressure. The respective boundary conditions, necessary to solve the equations in (10.43), are

$$H_B \equiv H(x_B) = 2(y_0 - \sqrt{a_s^2 - x_B^2})$$

$$\tan \phi(x_B) = \frac{x_B}{\sqrt{a_s^2 - x_B^2}} \qquad (10.44)$$

$$\gamma(x_B) = \gamma^l + \frac{1}{2}f(H_B)$$

where the interaction energy per unit area, $f(h)$, in an infinite flat film of thickness h is related to the disjoining pressure, $\Pi(H)$, through the expression

$$f(h) = \int_h^\infty \Pi(H)\,\mathrm{d}H \qquad (10.45)$$

The geometrical parameters y_0 and a_s are determined from the conditions for constant droplet volume (details of the computational procedure are given in ref. 49)

$$V = \frac{4}{3}\pi a^3 = \frac{1}{2}\int_{h_0}^{H_B} \pi x^2(H)\,\mathrm{d}H + \frac{\pi p^2(3a_s - p)}{3} \quad \text{where } p = a_s + \sqrt{a_s^2 - x_B^2}$$

$$(10.46)$$

and mechanical equilibrium

$$F = \int_{x=0}^{x_B} \Pi(x)2\pi x\,\mathrm{d}x = 0 \qquad (10.47)$$

Here, a is the radius of the non-deformed droplet, $x(H)$ is the droplet shape function in the region $h_0/2 \leqslant y \leqslant x_B$ and F is force. Knowing the shape of the droplet surfaces [obtained from the solution of eqns. (10.43), (10.44) (10.46) and (10.47)], one may proceed to calculate the contributions to the interdroplet energy. Thus we obtain the surface area extension energy

$$W^S = 2\gamma\left[2\pi\left(\int_0^{x_B} \mathrm{d}x\, x\sqrt{1 + \tan^2 \phi(x)}\right) + 2\pi a_s p - 4\pi a^2\right] \qquad (10.48)$$

the van der Waals energy

$$W^{VW} = 2\pi\int_0^{x_B} \mathrm{d}x\, x\sqrt{1 + \tan^2 \phi(x)}\left[-\frac{A_H}{12\pi H^2(x)}\right] \qquad (10.49)$$

and the electrostatic energy

$$W^{\mathrm{EL}} = 2\pi \int_0^{X_{\mathrm{B}}} \mathrm{d}x\, x\sqrt{1 + \tan^2\phi(x)}[64C_{\mathrm{el}}kT\kappa^{-1}\exp(-\kappa H)] \qquad (10.50)$$

The comparison of the results based on the Laplace equations (described above) to those obtained assuming the model shape of truncated spheres, showed that both approaches are in very good agreement within 7%;[49] see also Table 10.1. This (exact) approach allows a thermodynamic treatment of doublets of droplets by an analogy with macroscopic liquid foam and emulsion films,[59] involving the concept of line and transversal tensions, thermodynamic film radius and contact angle. Such an analysis is presented in ref. 49.

Table 10.1 *Comparison of the results obtained by the approach based on (i) the numerical integration of the augmented Laplace equation and (ii) the model approach (shape of truncated spheres)*

$a/\mu m$	W/kT		W^{VW}/kT		W^{EL}/kT		W^{S}/kT	
	(i)	*(ii)*	*(i)*	*(ii)*	*(i)*	*(ii)*	*(i)*	*(ii)*
1	−46.9	−39.6	−70.7	−65.0	12.4	11.6	11.3	10.8
2	−117.5	−109.7	−200.0	−192.1	37.6	36.8	44.8	45.6
3	−210.6	−200.4	−386.3	−379.9	75.4	75.5	100.3	103.9
4	−325.9	−314.7	−629.4	−628.2	125.8	127.8	177.8	185.8
5	−463.5	−452.4	−929.5	−947.0	188.7	195.7	277.2	298.9
6	−623.3	−613.7	−1286.3	−1315.1	264.3	274.6	398.8	396.7
7	−805.2	−798.6	−1700.1	−1739.5	352.5	366.1	539.3	574.8
8	−1009.4	−1006.9	−2170.6	−2229.5	453.3	472.3	708.0	750.3
9	−1235.7	−1238.6	−2698.0	−2748.3	566.6	584.7	895.7	925.0
10	−1484.2	−1493.6	−3282.4	−3327.5	692.8	710.7	1105.4	1123.2

[a]The droplet parameters are: $A_{\mathrm{H}} = 10^{-20}$ J, $\gamma = 1$ mN m^{-1}, $C_{\mathrm{EL}} = 0.2$ M, $\Psi_0 = 100$ mV.

10.3 Dense Systems

This section presents a theoretical study of more concentrated deformable emulsions and microemulsions where higher order interactions become important.[25] The purpose is to relate the microscopic droplet deformability to the structure of such systems and further to their macroscopic (thermodynamic) properties. The radial distribution function and static structure factor are calculated utilising an integral equation approach in an appropriate closure approximation.[60] This method allows us to obtain the virial equation of state as well. A semi-empirical equation of state, based on modifying the Carnahan–Starling expression,[36] as well as comparison with Brownian dynamics simulations are also presented.

10.3.1   Nonionic Emulsions and Microemulsions

We start with the simpler case of emulsions and/or microemulsions, stabilised with nonionic surfactant. This means that there are no charge effects present and in the extremely simplest situation (all long range forces are neglected, including van der Waals) we may account only for the surface bending. For a small deformation[25,26]

$$W^C(\tilde{z}) = 16\pi k_C \frac{1}{\tilde{R}_0}\left(1 - \frac{\tilde{R}_0}{2}\right)(1 - \tilde{z}) \quad \tilde{z} < 1$$

$$= 0, \quad \tilde{z} > 1 \tag{10.51}$$

or area extension

$$W^C(\tilde{z}) = 2\pi a^2 \gamma (1 - \tilde{z})^2 \quad \tilde{z} < 1$$

$$= 0 \quad \tilde{z} > 1 \tag{10.52}$$

energies in order to illustrate the effect of droplet deformation more clearly. In fact, these two contributions are often dominant for some experimental systems like emulsions or microemulsions stabilised by nonionic surfactants.[53]

10.3.1.1   *Percus–Yevick Approximation*

A convenient way to obtain the thermodynamic properties of dense colloidal suspensions is to employ the Ornstein–Zernike integral equation approach for calculating the radial distribution function of the particles (deformable fluid droplets in our case).[60–62] This equation reads

$$h(\tilde{z}) = c(\tilde{z}) + \frac{6\Phi}{\pi}\int_0^\infty d\tilde{z}' c(\tilde{z}')h(|\tilde{z} - \tilde{z}'|) \tag{10.53}$$

where $h(\tilde{z}) = g(\tilde{z}) - 1$ is the total correlation function, while the unknown direct correlation function, $c(\tilde{z})$, could be derived by means of the Percus–Yevick closure[60,62]

$$c(\tilde{z}) = \left[1 - \left(\frac{W(\tilde{z})}{kT}\right)\right]g(\tilde{z}) \tag{10.54}$$

where kT is the thermal energy. We chose the Percus–Yevick approximation since it is known to give accurate results for relatively short ranged interaction potentials.[62] The disperse medium is not taken into account explicitly. Eqns. (10.53) and (10.54) allow us to determine the radial distribution function in microemulsions or miniemulsions assuming that the pair interaction energy has the form of eqns. (10.51) or (10.52), respectively. The calculation of the thermodynamic properties using the radial distribution function is a straightforward statistical mechanical problem. For example, the equation of state has, in general, the following form[60]

$$\frac{P_{OS}}{nkT} = 1 - \frac{\Phi}{\pi kT} \int_0^\infty d\tilde{z} \frac{dW(\tilde{z})}{d\tilde{z}} g(\tilde{z})\tilde{z}^3 \qquad (10.55)$$

where P_{OS} is the osmotic pressure and n is the particle number density. The particular properties of the system (due to the specific type of the interactions) are hidden in the radial distribution function $g(\tilde{z})$ and pair energy $W(\tilde{z})$.

Another useful quantity is the structure factor, $S(\tilde{q})$, defined by[60,63]

$$S(\tilde{q}) = 1 - 24\Phi \int_0^\infty d\tilde{z}\, \tilde{z}^2 \left(\frac{\sin \tilde{q}\tilde{z}}{\tilde{q}\tilde{z}} \right) [1 - g(\tilde{z})] \qquad (10.56)$$

where $\tilde{q} = 2qa$ is the dimensionless wave vector. The structure factor could be directly measured by light scattering techniques.[63]

Therefore the integral equation approach allows us to obtain all the information about the thermodynamic properties of mini- and microemulsions and to compare (in some cases) the results with relevant experimental data or computer simulations.

10.3.1.2 Modified Carnahan–Starling Equation for Dispersions of Deformable Droplets

The Carnahan–Starling expression[36,60,61] was originally derived for hard spheres and represents a semi-empirical relationship which, however, remains the most accurate equation of state in analytical form for such systems

$$\frac{P_{OS}}{nkT} = \frac{1 + \Phi + \Phi^2 - \Phi^3}{(1 - \Phi)^3} \qquad (10.57)$$

It was later modified to include an attractive contribution,[60-62] regarding the latter as a small perturbation. It is therefore tempting to modify the Carnahan–Starling equation of state in such a way that it would become applicable to deformable fluid droplets. Unfortunately the perturbation approach is not relevant to the case of deformable fluid droplets. This becomes clear if one writes the perturbation term, P_{OS}^1, for the osmotic pressure[60-62]

$$\frac{P_{OS}^1}{nkT} = 12\Phi \int_0^\infty d\tilde{z} \frac{W^1(\tilde{z})}{kT} g^{HS}(\tilde{z})\tilde{z}^2 \qquad (10.58)$$

The quantity $g^{HS}(\tilde{z})$ is the hard sphere radial distribution function and it turns the integral to zero for the interparticle distance range below $2a$. On the other hand, this is exactly the range where the perturbation interactions, $W^1(\tilde{z})$, due to the droplet deformability are supposed to take place [see expressions (10.1) and (10.2) above]. Hence the perturbation approach in its original form[61,62] is inconsistent with the basic property of the system we are interested in: deformability of the fluid particles (droplets). We overcome this problem by suggesting an alternative approach.[25] It is based on the introduction of an effective droplet volume fraction

$$\Phi^{\mathrm{EFF}} = \frac{B_2}{B_2{}^{\mathrm{HS}}} \, \Phi \tag{10.59}$$

with B_2 being the second osmotic virial coefficient

$$B_2 = 12 \int_0^\infty \mathrm{d}\tilde{z} \left[1 - \exp\left(\frac{W(\tilde{z})}{kT} \right) \right] \tilde{z}^2 \tag{10.60}$$

and the energy $W(\tilde{z})$ is given by the expressions (10.1) or (10.2). $B_2{}^{\mathrm{HS}}$ is the hard sphere virial coefficient. The effective volume fraction, defined by eqn. (10.59), is then introduced into the Carnahan–Starling formula (10.57) and we obtain an expression relating the emulsion (or microemulsion) osmotic pressure to the droplet volume fraction and their interfacial properties: interfacial tension and bending elasticity. This approach is less rigorous than that based on solving the integral equation (10.53), but is much simpler and does not require substantial numerical effort.[25]

10.3.1.3 Thermodynamic Properties of Dense Micro- and Mini-emulsions—Radial Distribution Functions, Structure Factors and Osmotic Pressure

As we discussed above, knowing the radial distribution function of a multiparticle system, one may obtain all its thermodynamic properties. The radial distribution function for a microemulsion is shown in Figure 10.10a. The pair energy has the form of eqn. (10.51) with parameters: $k_C = kT$, $\tilde{R}_0 = 1$ and $\Phi = 0.42$ (see ref. 25). The droplet deformability is clearly manifested by the non-zero values of the radial distribution function (solid curve) for separations less than the droplet diameter ($\tilde{z} < 1$). This means that the droplet centres could approach closer due to the deformation, which is certainly impossible for hard spheres with the same size and volume fraction (dotted curve). The height of the peaks is lower for microemulsions, which means that the local ordering decays more rapidly if the droplets are prone to deformation upon contact. Hence, the deformation of the droplets changes the form of the radial distribution, compared to the respective hard sphere system, and must therefore affect all the properties of the microemulsion. An obvious example is the structure factor of the microemulsion shown in Figure 10.10b with the solid curve and compared to a hard sphere suspension with the same volume fraction and particle size (dotted curve). It also illustrates that the local ordering of microemulsions is suppressed because of the droplet deformability in comparison with its hard sphere reference system. The structure factor is a quantity that could be directly measured by light scattering experiments.[6,63]

A similar situation occurs with larger miniemulsions, where the droplet deformability could be due to the surface extension only [*i.e.* to the interfacial tension,

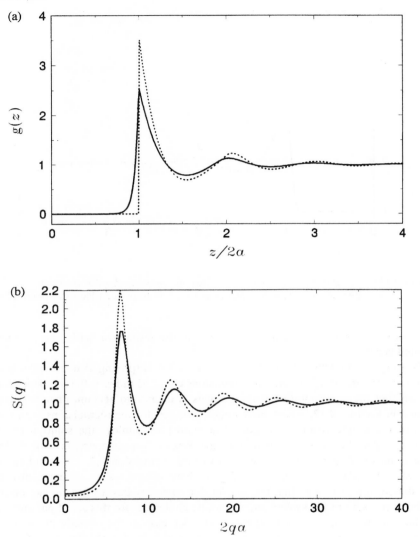

Figure 10.10 (a) *Radial distribution function and* (b) *structure factor for nonionic micro-emulsions. More details are given in the text*
(Reproduced by permission of the American Institute of Physics from ref. 25)

see eqn. (10.52)]. Figure 10.11 depicts the radial distribution function for emulsion drops at volume fraction $\Phi = 0.42$ and $\pi a^2 \gamma / kT = 760$. An example of such an emulsion is one with droplet radius $a = 100$ nm and interfacial tension $\gamma = 0.1$ mN m^{-1} at a temperature equal to 25 °C. Despite the different origin of the drop deformation, the trends in the radial distribution function form are very similar to the microemulsion case. Again it obtains non-zero values for $\tilde{z} < 1$ and

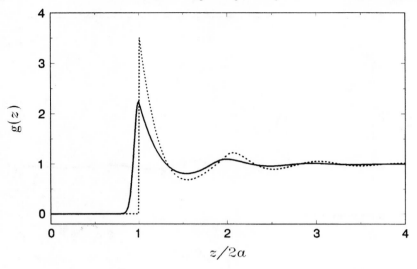

Figure 10.11 *Radial distribution function for nonionic emulsion droplets (see text)*
(Reproduced by permission of the American Institute of Physics from ref. 25)

the peaks are lower (solid curve) compared to the respective hard sphere system (dotted curve).

Bearing in mind the above discussion, it is not surprising that the calculated osmotic pressure of micro- and miniemulsions could be rather different, depending on whether the deformability of the droplets is taken into account or not. This is shown in Figure 10.12, where the osmotic pressure for microemulsions is plotted against the droplet volume fraction. The main plot employs the solution of the Ornstein–Zernike integral equation in the Percus–Yevick approximation.[25] The calculations were performed for two types of microemulsions: $\tilde{R}_0 = 1$ (solid curve) and $\tilde{R}_0 = 1.2$ (dashed curve). The reference hard sphere system is represented by the dotted line. It is seen that the droplet deformability has a great impact on the osmotic pressure of the system and cannot be neglected (for this set of parameters) without introducing a substantial error. The decrease in the osmotic pressure (in comparison with hard spheres) is about 30% for the 'harder' droplets ($\tilde{R}_0 = 1$) and about 40% for the 'softer' ones ($\tilde{R}_0 = 1.2$). This conclusion represents not only fundamental but also practical importance. The osmotic pressure of suspensions is a basic parameter in many applications and processes such as sedimentation (creaming), rheological behaviour, *etc.*[6] The inset in Figure 10.12 shows a comparison between the equations of state, derived by the virial (10.55) and the modified Carnahan–Starling (10.57), (10.59) and (10.60) approaches. The modified Carnahan–Starling equation gives slightly higher pressures for both deformable droplets and hard spheres. A more definite evaluation of both methods could be obtained by comparison with results derived by more elaborate approaches such as Monte Carlo or Brownian dynamics simulations (see below).

Assuming that the modified Carnahan–Starling approach gives reasonable

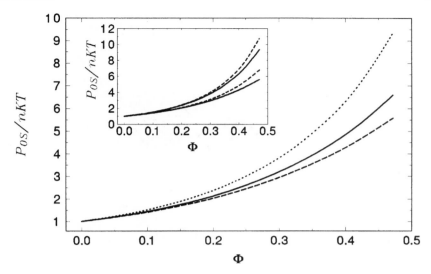

Figure 10.12 *Osmotic pressure vs. volume fraction for nonionic microemulsion with* $\tilde{R}_0 = 1$ *(solid curve),* $\tilde{R}_0 = 1.2$ *(dashed curve), and hard spheres (dotted curve). The inset represents a comparison between the Percus–Yevick (solid curves) approximation used and the Carnahan–Starling (dashed curves) results for hard spheres (upper pair) and* $\tilde{R}_0 = 1.2$ *(lower pair);* $k_C = kT$ (Reproduced by permission of the American Institute of Physics from ref. 25)

results for the osmotic pressure of liquid/liquid dispersions of deformable droplets, we derived three-dimensional plots for the osmotic pressure of microemulsions (Figure 10.13a) and miniemulsions (Figure 10.13b) as functions of the volume fraction of the droplets and the value of the deformation-determining parameters k_C (for microemulsions) and γ (for miniemulsions). As expected, the deformation is most important for a high volume fraction of droplets and low values of the constants k_C and γ.

10.3.1.4 Comparison between Ornstein–Zernike Integral Equation Approach and Brownian Dynamics Simulations for Deformable Droplets

In order to check the accuracy of the results presented above, we performed Brownian dynamics simulations, using an algorithm for a constant number of particles (108 in our case), volume and temperature.[64] The initial temperature was adjusted to give a final equilibrium value corresponding to 25 °C. Figure 10.14 shows the radial distribution functions for a microemulsion with the same parameters as in Figure 10.10a, except for the volume fractions which here are 0.16 (lower curves), 0.26 (middle curves) and 0.42 (upper curves).

The full curves correspond to the Ornstein–Zernike solutions in the Percus–Yevick approximation (*cf.* Figure 10.10a), while the dotted curves were obtained by

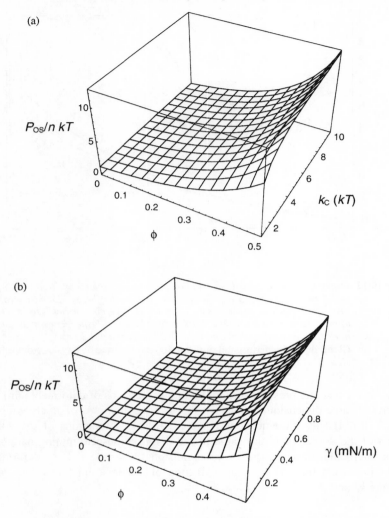

Figure 10.13 *3D plot of the osmotic pressure for microemulsion* (a) *and miniemulsion* (b) *as a function of the volume fraction and the bending elasticity constant* (a) *or interfacial tension* (b), *respectively*
(Reproduced by permission of the American Institute of Physics from ref. 25)

means of Brownian dynamics simulations. As seen from the figure, the agreement between the theoretical approach based on the integral equations and the simulations is excellent. We also compared the virial equations of state (10.55) with the osmotic pressure *vs.* volume fraction dependence for the same type of microemulsion (Figure 10.15). The agreement again is very good, which suggests that the calculations based on solving the Ornstein–Zernike equation are most reliable for investigating the properties of dense systems of deformable droplets.

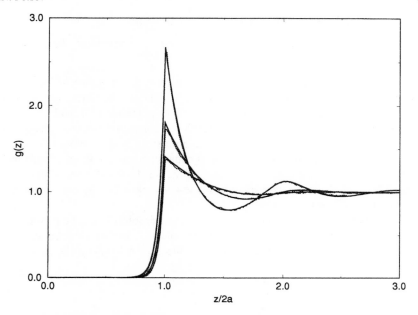

Figure 10.14 *Comparison between radial distribution functions for microemulsions ob-tained by means of the Ornstein–Zernike integral equation and Brownian dynamics simulations. The parameters are as in Figure 10.10a (see text). The volume fractions are: 0.16 (lower curves), 0.26 (middle curves) and 0.42 (upper curves)*

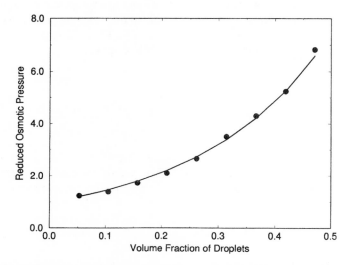

Figure 10.15 *Comparison between the reduced osmotic pressure, P_{OS}/nKT, of microemul-sion obtained by the Ornstein–Zernike integral equation approach (curve) and Brownian dynamics simulations (points)*

10.3.2 Ionic Emulsions and Microemulsions

10.3.2.1 Averaging of the Pair Interaction Energy over the Droplet Deformation

As already discussed in Section 10.2 of this chapter, the presence of long ranged repulsive interactions between the droplets may lead to the formation of a thin liquid film (see Figure 10.1) which is more or less plane parallel, and has a given thickness and radius. This fact leads to additional complexity in the consideration of the macroscopic properties of charged emulsions and microemulsions. An application of the Ornstein–Zernike integral equation approach to ionic deformable droplets was recently suggested[65] and some of the results are illustrated below. The van der Waals as well as surface extension and/or bending energy contributions are also taken into account.

The configuration of a doublet of two deformed droplets could be characterised by the distance between the centres of the droplets, z, the film thickness, h, and radius, r. It is obvious that these three variables are related by the geometrical relationship (see Figure 10.1)

$$z = h + 2a\sqrt{1 - \frac{r^2}{a^2}} \tag{10.61}$$

and, hence, only two are independent. A natural choice for one of them is the distance z, since it is consistent with the statistical mechanical distribution function formalism. We choose the other independent variable, characterising the doublet configuration, to be the film radius, r, although it could easily be the thickness, h. Hence, we have an additional configurational option – a variable film radius, besides the positions of the droplets in space. This implies that an appropriate averaging over the radius is needed in order to relate microscopic to macroscopic properties, which is our main goal. The configurational integral, $Z_N(V, T)$, for a system of charged and deformable droplets could be written in the following way[65]

$$Z_N(V, T) =$$

$$\int \cdots \int \int \cdots \int dr_1 \ldots dr_M \, dz_1 \ldots dz_N \exp\left[-\frac{W(r_1, \ldots, r_M; z_1, \ldots, z_N)}{kT} \right] \tag{10.62}$$

$W(r_1, \ldots, r_M; z_1, \ldots, z_N)$ is the energy of N interacting droplets which may form M films. This configurational integral leads to a pair correlation function of the form

$$g^{(2)}(r_1; z_1, z_2) =$$

$$\frac{V^2}{Z_N} \int \cdots \int \int \cdots \int dr_2 \ldots dr_M \, dz_3 \ldots dz_N \exp\left[-\frac{W(r_1, \ldots, r_M; z_1, \ldots, z_N)}{kT} \right] \tag{10.63}$$

Further we may assume pairwise additivity of the interaction energy

$$W(r_1, \ldots, r_M; z_1, \ldots, z_N) = \sum W(r_{ij}; z_{ij}) \qquad (10.64)$$

and liquid-like ordering of the droplets (which means that only the magnitude of the relative distance between each two of them matters for the energy). For vanishing volume fractions of droplets ($\Phi \to 0$) we obtain

$$g^0(r; z) = \exp\left[-\frac{W^0(r; z)}{kT}\right] \qquad (10.65)$$

which is the probability of finding a doublet of droplets at distance z, and forming a film with radius r. Averaging the above expression over all possible deformations (film radii) allowed, one arrives at the radial distribution function at infinite dilution (see also ref. 26)

$$g^0(z) = const \times \int dr g^0(r; z) \qquad (10.66)$$

The normalising constant is determined by the requirement $g(z \to \infty) = 1$. For infinitely dilute samples it depends only on the surface extension and bending properties. In the emulsion case (where the interfacial tension dominates) one obtains $const = 4(\pi a^2 \gamma / 2kT)^{1/4} / \Gamma(1/4)$, while in the microemulsion case (where the bending rigidity dominates) the result is $const = 4[(2k_C/kT\tilde{R}_0)(1 - \tilde{R}_0/2)]^{1/2}$.

Now the pair interaction energy, $w(z)$, as a function of the interdroplet distance only, has the following form

$$w(z) = kT \ln[g^0(z)] \qquad (10.67)$$

Introducing eqn. (10.67) into (10.54) allows us to treat deformable charged droplets in the framework of the integral equation approach. The energy $W^0(r; z)$ could be determined, using the results derived in Section 10.2. The film thickness, h, could be replaced by the distance, z, using eqn. (10.61).

10.3.2.2   *Entropy Driven Film Formation*

An interesting result, generated by the present method, is that doublets of deformed droplets with films between them could be formed, and probably detected in some experiments, even in cases in which the pair energy is not sufficiently attractive to produce a minimum in the $h-r$ energy contour diagram (see Section 10.2). The reason for this fact is that although the deformed states are not energetically favourable, they are still allowed and often the energy of some of them is only slightly higher than that of the non-deformed state. In other words, in the statistical ensemble, representing the emulsion and/or microemulsion, there are systems containing deformed droplets, since this is allowed and is entropically favourable, providing additional states. When taking the appropriate averages, this system also contributes to the respective quantity of interest. This could be illustrated by examining the average film radius as a function of the interdroplet distance. For infinitely dilute samples this is

$$\langle r(z) \rangle = const \times \int \mathrm{d}r \; r \exp\left[-\frac{W^0(r; z)}{kT}\right] \tag{10.68}$$

In Figure 10.16 the average film radius *vs.* interdroplet distance is shown for emulsions with size $a = 0.4$ μm, $\gamma = 0.1$ mN m^{-1}, $A_H = 5 \times 10^{-21}$ J and $\Psi_0 = 100$ mV. Expressions (10.3), (10.9), (10.10), (10.29) and (10.62) were used to derive $W^0(r; z)$. The full (upper) curve is for an electrolyte concentration equal to 0.01 M monovalent electrolyte, while the dotted one (lower) is for 0.005 M. In both cases the interaction energy is lowest in the absence of any deformation ($r = 0$).[65] Still, a well defined peak is observed at small distances which is due to the reasons discussed above. It can also be seen that the film radius seems to have non-zero values at separations where the interdroplet interactions vanish. This is due to the fact that the surface of a flexible droplet will fluctuate even if there is no second droplet to interact with. If one examines expressions (10.29) and (10.31) it becomes evident that they do not depend on the droplet positions and therefore do not decay with the distance. This reflects in a simple way the fact that due to fluctuations the droplet surface could be greater than that of a sphere with the same volume (see ref. 66 for more detailed discussion).

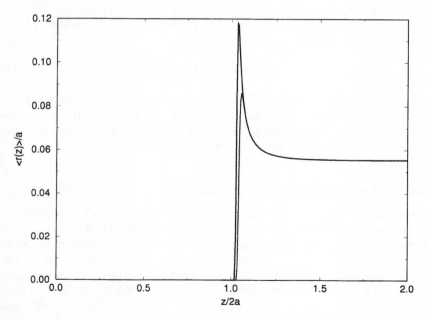

Figure 10.16 *Averaged film radius* vs. *interdroplet distance in an ionic emulsion. See text for the parameters*

10.4 Concluding Remarks

Understanding of the stability and the macroscopic properties of emulsions and microemulsions is continuously increasing. An important contribution to this fact is the great number of experimental studies on a variety of different systems and conditions. The theoretical interpretation of the results obtained is based on the assumption that either the droplets behave as hard spheres, or (as in the case of very large, above 10 μm, droplets) all the interactions take place within a large plane parallel film, ignoring the adjacent curved regions. In this review we present a summary of the recent efforts to formulate a uniform theoretical approach, which accounts for both the film and the surrounding curved portions of the interacting droplets, thus giving more realistic estimates for the energy, especially for droplets $\leqslant 1$ μm. Often the balance between the interaction contributions of these two regions is responsible for the formation of a doublet of deformed droplets with a liquid film separating them. The approach also allows us to find the conditions needed for deformation to take place, and is a first and necessary step to a further exploration of more dense systems.

However, we believe that the effects of the deformability are not studied enough yet. This seems to be mostly in the cases of emulsion and microemulsion dynamics. Since the dynamics and coalescence were not discussed above, here we state a brief view on the current status of the problem. A recent study[24] of the effect of the droplet deformability on the rate of coalescence in emulsions and microemulsions shows that it is negligible for the former although might be important for the latter. This is, however, true for very dilute droplets, where the diffusion approach of any two of them is the time limiting step. The picture could be rather different if energy barriers are present. For example, in concentrated systems, the diffusion time becomes less important and the eventual presence of an energy barrier in the interaction energy could determine the entire kinetics of coalescence. In cases of a very complex shape of the interaction energy surface (as a function of the film radius and thickness), like those in Figures 10.5 and 10.7, the diffusion approach of the droplets might determine the rate of flocculation in the far minimum only, while the flocculation in the primary minimum or coalescence rates could depend on the transition time the systems need to pass from one energy minimum to another. This is related to the phenomenon of reversible flocculation as well. All these problems are very important from a practical viewpoint and at the same time present rather fascinating and challenging questions for fundamental colloid science. Therefore we believe it is worth the effort one may put into such studies.

10.5 References

1. E.J.W. Verwey and J.Th.G. Overbeek, 'Theory and Stability of Lyophobic Colloids', Elsevier, Amsterdam, 1948.
2. B.V. Derjaguin, N.V. Churaev and V.M. Muller, 'Surfaces Forces', Plenum Press, New York, 1987.
3. B.V. Derjaguin, 'Theory of Stability of Colloids and Thin Liquid Films', Plenum Press, New York, 1989.

4. J.N. Israelachvili, 'Intermolecular and Surface Forces', 2nd edn., Academic Press, London, 1991.
5. P.A. Kralchevsky, K.D. Danov and I.B. Ivanov, in 'Foams: Theory, Measurements and Applications', ed. R.K. Prud'homme, Elsevier, Amsterdam, 1996, p. 1.
6. W.B. Russel, D.A. Saville and W.R. Schowalter, 'Colloidal Dispersions', Cambridge University Press, Cambridge, 1989.
7. G. Hadziioannou, S. Patel, S. Cranick and M. Tirrell, *J. Am. Chem. Soc.*, 1986, **108**, 2869.
8. S. Asakura and F. Oosawa, *J. Chem Phys.*, 1954, **22**, 1255; *J. Polym. Sci.*, 1958, **33**, 183.
9. P. Richetti and P. Kekicheff, *Phys. Rev. Lett.*, 1992, **68**, 1951; J.L. Parker, P. Richetti, P. Kekicheff and S. Sarman, *Phys. Rev. Lett.*, 1992, **68**, 1955.
10. M.P. Aronson, *Langmuir*, 1989, **5**, 494.
11. I.B. Ivanov and D.S. Dimitrov, in 'Thin Liquid Films', ed. I.B. Ivanov, Surfactant Science Series, vol. 29, Dekker, New York, 1988, ch. 8.
12. I.B. Ivanov, *Pure Appl. Chem.*, 1980, **52**, 124.
13. I.B. Ivanov, D.S. Dimitrov, P. Somasundaran and R.K. Jain, *Chem. Eng. Sci.*, 1985, **44**, 137.
14. T.T. Traykov and I.B. Ivanov, *Int. J. Multiphase Flow*, 1977, **3**, 471; T.T. Traykov, E.D. Manev and I.B. Ivanov, *Int. J. Multiphase Flow*, 1977, **3**, 485.
15. X. Zhang and R.H. Davis, *J. Fluid Mech.*, 1991, **230**, 479.
16. S.G. Yiantsios and R.H. Davis, *J. Colloid Interface Sci.*, 1991, **144**, 412.
17. M.P. Aronson and H. Princen, *Nature*, 1980, **286**, 370; M.P. Aronson and H. Princen, *Colloids Surf.*, 1982, **4**, 173.
18. J.A.M.H. Hofman and H.N. Stein, *J. Colloid Interface Sci.*, 1991, **147**, 508.
19. N.D. Denkov, P.A. Kralchevsky, I.B. Ivanov and C.S. Vassilieff, *J. Colloid Interface Sci.*, 1991, **143**, 157.
20. K.D. Danov, D.N. Petsev, N.D. Denkov and R. Borwankar, *J. Chem. Phys.*, 1993, **99**, 7179; *J. Chem. Phys.*, 1994, **100**, 6104.
21. J.K. Klahn, W.G.M. Agterof, F. van Voorst Vader, R.D. Groot and F. Groenweg, *Colloids Surf.*, 1992, **65**, 161.
22. H.C. Hamaker, *Physica*, 1937, **4**, 1058.
23. B.V. Derjaguin, *Kolloid-Z.*, 1934, **69**, 155.
24. K.D. Danov, N.D. Denkov, D.N. Petsev, I.B. Ivanov and R. Borwankar, *Langmuir*, 1993, **9**, 1731.
25. D.N. Petsev and P. Linse, *Phys. Rev. E*, 1997, **55**, 586.
26. N.D. Denkov, D.N. Petsev and K.D. Danov, *Phys. Rev. Lett.*, 1993, **71**, 3226.
27. D.N. Petsev, N.D. Denkov and P.A. Kralchevsky, *J. Colloid Interface Sci.*, 1995, **176**, 201.
28. V. Krachunov, B.Sc. Thesis, University of Sofia, Bulgaria, 1996.
29. D.H. Napper, 'Polymeric Stabilization of Colloidal Dispersions', Academic Press, New York, 1983.
30. Th.F. Tadros, in 'Thin Liquid Films', ed. I.B. Ivanov, Surfactant Science Series, vol. 29, Dekker, New York, 1988, ch. 6.
31. A.K. Dolan and S.F. Edwards, *Proc. R. Soc. London Ser. A*, 1974, **337**, 509.
32. S.J. Alexander, *Physique*, 1977, **38**, 983; P. G. de Gennes, *Adv. Colloid Interface Sci.*, 1987, **27**, 189.
33. H.J. Taunton, C. Toprakciaglu, L.J. Fetters and J. Klein, *Macromolecules*, 1990, **23**, 571.
34. H.J. Ploehn and W.B. Russel, *Adv. Chem. Eng.*, 1990, **15**, 137.
35. J. Bibette, D. Roux and F. Nallet, *Phys. Rev. Lett.*, 1990, **65**, 2470; J. Bibette, D. Roux and B. Pouligny, *J. Phys. II*, 1992, **2**, 401.

36. N.F. Carnahan and K.E. Starling, *J. Chem. Phys.*, 1969, **51**, 636.

37. A.D. Nikolov and D.T. Wasan, *J. Colloid Interface Sci.*, 1989, **134**, 1; A.D. Nikolov, P.A. Kralchevsky, I.B. Ivanov and D.T. Wasan, *J. Colloid Interface Sci.*, 1989, **134**, 13; A.D. Nikolov, D.T. Wasan, N.D. Denkov, P.A. Kralchevsky and I.B. Ivanov, *Prog. Colloid Polymer Sci.*, 1990, **82**, 87; D.T. Wasan, A.D. Nikolov, P.A. Kralchevsky and I.B. Ivanov, *Colloids Surf.*, 1992, **67**, 139.

38. P.A. Kralchevsky and N.D. Denkov, *Chem. Phys. Lett.*, 1995, **240**, 385.

39. W.Z. Helfrich, *Z. Naturforsch. C*, 1974, **30**, 510.

40. P.G. de Gennes and C. Taupin, *J. Phys. Chem.*, 1982, **86**, 2304; A.G. Petrov and I. Bivas, *Prog. Surface Sci.*, 1984, **16**, 389.

41. P.A. Kralchevsky, J.C. Eriksson and S. Ljunggren, *Adv. Colloid Interface Sci.*, 1994, **48**, 19.

42. B.P. Binks, H. Kellay and J. Meunier, *Europhys. Lett.*, 1991, **16**, 53; B.P. Binks, J. Meunier and D. Langevin, *Prog. Colloid Polymer Sci.*, 1989, **79**, 208; B.P. Binks, J. Meunier, O. Abillon and D. Langevin, *Langmuir*, 1989, **5**, 415.

43. T.D. Gurkov, P.A. Kralchevsky and I.B. Ivanov, *Colloids Surf.*, 1991, **56**, 119; P.A. Kralchevsky, T.D. Gurkov and I.B. Ivanov, *Colloids Surf.*, 1991, **56**, 149.

44. D. Henderson, D.-M. Duh, X. Chu and D. Wasan, *J. Colloid Interface Sci.*, 1997, **185**, 265.

45. P. Poulin, F. Nallet, B. Cabane and J. Bibette, *Phys. Rev. Lett.*, 1996, **77**, 3248; J. Bibette, T.G. Mason, H. Gang, D.A. Weitz and P. Poulin, *Langmuir*, 1993, **9**, 3352; J. Bibette, T.G. Mason, H. Gang and D.A. Weitz, *Phys. Rev. Lett.*, 1992, **69**, 981.

46. D.N. Petsev and J. Bibette, *Langmuir*, 1995, **11**, 1075.

47. V.N. Paunov, R.I. Dimova, P.A. Kralchevsky, G. Broze and A. Mehreteab, *J. Colloid Interface Sci.*, 1996, **182**, 239.

48. A. Scheludko, *Adv. Colloid Interface Sci.*, 1967, **1**, 391; S.S. Dukhin, N.N. Rulev and D.S. Dimitrov, 'Coagulation and Dynamics of Thin Films', Dumka, Kiev, 1986.

49. N.D. Denkov, D.N. Petsev and K.D. Danov, *J. Colloid Interface Sci.*, 1995, **176**, 189.

50. J. Bibette, *J. Colloid Interface Sci.*, 1991, **147**, 474.

51. B.P. Binks, W.-G. Cho and D.N. Petsev, to be submitted to *J. Chem. Soc., Faraday Trans.*

52. U. Olsson and H. Wennerstrom, *Adv. Colloid Interface Sci.*, 1994, **49**, 113; R. Aveyard, B.P. Binks and P.D.I. Fletcher, *Langmuir*, 1989, **5**, 1210.

53. P.D.I. Fletcher and D.N. Petsev, *J. Chem. Soc., Faraday Trans.*, 1997, **93**, 1383.

54. P.D.I. Fletcher and J. Morris, *Colloids Surf. A*, 1995, **98**, 147; P.D.I. Fletcher and J.F. Holzwarth, *J. Phys. Chem.*, 1991, **95**, 2550.

55. M.S. Leaver, U. Olsson, H. Wennerstrom and R. Strey, *J. Phys. II*, 1994, **4**, 515; M.S. Leaver, I. Furo and U. Olsson, *Langmuir*, 1995, **11**, 1524.

56. J.S. Huang, M. Kotlyarchyk and S.-H. Chen, in 'Micellar Solutions and Microemulsions', eds. S.-H. Chen and R. Rajagopalan, Springer, Berlin, 1990; G.J.M. Koper, W.F.C. Sager, J. Smeets and D. Bedeaux, *J. Phys. Chem.*, 1995, **99**, 13291.

57. T.L. Hill, 'An Introduction to Statistical Thermodynamics', Addison-Wesley, Massachusetts, 1960.

58. P.W. Atkins, 'Physical Chemistry', Freeman, New York, 1994.

59. P.A. Kralchevsky and I.B. Ivanov, *Chem. Phys. Lett.*, 1985, **121**, 111; P.A. Kralchevsky and I.B. Ivanov, *Chem. Phys. Lett.*, 1985, **121**, 116; I.B. Ivanov and P.A. Kralchevsky, in 'Thin Liquid Films', ed. I.B. Ivanov, Surfactant Science Series, vol. 29, Dekker, New York, 1988, ch. 2.

60. R. Balescu, 'Equilibrium and Nonequilibrium Statistical Mechanics', Wiley, New York, 1975.

61. R. Brout, 'Phase Transitions', Benjamin, New York, 1965.

62. J.A. Barker and D. Henderson, *Rev. Mod. Phys.*, 1976, **48**, 587.

63. J.W. Goodwin and R.H. Ottewill, *J. Chem. Soc., Faraday Trans.*, 1991, **87**, 357.

64. D. Frenkel and B. Smit, 'Understanding Molecular Simulations', Academic Press, New York, 1996.

65. D.N. Petsev, *Physica A*, 1998, **250**, 115.

66. J.R. Henderson and P. Schofield, *Proc. R. Soc. London, Ser. A*, 1982, **380**, 211; S.A. Safran, *J. Chem. Phys.*, 1983, **78**, 2073; L.C. Sparling and J.E. Sedlak, *Phys. Rev. A*, 1989, **39**, 1351; H. Gang, A.H. Krall and D.A. Weitz, *Phys. Rev. Lett.*, 1994, **73**, 3435.

CHAPTER 11

Gel Emulsions—Relationship between Phase Behaviour and Formation

CONXITA SOLANS[1], RAMON PONS[1] AND HIRONOBU KUNIEDA[2]

[1]Departament de Tecnologia de Tensioactius, CID/CSIC Jordi Girona, 18–26 08034 Barcelona, Spain
[2]Graduate School of Engineering, Yokohama National University, Tokiwadai 156, Hodogaya-ku, Yokohama 240, Japan

11.1 Introduction

The emulsions described in this chapter are highly concentrated emulsions which form in either water- or oil-rich regions of water/surfactant/oil systems.[1–12] These emulsions are characterised by their large internal phase volume fraction (as large as 0.99), low surfactant content (as low as 0.5 wt%) and gel-like consistency (the reason why we refer to them as gel emulsions). By appropriate selection of composition variables and temperature, optically transparent gel emulsions can be obtained. All these features make them of particular interest for theoretical studies (*i.e.* as models for foams) and for specific applications as formulation and novel delivery systems in, say, the cosmetic and pharmaceutical fields. In recent years, highly concentrated emulsions have received a great deal of attention as novel reaction media for polymerisation reactions in the continuous and/or dispersed phases to obtain materials with improved properties such as solid foams, composites, *etc.*[13,14]

Highly concentrated emulsions were described and started being the object of study long ago.[15–20] The initial research focused on the geometrical packing[16,17] and rheology of oil-in-water (O/W) type emulsions.[17–19] Lately, systematic studies about formation, stability, structure and interfacial properties of both emulsion types (O/W and W/O) have been undertaken.[1–12,21–25]

In this chapter, a summary of the structural aspects and the most relevant properties (stability, rheology and diffusion) of gel emulsions is given in section 11.2. It will be followed by a description of gel emulsion formation (section 11.3) and the relationship between spontaneous formation and phase behaviour of the corresponding systems (section 11.4).

11.2 Gel Emulsions: Structural Aspects and Properties

11.2.1   Structure

The internal phase volume fraction in a gel emulsion, ϕ, is larger than the critical value of the most compact arrangement of uniform, undeformed spherical droplets $\phi_c = 0.74$.[15] Consequently, the droplets in gel emulsions are deformed and/or polydisperse, as shown in Figure 11.1. This figure shows the foam-like structure of a gel emulsion with an internal phase volume ratio close to unity when viewed by optical microscopy. Typically, the droplet size of gel emulsions with water volume fractions above 0.95 range from sub-micrometre to about 10 μm, while the upper limit of those with volume fractions of the order of 0.90 is of about 1 μm.[11] However, depending on the gel emulsion-forming system and/or the method of preparation, both droplet size and polydispersity can be sufficiently low. The internal or dispersed phase in gel emulsions can be either polar or non-polar and, as ordinary emulsions, they are classified in two types: oil-in-water (O/W) and water-in-oil (W/O). Phase behaviour studies of gel emulsion-forming systems, *i.e.* water/polyoxyethylene nonionic surfactant/oil systems, indicated that at equilibrium they

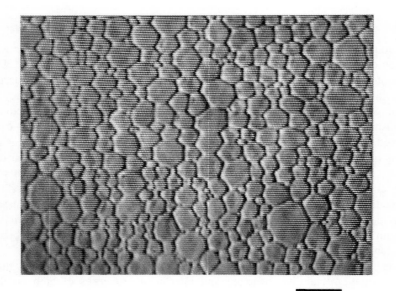

10μm

Figure 11.1 *Micrograph of a gel emulsion with a dispersed phase volume fraction close to unity*

separate into two isotropic liquid phases.[2,6] In W/O gel emulsions the nature of these phases are an aqueous phase and a swollen reverse micellar solution phase (or W/O microemulsion).[2] In O/W gel emulsions the equilibrium phases are an oil phase and an oil-swollen aqueous micellar solution (or O/W microemulsion).[6] Considering the nature of the phases at equilibrium, the emulsions are water (or oil)-in-microemulsion emulsions.

The structure of the equilibrium oil phase of W/O gel emulsions was confirmed by Fourier transform pulse gradient spin-echo (FT-PGSE) NMR determinations.[8] It was found that at high oil/surfactant weight ratios the values of the self-diffusion coefficient for water and surfactant are very low, whereas the corresponding values for oil are high. This means that the continuous phases are non-aqueous and water is confined. Therefore the structure corresponds to a reverse micellar solution or W/O microemulsion. At low oil/surfactant weight ratios, when gel emulsions are extremely unstable or cannot be formed, the self-diffusion coefficient of water was found to approach that of the oil, indicating a bicontinuous structure. From these results it was concluded that the continuous phase of W/O gel emulsions should be a reverse micellar solution or W/O microemulsion.

The equilibrium and also the non-equilibrium structure of W/O gel emulsions was also determined by small angle X-ray scattering (SAXS).[11] The SAXS spectra obtained were explained by the superposition of two spectra: a peak at values of the scattering vector, q, of the order of 0.07 Å^{-1} corresponding to the continuous phase, and an asymptotic part at lower values of q, proportional to q^{-4}, corresponding to the scattering due to the interface between water droplets and the continuous phase.[26] Comparison of SAXS spectra of gel emulsions with those of the continuous phase separated out by centrifugation, indicated that no significant differences exist between the structures of the microemulsion under equilibrium and non-equilibrium conditions.[11] However, theoretical considerations[11,27] suggest that the number of microemulsion droplets is decreased as the volume fraction of disperse phase increases. Consequently, at values of water volume fraction close to unity, most of the surfactant molecules would be adsorbed at the interface of water droplets. By comparison of calculated intensities, determined according to the Porod–Auvray equation[26] with experimental absolute intensities, the specific surface and mean droplet radius of gel emulsions was calculated. The droplet size was found to increase on increasing the volume fraction of dispersed phase, oil/surfactant ratio, salinity and temperature. These results were interpreted on the basis of surfactant availability and interfacial tension.[11]

The non-equilibrium structure of W/O gel emulsions was also investigated by electron spin resonance (ESR).[27,28] The change in apparent order parameter, S, and the isotropic hyperfine splitting constant, a_N, of two amphiphilic and water insoluble spin probes (5- and 16-doxylstearic acids) were determined in gel emulsions, middle- or low-internal-phase ratio emulsions, and single oil phases (microemulsion). This type of spin-probe molecule is incorporated with the surfactant aggregates or in the surfactant molecular layer (water–oil interface). The order parameters provide information regarding the degree of organisation of the surfactant molecules and the hyperfine splitting constant, a_N, is used as a measure of the polarity of the spin label environment.

The variations of S and a_N with water concentration in compositions of the systems water/$C_{12}EO_3$/cyclohexane and water/$C_{16}EO_4$/cyclohexane with a constant oil/surfactant weight ratio are shown in Figures 11.2 and 11.3.[27] Up to the solubilisation limit (points a and b of Figures 11.2 and 11.3), S and a_N increase with increasing water content, indicating that the environment of the spin label changes from an apolar one (surfactant and oil mixtures) to a more polar one (existence of reverse micelles). At water contents higher than the solubilisation limit, W/O type emulsions are formed due to the separation of an excess water phase. If the spin probe is in the same state in a single reverse-micellar solution phase or in a gel emulsion, the order parameter should be equal. The apparent order parameter, S, and the isotropic hyperfine splitting constant, a_N, of 5-doxylstearic acid (Figure 11.2) increase slowly but continuously in an ordinary emulsion region beyond the solubilisation limit and then increase abruptly in the gel emulsion region. However, in the system with 16-doxylstearic acid (Figure 11.3), both S and a_N remain unchanged over a wide range of water concentration except in the gel emulsion region.

The difference in order parameters at the water droplet film in gel emulsions and the inside of the reverse micelles in the single isotropic phase indicated that the surfactant molecules at the water droplet film are more tightly packed compared to those in reverse micelles. The observed difference in the apparent order parameter between the ordinary emulsions and the gel emulsions suggested that, when the fraction of water in the system is around 0.99, most of the surfactant molecules are adsorbed at the oil–water interface and no reverse micelles are present in the continuous oil phase.[27]

In the water-rich region, as the water content increases, the water–oil interfacial area increases and the surfactant molecules are preferentially adsorbed at the water–oil interface. Consequently, the number of micelles would decrease and the spin probe would move from reverse micelles to the water–oil interface with the increase in water concentration. The ESR studies[27] suggested that the positions of the spin probes are different in ordinary and highly concentrated emulsions. The surfactant molecules are distributed between the interface and reverse micelles. In the non-equilibrium state, the number of reverse micelles decreases with increasing water content and finally no reverse micelles are present in the continuous media at very high water content.

With 5-doxylstearic acid as the spin probe, the order parameter for the longer chain surfactant, $C_{16}EO_4$, is higher than that of the shorter chain surfactant, $C_{12}EO_3$, (Figure 11.2) an indication that it is more tightly packed at the interface. This was confirmed by stability studies.[27]

11.2.2 Stability

Gel emulsions are kinetically stable systems, which break into two liquid phases with time. The time taken for separation, which may vary from minutes to months, can be significantly retarded by the appropriate selection of composition variables and temperature.

A relationship between the stability of gel emulsions and the HLB (hydrophile

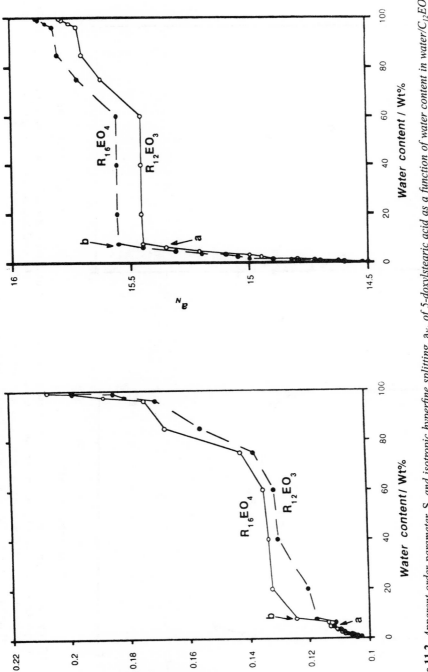

Figure 11.2 *Apparent order parameter, S, and isotropic hyperfine splitting, a_N, of 5-doxylstearic acid as a function of water content in water/$C_{12}EO_3$/cyclohexane and water/$C_{16}EO_4$/cyclohexane systems. 'a' and 'b' denote the solubilization limits in $C_{12}EO_3$ and $C_{16}EO_4$ systems, respectively. The oil/surfactant ratio in the $C_{12}EO_3$ system is 1.5 and in the $C_{16}EO_4$ system is 1.63* (Reproduced by permission of the American Chemical Society from ref. 27)

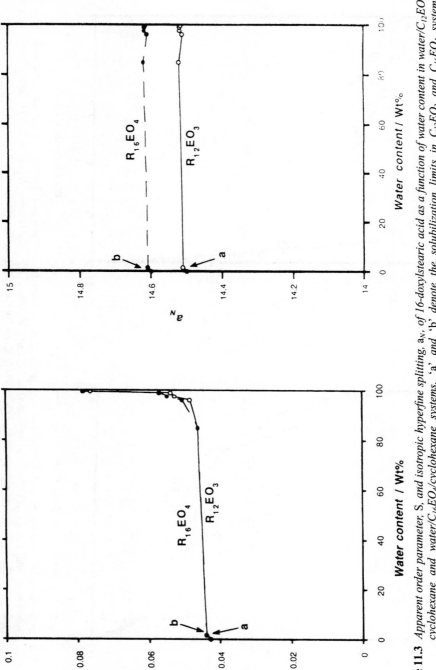

Figure 11.3 *Apparent order parameter, S, and isotropic hyperfine splitting, a_N, of 16-doxylstearic acid as a function of water content in water/$C_{12}EO_3$/ cyclohexane and water/$C_{16}EO_4$/cyclohexane systems. 'a' and 'b' denote the solubilization limits in $C_{12}EO_3$ and $C_{16}EO_4$ systems, respectively. The oil/surfactant ratio in the $C_{12}EO_3$ system is 1.5 and in the $C_{16}EO_4$ system is 1.63 (Reproduced by permission of the American Chemical Society from ref. 27)*

lipophile balance) temperature of the corresponding water/polyoxyethylene non-ionic surfactant/aliphatic hydrocarbon system was observed.[2-5] The stability of O/W gel emulsions decreases with increasing temperature, experiencing a pronounced coalescence rate at temperatures approximately 25 °C lower than the HLB temperature.[5] The stability of W/O gel emulsions first increases with temperature, reaching a maximum at 25–30 °C above the HLB temperature, and then decreases.[2,5] This fact was explained on the basis of FT-PGSE NMR studies.[8] Self-diffusion measurements of the continuous phase of W/O gel emulsions of the water/$C_{12}EO_4$/decane system as a function of temperature showed that the microemulsion structure changed from bicontinuous, at temperatures close to the HLB temperature of the system, to W/O-type at higher temperatures. Therefore, the changes in structure of the continuous phase strongly affect gel emulsion stability. As described above, a change in microemulsion structure from bicontinuous to W/O-type was observed in the same system with an increase in the oil/surfactant ratio.[8] Since the bicontinuous structure is highly permeable to the dispersed phase, it cannot stabilise the emulsion. However, when the structure of the continuous phase is of the droplet type, the stability can reach optimum values.

Stability studies of gel emulsions formed with the surfactants $C_{16}EO_4$ and $C_{12}EO_3$ (which have similar HLB values) and the same oil, cyclohexane (therefore showing the same HLB temperature), revealed that for low- and middle-internal phase emulsions the stability is higher in systems with the longer chain surfactant.[27] ESR studies showed that the order parameter values were higher in gel emulsions with the longer chain surfactant, $C_{16}EO_4$, than in those with the shorter chain surfactant, $C_{12}EO_3$, as described in the previous section. This is an indication of a tighter packing of the $C_{16}EO_4$ surfactant molecules at the interface.

The presence of additives may influence emulsion stability. In the context of gel emulsions, the effect of electrolytes on gel emulsion stability has been studied in some detail.[5,7,28] In general, a considerable increase in the stability and the viscosity of gel emulsions is produced with the addition of inorganic salts. Those with a large salting out effect, decreasing the cloud point in binary water/nonionic surfactant systems and consequently the HLB temperature in water/nonionic surfactant/oil systems, are the most effective stabilisers. The stability of emulsions as a function of $NaSO_4$, $CaCl_2$, NaCl and KI concentration, as determined by visual observation, showed that Na_2SO_4, $CaCl_2$ and NaCl stabilised the gel emulsions significantly while KI did not improve gel emulsion stability.[5] The former electrolytes depress the HLB temperature of the system while KI slightly increases it. ESR measurements[28] showed that the apparent order parameter in gel emulsions increased with the addition of electrolyte (Figure 11.4).

The increase in the apparent order parameter means that the packing of the surfactant molecules at the interface changes significantly with the addition of salt. The enhanced stability produced by electrolytes might be due to the dehydration of the hydrophilic part of the surfactant.[28] The surfactant–surfactant interactions would increase and consequently the interfacial films become more rigid. Results of rheological studies as well as interfacial tension measurements agreed with these hypotheses.[9]

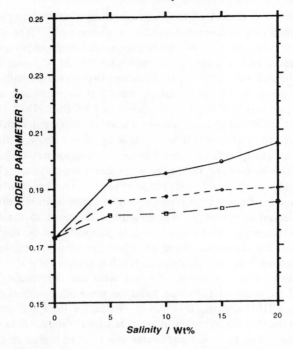

Figure 11.4 *Apparent order parameter, S, in gel emulsions as a function of electrolyte concentration in water: (☐) NaCl, (◇) CaCl₂ and (○) Na₂SO₄. Composition: $C_{12}EO_4$ + heptane (1.2 wt%) and water or aqueous solution (98.8 wt%)* (Reproduced by permission of Steinkopff from ref. 28)

11.2.3 Rheological Properties

The rheology of highly concentrated emulsions was studied from a theoretical and experimental point of view by Princen and co-workers in the 1980s.[18-20] They showed theoretically, for ordered structures, that the shear modulus of a concentrated system should depend directly on the interfacial tension and inversely on the radius (directly on the total surface area). In their bidimensional array of cylinders they obtained a volume fraction that was not directly extrapolable to truly tridimensional systems. This is due to the increase in interfacial area produced by the shear. Their model of deformation is shown in Figure 11.5.

They adopted a semi-empirical equation with two fitting parameters that were determined from experiment on highly concentrated oil-in-water emulsions of varying radius, interfacial tension and volume fraction. The equation they obtained is as follows:

$$G_0 = a \frac{\gamma}{R_{32}} \phi^3 (\phi - b) \qquad (11.1)$$

where G_0 is the static shear modulus, γ the interfacial tension, ϕ the dispersed

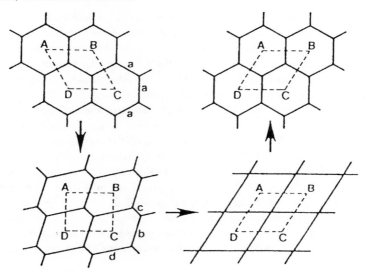

Figure 11.5 *Deformation of a unit hexagonal cell. The total interfacial area increases with strain, attaining a maximum (third stage). The configuration is unstable at the maximum, and it reverts to a new minimum*
(Reproduced by permission of Academic Press from ref. 19)

phase volume fraction and R_{32} the Sauter mean drop radius, with $a = 1.769$ and $b = 0.712$ as adjustable parameters.

Rheological measurements performed on highly concentrated W/O emulsions (gel emulsions) showed that the equation proposed by Princen also holds for these systems.[7,9] Measurements of the viscoelastic properties showed that these materials have a viscoelastic response that can be fitted to a Maxwell liquid element. The elastic and viscous terms of this model vary as a function of frequency as:

$$G' = G_0 \frac{(\omega\theta)^2}{1 + (\omega\theta)^2} \tag{11.2}$$

$$G'' = G_0 \frac{(\omega\theta)}{1 + (\omega\theta)^2} \tag{11.3}$$

where G' is the elastic modulus, G'' is the viscous modulus, G_0 the shear modulus, ω is the pulsation and θ is the relaxation time that has the form:

$$\theta = \frac{\eta}{G_0} \tag{11.4}$$

These parameters were also shown to depend on the system variables such as composition and temperature. The elastic modulus at high frequency is equivalent in these systems to the shear modulus and depends on the interfacial tension, droplet radius and volume fraction in the way predicted by eqn. (11.1) within experimental error.[9] Owing to the specific dependence of interfacial tension and droplet size in gel emulsions, a maximum in shear modulus is found as a function of temperature. This is shown in Figure 11.6 as G' (elastic modulus at high frequency G_0) as a function of temperature both for the experimental points and calculated values obtained from eqn. (11.1) by substitution of independently determined interfacial tension and and droplet radius values.[9]

Both data sets show this maximum, although the experimental agreement is not very good. The appearance of these maxima is due to the compensation of the effect of temperature on the interfacial tension and on the droplet size. As temperature increases the interfacial tension in nonionic surfactant systems above the HLB temperature increases; this implies an increase in the elastic modulus. At the same time, an increase in temperature produces an increase in droplet size (not only related to coalescence but to the interfacial tension as well[9,11]). An increase of droplet size produces a decrease of the elastic modulus. At low temperatures the interfacial tension dominates, at an intermediate point both factors compensate and at high temperatures the droplet size finally dominates. The dependence of the

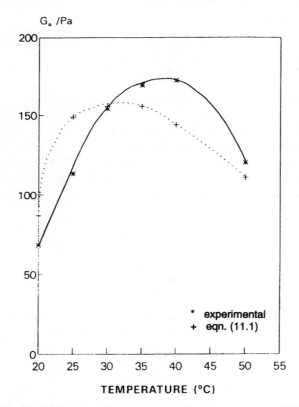

Figure 11.6 G_0 *as a function of temperature: experimental and calculated values (eqn. 11.1). The system is $H_2O/C_{16}EO_4/C_{10}H_{22}$ with oil/surfactant = 1.5 and 99% w/w water content. The lines are drawn as guide for the eye*
(Reproduced by permission of the American Chemical Society from ref. 9)

elastic modulus with volume fraction agrees with the predictions of eqn. (11.1) also. Usually a maximum is found, due to the increase of droplet size of the emulsions when reaching high volume fractions (droplet size approximately constant up to about $\phi = 0.95$, increasing steeply for higher volume fractions[11]). The effect of surfactant and salt concentration of the aqueous component was also checked and good agreement with eqn. (11.1) was always obtained through the study of the influence of these parameters on droplet size and interfacial tension.[9] Notable is the increase of elastic modulus produced by salt addition. For systems close to the HLB temperature, relatively small additions of salt produce a big increase in interfacial tension and thus in the elastic modulus.

Concerning the relaxation time of the system, it was found that a simple model could account for the observed behaviour. The system is considered as a two-element body: an elastic modulus produced by interfacial area increase and a viscous modulus produced by the loss caused by slippage of droplets against droplets. Putting these elements in series means that the systems behave as a Maxwell liquid and the value of the relaxation time can be calculated, taking into account the viscosity of the continuous phase and its participation in the global volume.[9]

$$\theta = c \frac{\eta}{(1 - \phi)G_0} \tag{11.5}$$

where η is the continuous phase viscosity and c an adjustable parameter. The dependence of the relaxation time on temperature is shown in Figure 11.7. G_0 is taken as the value of the high frequency elastic modulus. Good agreement is obtained by using the adjustable parameter c. Although no physical meaning is attached to this parameter, it can be taken as an effective friction between the droplets.

11.2.4 Diffusion

Diffusion of molecules within emulsions is a subject that is relevant for emulsion applications.[29] The use of emulsions as a form of drug and cosmetic delivery system is one in which the diffusion within the emulsion and to a receptor solution is of the utmost importance. Diffusion of probe molecules has been studied in gel emulsions with three different methods.[23–25,30] The first two methods are similar in principle: the study of the release of a probe molecule from a gel emulsion to a receptor solution in contact with the emulsion. Owing to the high viscosity and W/O nature of gel emulsions, this contact can be made directly with the receptor solution (provided the composition is right[23,30]). For gel emulsions whose viscosity is not high enough to ensure that the emulsion is not dispersed in the medium, a permeable barrier has been used.[24] Alternatively, a method based on the determination of the diffusion within the emulsion has been proposed.[23] In this method a gel emulsion containing a probe molecule is contacted with an emulsion of the same composition except for the probe molecule. After a given time the system is finely sliced in planes parallel to the contacting plane. Analysis of the probe molecule content in these slices produces concentration as a function

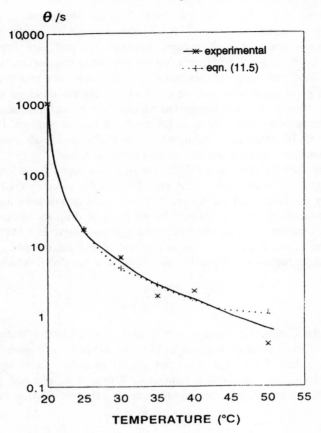

Figure 11.7 *Relaxation time as a function of temperature. Experimental and calculated values from eqn. 11.5 for the same system as in Figure 11.6. The lines are drawn as a guide for the eye*
(Reproduced by permission of the American Chemical Society from ref. 9)

of distance data. These data can be fitted to the exact solution for a semi-infinite slab:[23,31]

$$C = \frac{C_0}{2}\left[1 - \mathrm{erf}\left(\frac{x}{\sqrt{4Dt}}\right)\right] \tag{11.6}$$

(where C is the concentration at a distance x, C_0 is the initial concentration, D is the mutual diffusion coefficient and t is time) or to numerical solutions for Fick's second law if the semi-infinite boundary conditions are not accomplished. Typical results are shown in Figure 11.8 for gel emulsions of varying volume fraction.

Several factors that modulate the release and diffusion in gel emulsions have been studied, namely dispersed phase volume fraction, oil/surfactant ratio, electro-

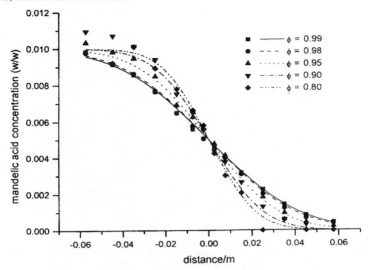

Figure 11.8 *Mandelic acid concentration profiles in gel emulsions of different dispersed phase content after nine days of diffusion. The surfactant was hydrogenated castor oil with 7 ethylene oxide groups, and the oil was decane. The curves are fits of eqn. (11.6), which corresponds to Fick's law*
(Reproduced by permission of Dekker from ref. 10)

lyte concentration, surfactant ethylene oxide length, surfactant hydrophobic chain length, hydrocarbon chain length, temperature and pH.[23-25,30,32] Quite interesting is the effect of dispersed phase volume fraction. As the dispersed phase volume fraction increases, the diffusion coefficient (obtained from the data of Figure 11.8) increases. These diffusion coefficients are shown in Figure 11.9, together with fits to the following equation:

$$\frac{1}{D} = \frac{\phi}{D_1} + \frac{(1-\phi)}{D_2} \tag{11.7}$$

In this equation, D is the experimental diffusion coefficient and ϕ is the dispersed phase volume fraction. D_1 corresponds to the extrapolation to the dispersed phase volume fraction equal to 1. In this situation only the dispersed phase and the interfaces contribute to the diffusion; therefore D_1 corresponds to this contribution. D_2 corresponds to the zero dispersed volume fraction situation where only the continuous phase contributes to the diffusion. This simple model corresponds to adding the resistance of the continuous phase to the resistance of the dispersed phase and the interfaces to obtain the total resistance. Results obtained in several systems showed a dependence of D_1 on the nature of the surfactant,[24,30] supporting the idea of a mixed contribution of the dispersed phase diffusion coefficient of the probe molecule reduced by the resistance to pass through the interfaces. The values obtained for D_2 in several systems correspond to the diffusion coefficient of

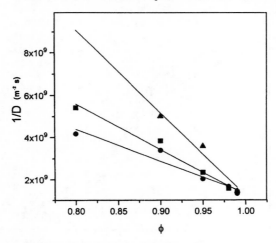

Figure 11.9 *Reciprocal diffusion coefficient as a function of volume fraction:* ■, *release from a gel emulsion 1 (analysed by HPLC);* ●, *release from a gel emulsion 1 (analysed by UV);* ▲, *diffusion within the emulsion (UV and HPLC analysis). The straight line corresponds to the best fit of eqn. (11.7)*
(Reproduced by permission of Steinkopff from Pons et al.[23])

microemulsion droplets, with good agreement in the trends observed for the microemulsion droplet size when changing the surfactant.[24,32]

An important contribution to the diffusion coefficient is that of the partition coefficient P of the probe molecule between the continuous and dispersed phases of the emulsion. Changing the partition coefficient by adding salt to the aqueous phase caused a linear variation in D shown in Figure 11.10.[24] This linear trend would be valid for relatively low values of the partition coefficient for which a change of partition coefficient scales linearly with the probe molecule concentration in the continuous phase. However, for large partition coefficient values the increase in continuous phase concentration of the probe molecule approaches a constant value and, in this situation, as most of the probe molecule is already in the continuous phase, no effect of the partition coefficient can be observed.[33]

Some of the experimental results cannot be explained solely on the basis of partition coefficient changes, so we need to consider the stability of the whole system. If the structure of the gel emulsion is lost, an increase in the release rate of the probe molecule can be observed because the interfacial barrier produced by the emulsion thin films is no longer present as a barrier for the diffusion. This has been observed for mixed gel emulsions of hydrogenated and fluorinated surfactants. Apparently the mixtures are somewhat less stable than the pure systems and the release is increased after some time of diffusion, producing a release that approaches a zero order release rate.[33]

The effect of temperature is strong, as could be expected for diffusion processes. An exponential growth has been observed. This must be a consequence of the effect of temperature on the diffusion coefficient within the continuous phase, the dispersed phase and the passage through the interfacial barrier (and, possibly, some

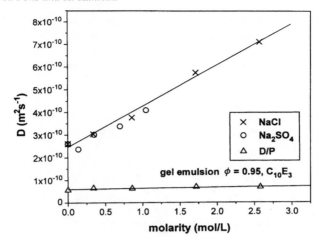

Figure 11.10 *Diffusion coefficients of mandelic acid from gel emulsions as a function of added electrolyte. Diffusion coefficients (D) over partition coefficients (P) are also plotted*
(Reproduced by permission of the American Chemical Society from ref. 24)

effect on the partition coefficient). Stokes–Einstein evaluation of the diffusion coefficient of the microemulsion droplets, taking into account their size as obtained by SAXS, could account for the main part of the observed increase.[32]

The pH of the dispersed phase has been shown to greatly influence the diffusion. This is probably related to two effects. On the one hand, the partition coefficient changes due to the fact that the ionised species partition more favourably in the water than in the oil phases; therefore the partition coefficient is reduced when the pH is increased. A reduction in diffusion coefficient of one order of magnitude is found for high pH (12.5). The effect of lowering the pH below that produced by the mandelic acid is far less important, because in the unbuffered mandelic acid solution only about 5% is in the ionised form. If an emulsion is contacted with another emulsion of different pH, several combinations are possible. Contacting an emulsion containing mandelic acid with an emulsion containing NaOH produces a complex concentration–distance profile that shows a peak at the basic emulsion side where the mandelic acid is willing to diffuse but is trapped in the high pH water droplets.[32] This profile can be fitted to a Fick's law model that takes into account a global partition coefficient between the two emulsions.[32]

11.3 Gel Emulsion Formation

11.3.1 Phase Behaviour of Water/Polyoxyethylene Nonionic Surfactant/Oil systems

High water content W/O gel emulsions form in the water-rich region whereas O/W gel emulsions form in the oil-rich region of ternary water/polyoxyethylene nonionic surfactant/oil systems.[1-6] These types of surfactant change from water soluble to oil soluble with an increase to temperature, as depicted in Figure 11.11 which is a schematic representation of the phase behaviour of a ternary water/ polyoxyethylene-type surfactant/oil system at increasing temperatures from T_1 to T_7.[34,35]

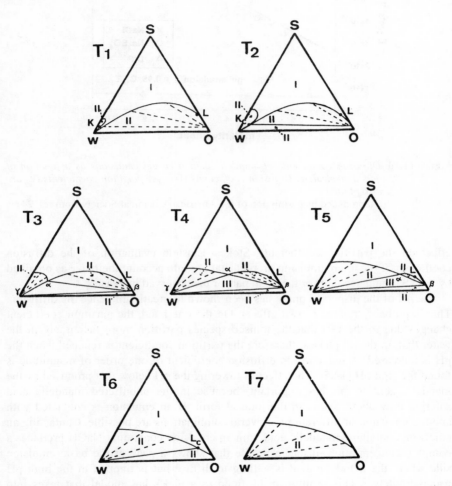

Figure 11.11 *Schematic phase diagrams of a water (W)/polyoxyethylene nonionic surfactant (S)/oil (O) system at various temperatures. I, II, and III designate one-, two- and three-phase regions*
(Adapted with permission of Academic Press from ref. 35)

The two-phase region (extending from the water to the oil axis) consists of aqueous micellar (or O/W microemulsion) and excess oil phases, at low temperatures, T_1 and T_2, as indicated by the tie lines that converge near the oil apex. At higher temperatures, T_6 and T_7, the tie lines in the two-phase region converge towards the water apex, the surfactant is mainly dissolved in the oil and the constituting phases are a reverse micellar solution (or W/O microemulsion) and excess water phase. At intermediate temperatures T_3-T_5, the surfactant does not show preferential solubility for oil or water and a three-phase region consisting of water, surfactant (middle phase or bicontinuous microemulsion) and oil phases appears. The intermediate temperature, T_4, at which equal amounts of oil and water are solubilized with the minimum amount of surfactant, is the HLB temperature of the system, also called the PIT because inversion from oil in water (O/W) to water in oil (W/O) emulsions or *vice versa* is produced at this temperature.[36]

W/O gel emulsions separate into two isotropic liquid phases at equilibrium: one phase is a submicellar surfactant solution in water and the other phase is a swollen reverse micellar solution (or W/O microemulsion).[5] This phase equilibrium is represented in Figure 11.11 by the ternary diagrams at T_6 and T_7. Similarly, O/W gel emulsions separate into two isotropic liquid phases at equilibrium: an oil phase and an aqueous micellar solution or O/W microemulsion. The phase equilibrium is represented by the ternary diagrams at T_1 and T_2 of Figure 11.11. Gel emulsions exist only in limited regions of the miscibility gap (the two-phase region). The boundaries of the gel emulsion regions depend on the system and also on the method of preparation.

11.3.2   Methods of Preparation

Gel emulsions can be prepared by the usual preparation method for W/O emulsions, dissolving a suitable emulsifier in the component that will constitute the continuous phase followed by addition of the component which will constitute the dispersed phase, with continuous stirring. However, they can be prepared according to several other methods that have been proposed specifically for highly concentrated emulsions.[1,2,12,37–39] In the following, attention will be focused on the multiple emulsion method[1,2,10] and spontaneous formation method,[12,40,41] because they constitute interesting novel methods of preparation.

The multiple emulsion method consists of weighing all the components at the final composition, followed by shaking or stirring the sample. The process of emulsification is facilitated by addition of small glass beads or porous materials such as textile fabrics.[1,2,12] The size of the container, the agitation method and the intensity of agitation also play an important role in determining the time taken to achieve complete emulsification. The time taken for gel emulsion formation may range from a few minutes to about one hour. It has been interpreted that the role of glass beads is of a mechanical nature, increasing the efficiency of agitation which is especially important in the final stages of the process. In contrast, the role of a porous material could be to facilitate the aggregation of oil droplets in the earlier stages of the emulsification process.[12] The macroscopic aspect during the preparation of a W/O gel emulsion consisting of 99 wt% water and equal weight ratios of oil and surfactant at various stages of the manual agitation process is depicted in Figure 11.12.

After adding the weighed components and with slight agitation, the mixture

Figure 11.12 *Steps in the emulsification process by the multiple emulsion method*

becomes turbid: if let to rest for a while, some creaming is observed (Figure 11.12a). By optical microscopy it was detected that after slight agitation, the mixture consists of oil droplets dispersed in water (O/W emulsion). As agitation proceeds, the mixture becomes inhomogeneous; if agitation is stopped, big white flocs rapidly reach the surface (Figure 11.12b). Since the surfactant has lipophilic properties at the temperature of preparation, it tends to stabilise W/O emulsions. Therefore, with further agitation, coalescence of oil droplets is observed. Moreover, the oil droplets experience growth by incorporation of water. This causes an increase in the viscosity of the system. At this step the mixture consists of a multiple W/O/W emulsion. The process of water in oil emulsification inside the oil droplets can be considered analogous to that of the preparation of an emulsion by the classical method of stepwise addition of the disperse phase to the continuous phase. The final step is reached with continued agitation, when all the water is emulsified resulting in a W/O gel emulsion (Figure 11.12c). Emulsification by this method results in rather polydisperse gel emulsions, like that shown in Figure 11.1.

In the spontaneous formation method, emulsification is achieved by a rapid temperature change of a micellar solution or microemulsion without the need for mechanical stirring.[12,40,41] It should be noted that for W/O gel emulsions the system is below the HLB temperature at the start and emulsification takes place when an O/W microemulsion is rapidly heated from temperatures below to above the HLB temperature. Formation of O/W gel emulsions, by this method, is achieved by quickly cooling a water-in-oil microemulsion from a temperature higher than the HLB temperature of the system to a temperature below it, at which two phases, a micellar solution phase and an oil phase, appear. The emulsification process for a gel emulsion of the W/O type is depicted in Figure 11.13.

Figure 11.13a shows a microemulsion of the 0.1 M NaCl aqueous solution/$C_{12}E_4$/ decane system with 90% aqueous solution and an oil/surfactant ratio of 2.33, at 7 °C. The HLB temperature of this system is approximately 18 °C. When the sample is rapidly brought to a higher temperature, 40 °C in the experiment shown in Figure 11.13b, the sample becomes milky in less than 40 s. If the test tube is inverted, no flow is observed, an indication that complete emulsification has been achieved. The emulsions produced by this emulsification method have finer and narrower droplet size distributions (Figure 11.14) than those obtained by the usual methods.

11.4 Relationship between Phase Behaviour and Spontaneous Gel Emulsion Formation

11.4.1 Water-in-oil (W/O) Gel Emulsions

When the process of spontaneous gel emulsion formation, depicted in Figure 11.13, was followed by conductivity as a function of time,[12] a monotonic decrease in conductivity of about four orders of magnitude in a short period of time (less than one minute) was observed when the O/W microemulsion (Figure 11.13a) was

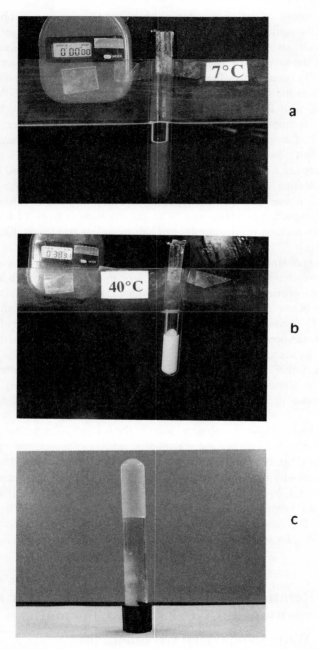

Figure 11.13 *The process of emulsification of a W/O gel emulsion by the spontaneous emulsification method. An oil-in-water (O/W) microemulsion at 7 °C (a) is rapidly heated to 40 °C and a gel emulsion is spontaneously formed after 38 s (b). At this stage, the emulsion does not flow if the container is turned upside down (c)*

10 μm

Figure 11.14 *Micrograph of a W/O gel emulsion with a 90 wt% 0.1 M NaCl solution, prepared by the spontaneous emulsification method*

transferred to a higher temperature (Fig 11.13b). In this emulsification process, no mechanical energy input was applied. However, when the gel emulsion was formed with stirring in addition to the temperature change, the conductivity did not decrease monotonically but showed a maximum.[40] Study of the phase behaviour of the system as a function of temperature (Figure 11.15) allowed interpretation of these results.

Figure 11.15b shows the detailed phase diagram in a narrower range of temperatures than Figure 11.15a, of the system 0.1 M aqueous $NaCl/C_{12}EO_4/$ decane as a function of decane concentration and temperature.[40] When a composition corresponding to, for instance, point A is heated from 7 °C to higher temperatures, the phase changes are the following: initially an O/W microemulsion (W_m) forms and as the temperature increases a narrow two-phase region consisting of bicontinuous microemulsion and aqueous phases $(D + W)$ is crossed and it is followed by another two-phase region composed of aqueous phase and lamellar liquid crystalline phases $(LC + W)$. The lamellar liquid crystal is dispersed as vesicles in this region. With increasing temperature, a single lamellar liquid crystalline phase is formed (LC). With further increase of temperature, the lamellar liquid crystal is melted and an isotropic phase L_3 appears in which the surfactant molecules form a bicontinuous network.[42] Above

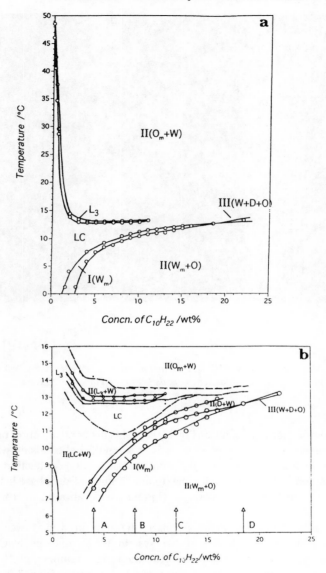

Figure 11.15 *Phase diagram of the 0.1 M aqueous NaCl/C$_{12}$EO$_4$/decane system as a function of temperature. Decane was added to 3 wt% C$_{12}$EO$_4$ aqueous solution, and the weight percent of decane in the system is plotted horizontally. (a) Phase diagram over a wide range of temperature. (b) Detailed phase diagram around the HLB temperature. W$_m$ oil-swollen micellar solution (O/W-type microemulsion); O$_m$, water-swollen reverse micellar solution (W/O-type microemulsion); D, surfactant phase (middle-phase microemulsion); L$_3$, bicontinuous surfactant phase; LC, lamellar liquid crystal; W and O, excess water and oil phases, respectively. I, II, and III indicate one-, two- and three-phase regions*
(Reproduced by permission of the American Chemical Society from ref. 40)

the single-phase L_3 region, excess water (or brine) is separated (region $L_3 + W$). With increase in temperature, this two-phase region is continuously changed to reverse micellar solution (or W/O microemulsion) and aqueous phase ($O_m + W$). Therefore, the spontaneous curvature or the surfactant aggregates change from convex to concave towards water. Phase inversion from water continuous O/W microemulsion to W/O gel emulsion occurs through a lamellar liquid crystalline phase and a bicontinuous surfactant L_3 phase. The change in self-organising structure during spontaneous formation is schematically represented in Figure 11.16.

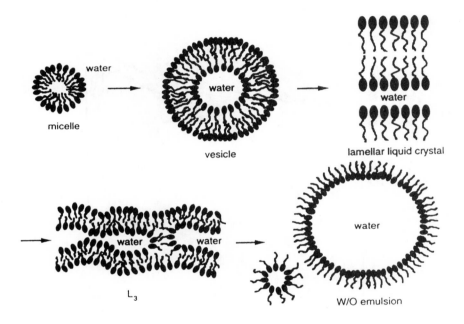

Figure 11.16 *Schematic change in the spontaneous curvature of surfactant layers in the process of spontaneous formation of W/O gel emulsions*
(Reproduced by permission of the American Chemical Society from ref. 40)

The maximum in conductivity observed when the gel emulsion was formed under stirring, or when the temperature change was very slow, was produced at the temperatures where the $L_3 + W$ region is found. In this temperature range the emulsions are very unstable and, in this system, separate into two isotropic phases within 15 min. The maximum in conductivity was attributed to the excess water separated when crossing this region. When the temperature change is slow the water droplet size in the gel emulsion is large because of coalescence of water droplets in the $L_3 + W$ region. In order to obtain fine droplet size gel emulsions, the temperature change should be very quick.

11.4.2   Oil-in-water (O/W) Gel Emulsions

The mechanism of O/W gel emulsion formation was determined[41] by studying the phase behaviour of gel emulsion-forming systems as a function of temperature and following the emulsification process conductimetrically, as for W/O gel emulsions. The phase diagram of 0.1 M NaCl aqueous solution/$C_{12}EO_6$/monolaurin/ *n*-decane as a function of temperature and brine concentration is shown in Figure 11.17. Monolaurin was added to the system to increase the lateral interactions of the surfactant layer and consequently to enhance the stability of the gel emulsions.[5]

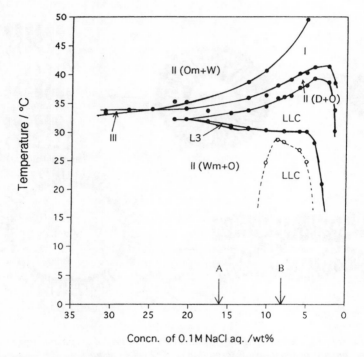

Figure 11.17 *Phase diagram of the 0.1 M aq. NaCl/$C_{12}EO_6$/monolaurin/*n-*decane system as a function of temperature. The $C_{12}EO_6$:monolaurin:*n-*decane ratio is kept constant at 3.5:1.5:95. The weight percentage of 0.1 M aq. NaCl in the system is plotted horizontally. I, II and III indicate one-, two- and three-phase regions. W, excess water phase; O, excess oil phase; W_m, oil-in-water micro-emulsion; O_m, water-in-oil microemulsion; L_3, reverse bicontinuous phase; D, surfactant phase; LLC, multiphase region including lamellar liquid crystal (Reproduced by permission of Academic Press from ref. 41)*

The surfactant changes from hydrophilic to lipophilic with an increase in temperature, as illustrated by following the changes experienced by composition A in Figure 11.17. At low temperatures, the surfactant dissolves as micelles in water and this phase coexists with excess oil ($W_m + O$). The excess oil phase is completely solubilised in the lower phase within a narrow temperature range at which the single isotropic L_3 phase appears. In contrast to the L_3 phase that appears in the water-rich region, the hydrocarbon parts of the surfactant layer are separated.[42] Above the L_3 region, a multiphase region with lamellar liquid crystal (LLC) is formed. With further increase of temperature a two-phase isotropic region (D + O) composed of a bicontinuous microemulsion phase (D) and an oil phase (O) is observed. In this two-phase region, emulsions are highly unstable and reach equilibrium within several minutes. At higher temperatures a single water-in-oil-microemulsion phase (I) appears and with further increase of the temperature an excess water phase is separated from the single-phase microemulsion. A W/O emulsion consisting of two liquid phases, a W/O microemulsion (Om) and an aqueous phase, exists in this (O_m + W) region.

When microemulsions in region I with compositions corresponding to points A and B in Figure 11.17 were quickly cooled to lower temperatures, gel emulsion formation was achieved in composition A while emulsification was not completed in composition B. At the final temperature, the phases existing for composition A were two isotropic phases (W_m + O) while composition B falls in a region including lamellar liquid crystals (LLC). It was observed that the faster the cooling rate of composition A, the smaller the emulsion droplets and the narrower the size distribution. This was attributed to the fact that at a temperature lower than the single-phase microemulsion (I), there is an extremely unstable emulsion region (D + O). If the cooling rate is fast, the system passes this region in a short time and the coalescence of oil droplets is very slow.

The emulsification process was also followed by conductivity and with simultaneous temperature measurement. The change in conductivity at the composition indicated by A in Figure 11.17 is shown in Figure 11.18. In Figure 11.18a, emulsification was carried out with stirring but in Figure 11.18b there was no stirring. Independent of the final temperature, 20 °C (Figure 11.18a) and 5 °C (Figure 11.18b), the conductivity increased with decreasing temperature. The shapes of the curves are similar, although in Figure 11.18a the conductivity changes more slowly (the cooling speed is slow). In Figure 11.18c, at the final stage of the emulsification process, the conductivity is much lower than in the former cases and the reproducibility is rather poor because oil-in-water emulsification is not completed and a bulk oil phase appears.

The phase behaviour studies represented in Figure 11.17 suggested that at composition A the spontaneous curvature of surfactant molecular layers continuously changes from convex to oil to convex to water while cooling. The change in self-organising structures during emulsification is schematically shown in Figure 11.19. The self-organising structures change from reverse micelles or water-in-oil (W/O) microemulsions to an oil-in-water (O/W) emulsion *via* the bicontinuous microemulsion D phase, the lamellar liquid crystal and the L_3 phase with a decrease in temperature. The D phase and LLC phase coexist with the oil (O) phase, but oil and surfactant form a bicontinuous structure in the L_3 phase. At composition B, a lamellar liquid crystalline phase is formed at lower temperatures, the curvature of surfactant molecular layer is flat to water and gel emulsions are not formed. Therefore, the existence of a single L_3 phase region seems to be

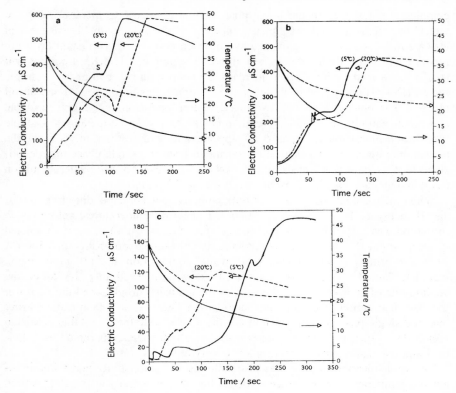

Figure 11.18 *Changes in conductivity and temperature as a function of time.* (a) *The composition of the system is at point A in Figure 11.17. The temperature is decreased from 37 to 20 °C (- - -) or 5 °C (—); the system is continuously agitated while cooling.* (b) *The composition of the system is at point A in Figure 11.17. The temperature is decreased from 37 to 20 °C (- - -) or 5 °C (—); the system is not agitated while cooling.* (c) *The composition of the system is at point B in Figure 11.17*
(Reproduced by permission of Academic Press from ref. 41)

very important for spontaneous emulsification. Similar to the process of emulsification in W/O gel emulsions, an unstable emulsion region is crossed during the process. Therefore the more rapidly the temperature is lowered, the better the emulsification.

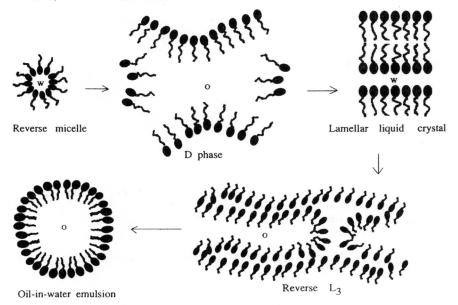

Figure 11.19 *Schematic change in self-organising structures during the spontaneous forma-tion of highly concentrated O/W gel emulsions*
(Reproduced by permission of Academic Press from ref. 41)

Acknowledgements

The authors gratefully acknowledge support by CICYT (Grant QUI 96-0454) and from Generalitat de Catalunya (Comissionat per a Universitat i Recerca Grant 1997SGR 00105).

11.5   References

1. C. Solans, F. Comelles, N. Azemar, J. Sanchez Leal and J.L. Parra, *J. Com. Esp. Deterg.*, 1986, **17**, 109.
2. H. Kunieda, C. Solans, N. Shida and J.L. Parra, *Colloids Surf.*, 1987, **24**, 225.
3. C. Solans, N. Azemar and J.L. Parra, *Prog. Colloid Polym. Sci.*, 1988, **76**, 224.
4. C. Solans, J.G. Dominguez, J.L. Parra, J. Heuser and S.E. Friberg, *Colloid Polym. Sci.*, 1988, **266**, 570.
5. H. Kunieda, N. Yano and C. Solans, *Colloids Surf.*, 1989, **36**, 313.
6. H. Kunieda, D.F. Evans, C. Solans and M. Yoshida, *Colloids Surf.*, 1990, **47**, 35.
7. R. Pons, C. Solans, M.J. Stébé, P. Erra and J.C. Ravey, *Prog. Colloid Polym. Sci.*, 1992, **89**, 110.
8. C. Solans, R. Pons, S. Zhu, H.T. Davis, D.F. Evans, K. Nakamura and H. Kunieda, *Langmuir*, 1993, **9**, 1479.
9. R. Pons, P. Erra, C. Solans, J.C. Ravey and M.J. Stebe, *J. Phys. Chem.*, 1993, **97**, 12320.
10. H. Kunieda, A. Cherian John, R. Pons, C. Solans, in 'Structure–Performance Relation-ships in Surfactants', eds. K. Eourni and M. Meno, Dekker, New York, 1997, p. 359.

11. R. Pons, J.C. Ravey, S. Sauvage, M.J. Stebe, P. Erra and C. Solans, *Colloids Surf.*, 1993, **76**, 171.
12. R. Pons, I. Carrera, P. Erra, H. Kunieda and C. Solans, *Colloids Surf. A*, 1994, **91**, 259.
13. E. Ruckenstein, *Adv. Polym. Sci.*, 1997, **127**, 3.
14. N.R. Cameron, D.C. Sherrington, *Adv. Polym. Sci.*, 1996, **126**, 165.
15. W. Ostwald, *Kolloid Z.*, 1910, **6**, 103; 1910, 7, 64.
16. K.J. Lissant, *J. Colloid Interface Sci.*, 1966, **22**, 462.
17. K.J. Lissant and K.G. Mayhan, *J. Colloid Interface Sci.*, 1973, **42**, 201.
18. H.M. Princen, *J. Colloid Interface Sci.*, 1979, **71**, 55.
19. H.M. Princen, *J. Colloid Interface Sci.*, 1983, **91** 160.
20. H.M. Princen and A.D. Kiss, *J. Colloid Interface Sci.*, 1986, **112**, 427.
21. J.C. Ravey and M.J. Stébé, *Physica B*, 1989, **394**, 156.
22. J.C. Ravey and M.J. Stébé, *Prog. Colloid Polym. Sci.*, 1990, **82**, 218.
23. R. Pons, G. Calderó, M.J. García, N. Azemar, I. Carrera and C. Solans, *Prog. Colloid Polym. Sci.*, 1996, **100**, 132.
24. G. Calderó, M.J. García-Celma, C. Solans, M. Plaza and R. Pons, *Langmuir*, 1997, **13**, 385.
25. G. Calderó, N. Azemar, I. Carrera, M.J. García-Celma, C. Solans and R. Pons, 'Proc. 4th World Surfactants Congress, Asociación Española de Productores de Sustancias para Aplicaciones Tensioactivas, Barcelona', 1996, vol. 2, p. 245.
26. L. Auvray, J.P. Cotton, R. Ober and C. Taupin, *J. Phys.*, 1984, **45**, 913.
27. H. Kunieda, V. Rajagopalan, E. Kimura and C. Solans, *Langmuir*, 1994, **10**, 2570.
28. V. Rajagopalan, C. Solans and H. Kunieda, *Colloid Polym. Sci.*, 1994, **272**, 1166.
29. S.S. Davis, J. Hadgraft, K.J. Palin, in 'Encyclopedia of Emulsion Technology', ed. P. Becher, Dekker, New York, 1985, vol. 2, p. 159.
30. G. Calderó, M.J. García-Celma, C. Solans, M.J. Stébé, J.C. Ravey, S. Rocca and R. Pons, *Langmuir*, 1998, **14**, 1580.
31. E.L. Cussler, in 'Diffusion: Mass Transfer in Fluid Systems', Cambridge University Press, Cambridge, ch. 3, 1994.
32. G. Calderó, to be published.
33. S. Rocca, M.J. García-Celma, G. Calderó, R. Pons, C. Solans and M.J. Stébé, *Langmuir*, 1998, **14**, in press.
34. H. Kunieda and K. Shinoda, *J. Disp. Sci. Technol.*, 1982, **3**, 233.
35. H. Kunieda and K. Shinoda, *J. Colloid Interface Sci.*, 1985, **107**, 107.
36. K. Shinoda and H. Saito, *J. Colloid Interface Sci.*, 1968, **26**, 70.
37. H.M. Princen, M.P. Aronson and J.C. Moser, *J. Colloid Interface Sci.*, 1980, **75**, 246.
38. R.D. Vold and R.C. Groot, *J. Phys. Chem.*, 1964, **68**, 3477.
39. J. Bibette, D. Roux and F. Nallet, *Phys. Rev. Lett.*, 1990, **65**, 2470.
40. H. Kunieda, Y. Fukui, H. Uchiyama and C. Solans, *Langmuir*, 1996, **12**, 2136.
41. K. Ozawa, C. Solans and H. Kunieda. *J Colloid Interface Sci.*, 1997, **188**, 275.
42. U. Olson, M. Jonströmer, K. Nagai, O. Söderman, H. Wennerström and G. Klose, *Prog. Colloid Polym. Sci.*, 1988, **76**, 75.

CHAPTER 12

Applications of Emulsions

THOMAS FÖRSTER AND WOLFGANG VON RYBINSKI

Henkel KGaA, Henkelstrasse 67, 40191 Düsseldorf 1, Germany

12.1 Introduction

Emulsions are formed when two immiscible liquids are mixed with each other. The most familiar forms are oil-in-water emulsions (O/W emulsions), which consist of oil droplets in water, and water-in-oil emulsions (W/O emulsions), where an aqueous solution is emulsified in an outer oil phase. In specific cases, two immiscible liquids form transparent systems, which are termed microemulsions.[1] According to IUPAC, a microemulsion is a thermodynamically stable emulsion.[2] This definition applies to a few systems made up of oil, water and emulsifiers, which spontaneously form transparent mixtures. Over and above this narrow definition, however, in patents and scientific literature the name microemulsion is applied to other transparent to translucent systems, which are not created sponta- neously but by special production procedures.

Emulsions are not a human invention. In living nature they play an especially important role in the absorption of fats with nutrients. The earliest known use of an emulsion by humans is certainly the exploitation of milk and milk products such as cream, butter and cheese for nutritional purposes. With increasing prosperity, the advanced civilizations of antiquity began to use emulsions for cosmetic purposes. Nowadays these ancient uses have been joined by countless others, primarily in a wide variety of technical processes, so that emulsions play a role in connection with almost all everyday products and processes. What is it that makes emulsions so attractive?

Emulsions are so attractive because they consist of at least two phases, an oil and a water phase, so that they are suitable solvents not only for water-soluble but also for hydrophobic substances. The good solubilizing properties of emulsions with regard to substances of different polarity is exploited mainly in the pharmaceutical and agricultural sectors, but also increasingly in the field of soil remediation. In other fields the interactions of emulsions with solid surfaces play a central role. Cooling lubricants, rolling oil emulsions, fiber and textile auxiliaries and other lubricant emulsions have the primary task of preventing undesirable frictional effects during machining processes, or at least keeping them within acceptable limits. The water phase, with its thermal capacity, dissipates the frictional heat generated, while oil, emulsifiers and other auxiliary substances are adsorbed on the treated materials, making the surface hydrophobic and therefore causing a lubricating effect.

In this overview a detailed explanation of the connection between the structure and use of an emulsion is given. A look at the various fields of application in the cosmetics and technology sectors shows how powerfully the specific profile of requirements concerning the technical properties of an emulsion determine physicochemical characteristics such as emulsion type, particle size, rheological properties and manufacturing process. The current state of development of emulsions in the various fields of application is described, as are trends and future developments.

12.2   Cosmetics

All skin creams and lotions and many other toiletry products are emulsions. The application of an aqueous solution on its own has no protective or caring effect on the outermost hydrophobic layer of the skin, the stratum corneum, whereas the application of an oil film makes a definite contribution to skin care but is experienced as too greasy in most cases. Only the skillful combination of water and oil in the right proportions, together with emulsifiers and lipids, imparts a pleasant sensation to skin care. The aqueous phase contains water-soluble active components such as moisturizers, for example glycerol, urea, amino acids and sugar, or cell growth stimulators such as α-hydroxy acids or special oligopeptides. The oil phase usually consists of a number of oil components and active substances such as oil-soluble vitamins, *etc.*[3]

12.2.1   Spreading Properties and Viscosity

A number of aspects have a crucial influence on the sensory properties of a cream. The various oil components differ considerably in terms of their molecular weight, their viscosity and their surface tension,[4] and this is reflected in widely differing spreading properties (Table 12.1). Cosmetic creams therefore

Table 12.1 *Spreading properties of various oils on human skin*

Oil component	Spreading area after 10 min/mm^2
Almond oil	197
Paraffin oil, high-viscosity	201
Caprylic/capric acid triglyceride	566
Octyldodecanol	596
Paraffin oil, low-viscosity	661
Decyl oleate	731
Diisooctylcyclohexane	805
Isopropyl myristate	1044
Dicaprylyl ether	1600

often contain suitable mixtures of strongly, moderately and weakly spreading oil components in order to achieve a pleasant skin sensation.[5] The other parameters that exert an influence are the emulsion type and the selected viscosity. Oil-in-water emulsions are usually experienced as lighter, and water-in-oil emulsions as heavier and greasier. The latter are therefore more widely used in northern countries, especially as a base for night creams with a pronounced caring effect.

The crucial factor determining the viscosity, and also the stability, of W/O emulsions is the so-called phase-volume ratio. Because cosmetic emulsions, irrespective of the emulsion type, usually contain between 65 and 80% water, in W/O emulsions the drops of the internal aqueous phase are more densely packed. According to the Krieger–Dougherty equation, the viscosity increases exponentially as the maximum drop density is approached (Figure 12.1).[6,7] If the maximum density is exceeded the W/O emulsion becomes unstable and the viscosity decreases sharply, because parts of the emulsion invert to O/W or multiple emulsions as a consequence of the geometrical restrictions (Figure 12.1). W/O emulsions are usually stable on the rising arm of the viscosity curve.

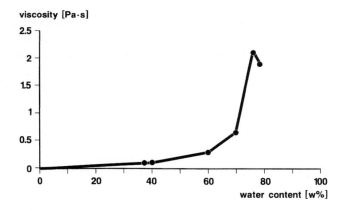

Figure 12.1 *Viscosity behavior of water-in-oil emulsions containing PEG7-hydrogenated castor oil/dicaprylyl ether/decyl oleate/glycerol/MgSO₄ = 3.5:7:7:5:0.7 as a function of water content*
(Reproduced by permission of Allured from ref. 7)

In cosmetic O/W emulsions the internal oil phase is not present in large enough quantities to make a significant contribution to the viscosity. The required viscosity is achieved by adding water-swelling polymers, layer silicates or gel forming lipids,[8–18] which can differ considerably in their viscosity properties and therefore also in their sensory properties.[19,20]

The differences in thickening properties are especially obvious when the viscosity of an emulsion concentrate with a defined particle size distribution is subsequently adjusted by means of a polymer or a lamellar fatty alcohol gel (Figure 12.2). Comparable absolute viscosity values are obtained after adding

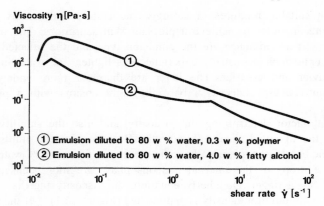

Figure 12.2 *Viscosity curves of thickened emulsions containing 80% water and 0.3 wt% polymer (1) or 4.0 wt% fatty alcohol (2) at 25 °C*
(Reproduced by permission of Steinkopff from ref. 20)

0.3% polyacrylate or 4.0% fatty alcohol. A close examination of the viscosity curves, however, reveals marked differences. The fatty alcohol gel soon breaks down under shear and therefore exhibits strongly shear thinning properties. Strain-sweep experiments performed for the purpose of studying viscoelastic gel properties confirm this finding. The critical strain at which a fatty alcohol gel breaks down is smaller by a factor of ten than that at which the polymer grid breaks down (Figure 12.3).

How do such differences in viscosity properties affect sensory properties? Studies show that the cream consistency preferred by customers correlates with the

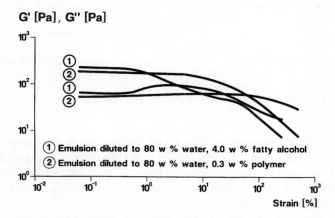

Figure 12.3 *Strain-sweep curves of thickened emulsions containing 80% water and 0.3 wt% polymer (1) or 4.0 wt% fatty alcohol (2) at 25 °C; upper set of curves shows G', lower set G"*
(Reproduced by permission of Steinkopff from ref. 20)

viscosity at a lower shear rate. By contrast, the viscosity at a higher shear rate is an indicator of whether a product can be distributed easily or not. On the basis of these correlations, cosmetic products can be quickly and easily ranked in a sensory-rheological grid and compared (Figure 12.4).

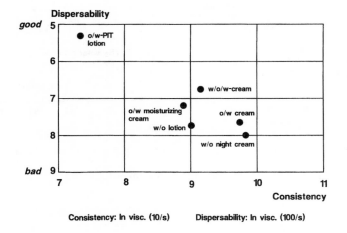

Figure 12.4 *Estimation of sensoric properties of creams by positioning in the rheological grid; consistency was obtained from natural logarithm of viscosity at 10 Pa s, dispersability from natural logarithm of viscosity at 100 Pa s, both at 25 °C*

12.2.2 New Types of Emulsions

In recent years, PIT emulsions have aroused considerable interest (PIT = phase inversion temperature). Emulsions consisting of oil, water and at least one ethoxylated nonionic surfactant undergo a temperature-induced phase inversion, during which a microemulsion is formed.[21,22] A system consisting of tetradecane, water and dodecyl polyethyleneglycol-(5) ether ($C_{12}E_5$) inverts from an oil-in-water to a water-in-oil emulsion in the temperature range between 45 and 55 °C (Figure 12.5). A microemulsion is formed in the phase inversion zone, resulting in the so-called Kahlweit fish in the phase diagram. At low emulsifier concentrations (below 15%) the microemulsion phase is in equilibrium with an oil phase and a water phase, and is therefore called three-phase microemulsion (W + D + O). Emulsifier concentrations of more than 15% are sufficient to solubilize the whole volume of water and oil in the form of a single-phase microemulsion (D) or a lamellar phase L_α. In the PIT zone the hydrophilic and lipophilic properties of the fatty alcohol ethoxylates are balanced, and this is manifested in a clear minimum in the interfacial tension curve (Figure 12.5).[23]

The phase-inversion temperature process has a number of advantages. Finely-dispersed emulsions can be created with a simple hot–cold process,[24] during which a microemulsion is passed through in the higher temperature range (Figure 12.6).

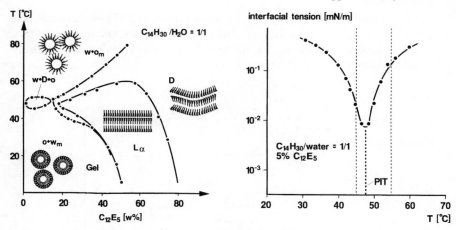

Figure 12.5 *Phase behavior (left) and interfacial tension (right) of water/tetradecane/$C_{12}E_5$ mixtures*
(Reproduced by permission of Dekker from ref. 22)

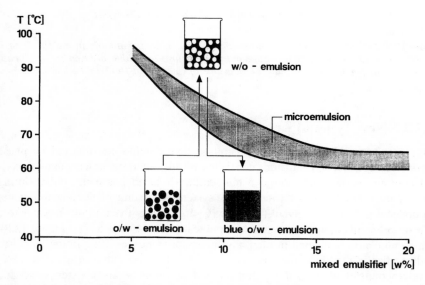

Figure 12.6 *Principle of the phase inversion temperature (PIT) method*
(Reproduced by permission of Dekker from ref. 22)

The finely-dispersed nature of the microemulsion is partially retained after cooling.[25] Emulsions with particle sizes of about 100 nm exhibit long-term stability simply as a consequence of the Brownian motion of the oil droplets,[26] so that low-viscosity, sprayable emulsions can also be produced. Pump-type milky deodorant emulsions are one of the cosmetic product innovations based on this process.

The PIT phenomenon is predictable by parameters. With the help of characteristic variables for oil and emulsifiers, new formulations with desired components can be calculated on a computer by means of the CAPICO procedure (calculation of phase inversion in concentrates), so that development times can be dramatically reduced.[27]

Besides W/O and O/W emulsions, a number of multiple emulsions of the type W/O/W are commercially available. W/O/W emulsions can be produced in a one-stage modified PIT process.[28] In principle, W/O/W emulsions with different, ideally separate, water compartments can be produced in a two-stage process by emulsifying a primary W/O emulsion in an external water phase.[29–31] These emulsions represent a promising basis for approaches to solving the problem of blending sensitive active components such as vitamin C derivatives or enzymes in a stable form.[32] Another approach to the stable formulation of sensitive active substances, which is still at the research stage, involves non-aqueous microemulsions, *e.g.* decane in ethylene glycol[33] or polyols in silicone oils.[34]

12.2.3 Microemulsions

Microemulsions have been the subject of intensive research efforts for many years. This research has been concentrated above all on microemulsions with ethoxylated nonionic surfactants.[21,35–40] However, these microemulsions have not yet been able to establish themselves on the market. The main reason is probably the very high concentrations of surfactants—often in combination with large amounts of short- or long-chain alcohols[38–40]—that are needed to obtain stable microemulsions. This results in high raw material costs and can have a negative effect on skin compatibility.

The alkyl polyglycosides (APGs) form a new class of nonionic surfactant that are characterized by their excellent environmental and dermatological compatibility.[41] Combinations of alkyl polyglycosides with certain hydrophobic cosurfactants are extremely effective in expediting the formation of stable microemulsions.[42,43]

Whereas emulsions consisting of oil, water and an ethoxylated nonionic surfactant pass through a temperature-induced phase inversion, emulsions containing oil, water and a mixture of alkyl polyglycosides with a hydrophobic coemulsifier undergo inversion when the ratio of the components in the emulsifying mixture is varied. A system consisting of dodecane, water and lauryl glycoside, with sorbitan monolaurate (SML) as a hydrophobic coemulsifier, forms microemulsions when the ratio of alkyl polyglycoside to SML is in the range 4:6 to 6:4 (Figure 12.7).[43,44] W/O emulsions are formed when a higher proportion of SML is present, and O/W emulsions are formed if the proportion of alkyl polyglycosides is higher. If the total emulsifier concentration is varied the Kahlweit fish appears in the phase diagram again, with three-phase microemulsions in its body and one-phase microemulsions in its tail. The greater emulsifying power of the alkyl polyglycoside/SML mixture in comparison with the fatty alcohol ethoxylate system (Figure 12.5) is demonstrated by the fact that only 10% of the mixed emulsifier suffices to form a one-phase microemulsion.

The similarity of both types of surfactant in the phase inversion zone is not

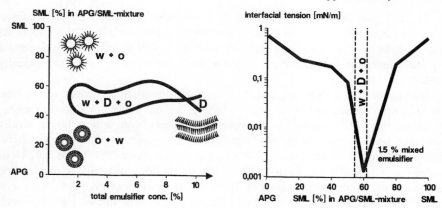

Figure 12.7 *Phase behavior (left) and interfacial tension (right) of water/dodecane/alkyl polyglycoside/sorbitan laurate mixtures with an oil/water ratio of 1:1 at 25 °C* (Reproduced by permission of Verlag für Chemische Industrie from ref. 44)

limited to their phase behavior but is also apparent in the interfacial tension of the mixed emulsifier. At an alkyl polyglycoside/SML ratio of 4:6 the hydrophilic–lipophilic properties of the mixed emulsifier are balanced and the interfacial tension is at a minimum. It is worth remarking that the alkyl polyglycoside/SML mixture has a very low minimum interfacial tension (about 10^{-3} mN m^{-1}), which is an order of magnitude lower than that of the comparable fatty alcohol ethoxylate system and much lower than expected from the slightly shorter alkane chain length (Figure 12.5).

In the case of the alkyl polyglycoside-containing microemulsion, the high level of interfacial activity is attributable to the fact that the hydrophilic alkyl polyglycoside with its large polyglycoside head group is present in exactly the correct ratio with a hydrophobic coemulsifier with a smaller head group at the oil–water interface. Unlike the ethoxylated nonionic surfactants, the hydration—and therefore the effective size of the head group—is almost independent of temperature,[42] so that a smaller temperature dependence is to be expected for the phase behavior of the emulsion, just as for the interfacial tension.[41] This is indeed the case, as Figure 12.8 illustrates for a system consisting of equal parts of dioctylcyclohexane and water as well as 15% of an emulsifying mixture of lauryl glycoside and glyceryl monooleate (GMO).[44] Irrespective of temperature, at alkyl polyglycoside/GMO ratios between 60:40 and 75:25 the system forms transparent microemulsions (ME) or very finely-dispersed blue emulsions (with particle sizes of about 100 nm and smaller) of the O/W type, which were determined by electrical conductivity measurements.

A well studied potential application for alkyl polyglycoside microemulsions in the toiletries sector is cleansing facial care products, which combine good cleansing performance with a refatting effect. Table 12.2 shows a model formulation for an alkyl polyglycoside microemulsion, which was studied as a 2-in-1 facial cleanser in

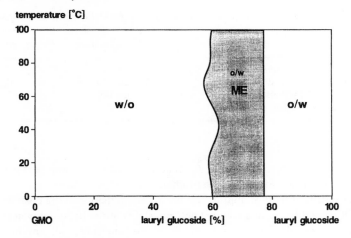

Figure 12.8 *Phase behavior of water/dioctylcyclohexane/lauryl glycoside/glyceryl oleate mixtures with fixed oil/water ratio of 1:1 and 15% mixed emulsifier* (Reproduced by permission of Verlag für Chemische Industrie from ref. 44)

Table 12.2 *Model facial cleanser formulations*[a]

	Microemulsion	Surfactant solution
Lauryl glycoside	1.7	1.7
Coco glycoside	1.1	1.1
Glyceryl oleate	0.7	–
Dicaprylyl ether	4.0	–
Octyldodecanol	1.0	–
Perfume oil	1.0	–
1,2-Propylene glycol	5.0	5.0
Methylparaben	0.2	0.2
Water, dist.	85.3	92.0

[a] in wt% active substance

comparison with an alkyl polyglycoside surfactant solution, which served as a standard.[44] The cleansing performance of both products in the standardized *in vitro* cleansing test on pig skin with black model soiling was excellent (Table 12.3). Even in the presence of oil, the cleansing performance of the microemulsion remained good. The refatting was measured in both cases (*in vitro* on the pig skin model and *in vivo* on the forearms of volunteers) by quantitative analysis of the marker substance dicaprylyl ether, which does not occur naturally in the skin.[44] In both models the microemulsion facial cleanser was associated with a clearly detectable release of oil. These tests therefore confirmed the 2-in-1 concept for the

Table 12.3 *Cleansing performance and refatting effect of the facial cleanser*

	Microemulsion	Surfactant solution
Cleansing performance/% *in vitro*	95 ± 3	93 ± 1
Refatting *in vitro* marker: dicaprylyl ether	300 µg/100 cm^2	0
Refatting *in vivo* (12 volunteers) marker: dicaprylyl ether	270 µg/100 cm^2	0

alkyl polyglycoside microemulsion, *i.e.* that very good cleansing performance can be combined with care effects in one product.

Another potential field of application for microemulsions in the cosmetics sector is low-VOC (volatile organic carbon) perfumes. Traditionally, toilet waters, shaving lotions, *etc.*, are alcoholic solutions of certain perfume oil mixtures. Alkyl polyglycoside microemulsions can now be used as a basis for creating non-alcoholic toilet waters with new property profiles.[45,46]

12.2.4 Skin Penetration

Modern skin care products contain active substances that have not only a physical effect on the skin but also a biological effect inside it. These products need to penetrate the uppermost layers of the skin to a sufficient degree before they can realize their full potential (Figure 12.9). Figure 12.10 illustrates this, using the example of the oil-soluble, antioxidative vitamin E to show how different galenic vehicles—in this case an oil solution, a W/O cream and two O/W creams with different droplet sizes—influence penetration into the uppermost layer of the skin. The penetration studies were carried out on the most recently developed *in vitro* skin model, *i.e.* perfused bovine udder skin.[47,48] In comparison to simple oil solutions, emulsions generally enhance penetration, because at the same total concentration the effective active substance concentration in the solvent phase—in this case in oil—is higher.[7] Moreover, it can be seen that smaller amounts penetrate the skin from the two O/W emulsions than from the W/O emulsion. The cause is the lamellar gel network, which is responsible for the viscosity of the external water phase in the two O/W emulsions but simultaneously hinders the free diffusion of the vitamin E from the oil droplets into the skin.[49]

As well as the direct influence of the emulsion type and structure on skin penetration, attention has recently been focused on effects which are associated with changes in the emulsion structure after topical application. When emulsions are applied to the skin, they are spread out over it in the form of a thin film. The emulsions heat up to skin temperature, and components with a high vapor pressure begin to evaporate.[7,50–52] Drying experiments on thin emulsion films show that—depending on atmospheric humidity and layer thickness—almost all of the water escapes from the emulsion within 5 to 10 minutes. In this window of time there is a change in the viscosity behavior of the emulsion, and in some cases also in its structure. Such phase changes may have dramatic effects on the thermodynamic

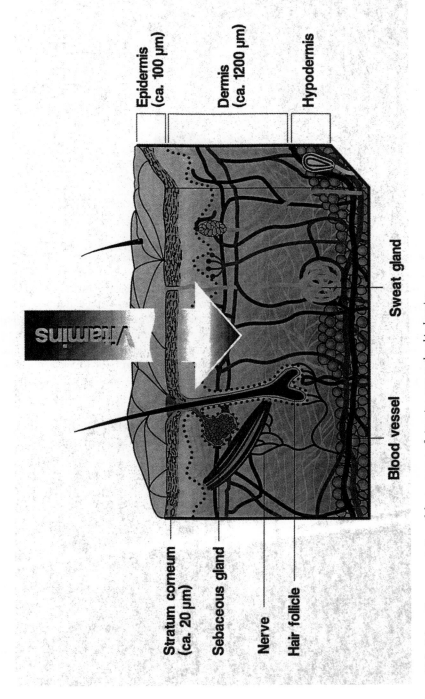

Figure 12.9 *Schematic representation of the transport of vitamins across the skin barrier*

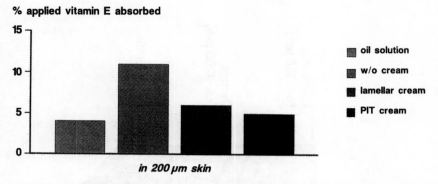

Figure 12.10 *Influence of galenic vehicles on vitamin E penetration into the epidermis for a constant applied amount of vitamin E (800 µg cm⁻² skin)*

activity of active substances, which can be exploited to expedite their liberation. This can be of considerable interest not only for the cosmetics but also for the pharmaceutical sector.[53]

So-called temperature/water-content maps give an impression of the changes in the galenic vehicles during the drying process. Emulsions with a defined water content are prepared for this purpose, and are characterized with regard to their phase behavior and viscosity.[7] In Figure 12.11 the PIT cream is used as an example to illustrate the viscosity and phase changes that can be expected in O/W emulsions. The PIT cream passes through a phase inversion from O/W to W/O when the water content is in the range from 20 to just below 50% and the temperature is

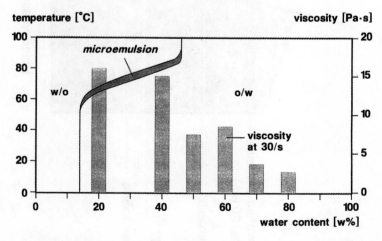

Figure 12.11 *Temperature–water content map of PIT emulsions containing ceteareth-20/ glyceryl stearate/cetearyl alcohol/dicaprylyl ether/decyl oleate/dimethicone/ glycerol = 2.1:2.2:6:8:7:0.5:5 with varying water content; bars show viscosity at 30/s and 25 °C*
(Reproduced by permission of Allured from ref. 7)

between 70 and 80 °C. A microemulsion forms in the phase inversion zone, as described above. After topical application, however, the cream only reaches the skin temperature of about 30 °C, so that the microemulsion zone is not reached. Just like the lamellar emulsion, the PIT emulsion is thickened by a lamellar fatty alcohol gel. During drying the viscosity increases as the lamellar fatty alcohol gel becomes more densely packed. When the water content is low (about 15% and lower) the O/W cream undergoes a phase inversion to a W/O emulsion, caused not by the temperature but solely by the geometrical packing relationships when the water content is so low.

The situation is different in the case of the W/O cream (Figure 12.1). Here the external phase is oil and the viscosity is adjusted by increasing the proportion of the internal phase. In contrast to the O/W systems, the viscosity therefore decreases during drying.

12.3 Metal Processing

In the cosmetics sector, emulsions are mainly employed as leave-on products. They remain on the skin after application. When emulsions are used as additives for technical purposes, *e.g.* metal processing, however, the leave-on situation is the exception. This means that they are applied to a surface before a specific process is carried out, change the surface properties in the required manner during the process, and are then removed to ensure that they do not interfere with the subsequent downstream processes or the final use of the product. The profile of requirements for technical emulsions in terms of their composition and physico-chemical properties therefore covers not only the original application but also the total process involved.

12.3.1 Rolling Oil Emulsions

A typical cold rolling mill produces more than one million metric tons of strip steel each year. While passing through a series of—usually four—roller stands, the strip thickness is reduced by a factor of about 4 to a final thickness of between 0.45 and 3 mm. The strip speeds are of the order of magnitude of 600 to 1000 m min^{-1}.[54] Rolling oil emulsions must perform two main tasks, *i.e.* dissipating the heat generated and providing sufficient lubrication, so that an even surface texture is imparted to the strip steel at these high strip speeds. The lubricant film on the strip surface would interfere with most of the subsequent processing steps, *e.g.* welding, electroplating or phosphatizing. Rolling oil emulsions therefore contain only a low concentration of oil ranging from approx. 1.5% for the first rolling stand down to 0.5% for the fourth stand.[54] In view of the large steel throughput the quantities of emulsion used are enormous. On average, between 4 and 5 m^3 min^{-1} are consumed per rolling stand, giving a consumption rate of 0.2 kg t^{-1} rolled steel.[54]

The lubricating emulsion passes through a cycle. Fresh rolling oil emulsion is applied at the stand. Some of the water evaporates in the roll gap and the emulsion and the oil form a lubricating film on the surface of the steel. The residual emulsion is run off. The run-off emulsion, whose composition now differs from that of the original emulsion, is purified, the water and oil content is replenished, and the emulsion is returned to the cycle. The total emulsion volume circulates several times each hour. The emulsion is not regularly replaced.

Development work in recent years has focused mainly on bringing the mutually contradictory requirements—good lubrication during the rolling process and the least possible amount of surface residues after the rolling process—into alignment. The surface residues that interfere with the further processing of the sheet steel include abrasion particles and the lubricant film, usually measured as the carbon coating.

The main components of cold rolling lubricants are mineral oils and ester oils. The higher aromatics that used to be included have now been prohibited in many countries on health grounds. Synthetic ester oils have proved superior to natural triglycerides on account of their better resistance to hydrolysis and their temperature stability. The addition of a so-called anti-wear package has made it possible to further reduce the degree of chemical breakdown of the rolling oil caused by oxidation or hydrolysis, and to prevent the catalytic decomposition effects of iron or trace metals. The rolling oil residue can then be thoroughly vaporized in the subsequent annealing process.[55]

Mixtures of alkyl ethoxylates are frequently employed for emulsifying purposes,[54] and in some cases protective colloids are also added to provide electrosteric stabilization.[55] It is of crucial importance for the rolling result that the emulsifier has an optimal composition. Not only the surface quality of the sheet steel but also the rolling speed are decisively influenced by the size distribution of the oil droplets. A narrow spread of droplet sizes of about 2−4 μm has been found to be optimal, and even remains stable after being fed back into the emulsion cycle.

The development of controlled-particle-size (CPS) emulsifier systems brought about an important breakthrough in cold rolling technology by enabling stable emulsions with such large droplets and narrow size distribution to be produced in the rolling process. The main advantage of these CPS rolling oil emulsions is the dramatically reduced amount of friction debris. This is also reflected in economic terms, because the tonnages between two roller changes are much higher.[55]

12.3.2   Lubricant Emulsions

The problems of heat dissipation and lubrication are common to many activities in the metal processing industry. Cooling lubricants are needed during drilling and cutting operations, for example to ensure that heat is dissipated quickly enough to protect the workpiece against thermal damage. Moreover, the cooling lubricant stream carries metal chips away from the processing zone and covers the freshly exposed metal surface with a protective film of emulsion.[56] Besides improvements in technical performance, which is quantified as the ratio of removed metal to tool wear,[56] toxicological properties[57] and recycling[58] have been at the heart of development efforts in recent years.

The lubricant emulsions used are of the O/W type, with a more or less substantial proportion of oil. Microemulsions are employed, as well as finely dispersed O/W emulsions, which can also be created with the PIT process.

12.4 Textiles

In contrast to the metal processing sector, where the emphasis is mainly on the use of emulsions, for lubrication purposes and to dissipate heat, the textile industry is characterized by the frequent use of microemulsions. Metal surfaces are usually large and smooth, and the surface energies are comparatively high, so that metals are readily wettable. Textile fibers and fabrics, on the other hand, consist of individual fibers with a small radius of curvature, and they are spun or woven, so that lots of small cavities are formed. In many fibrous materials, especially synthetics, these cavities are not readily susceptible to wetting. Textile auxiliaries therefore contain surfactants in higher concentrations, which serve not only as emulsifiers but also as wetting agents. Microemulsions, which are characterized by their pronounced interfacial activity, are especially suitable for wetting fibers quickly (Figures 12.5 and 12.7).

12.4.1 Textile Fiber Preparations

Thanks to the contribution of ever more effective textile fiber preparations, processing speeds in nylon and polyester factories are now of the order of up to 4000 m min^{-1}. The most important tasks of textile fiber preparations are to reduce friction at guide pulleys and machine parts, and to ensure that there is sufficient fiber/fiber interaction in the fiber strands after the spinning process, so that there will be adequate cohesion between the fibers. They also have the task of preventing electrostatic charge from building up on the fibers, so that there will be no frictionally induced electrostatic charge during subsequent processing stages, such as sewing.

Although textile fiber microemulsions used to contain roughly equal amounts of mineral oil and emulsifiers (sulfated triglycerides, alkanolamines and potassium soaps), more modern systems are based on thermally stable ester oils and mainly nonionic ethoxylated emulsifiers. Alkyl phosphate esters, sulfonated mineral oils and dialkyl sulfosuccinates are employed as antistatic agents.[59]

With the introduction of high-speed spinning mills, polyester partially oriented yarns (POY) became available. The orientation process requires the fibers freshly solidified from the melt to be stretched to a certain degree, and this is made possible by controlled fiber/metal friction in the texturing machine. For POY textile fiber preparations, therefore, a limited lubricating effect is now required rather than a maximum one. Water-soluble block copolymers or random ethylene oxide (EO)/propylene oxide (PO) copolymers ideally satisfy this modified requirement profile. Moreover, the EO/PO copolymers are characterized by their special thermal properties. After the fiber production process proper, the fibers are passed over heaters in the texturing machine for the purpose of aligning the polymers. This causes the EO/PO copolymers to break down as a consequence of thermal oxidation, so that hardly any spinning lubricant residues remain on the fibers. In combination with small amounts of wetting and antistatic agents, a new generation of textile fiber preparations has been created on the basis of the EO/PO copolymers.[59]

12.4.2   Textile Finishing

The main function of textiles used in articles of clothing is to protect the body by keeping heat in and moisture out. The wearer comfort of textiles depends not only on heat and moisture management but also on the softness and feel of the material. Finishing agents impart a pleasant, soft and supple feel to textiles, especially to cotton but also to cotton–polyester blended fabrics.

The principle of finishing is based on selective adjustment of the hydrophilic and hydrophobic properties of the surface. Because almost all fiber surfaces carry a negative charge, finishing aids often contain cationic surface-active substances, which are adsorbed on the textile surface due to electrostatic attraction. Besides quaternary alkylammonium compounds and the ecotoxicologically more compatible esterquats, modified silicone oils are also very popular softening agents in the field of textile finishing due to their versatility. By varying the functional amino groups and the degree of polymerization of the basic silicone backbone, an optimal combination of technically relevant properties such as feel, water repellency, water permeability, resistance to yellowing on drying, and soil release can be achieved.[60,61]

Textile finishing aids are used preferentially in the form of very finely distributed dispersions or emulsions, to ensure that the particles are deposited evenly over the whole surface of the textile. Microemulsions are especially advantageous. The small drop size and good wetting properties enable the product to penetrate into the textile and the fibers, thus ensuring excellent distribution of the microemulsion, combined with outstanding softness and smoothness. In practice, textile finishing aids are usually applied by means of the Foulard process, in which the textile is immersed in the strongly diluted microemulsion for a certain period of time, expressed as about 90% liquor pickup, and then dried at a high temperature. The finishing agent content of the finished textile is less than 0.5 wt%.[62]

Although silicone oil microemulsions used to be produced by high-pressure emulsification, modern methods exploit an emulsion inversion that can be induced by the oil/water ratio[63,64] or the pH-dependent degree of dissociation of amino-functional silicone oils.[62] A typical silicone oil microemulsion used for textile finishing consists of an amino-functional silicone oil, an ethoxylated nonionic surfactant and water. Owing to the functional amino group the silicone oil becomes surface active, with hydrophilic and lipophilic properties that vary with the degree of dissociation. The phase behavior of such a system is strongly dependent on the pH and water content (Figure 12.12).

If the water content is low, the silicone oil Q2-8166 (Dow Corning; viscosity 1500 mPa s, N-content 0.8%) forms a lamellar liquid crystalline phase. In an alkaline environment the lamellar phase is in close proximity to a microemulsion phase and the system inverts when water is added, forming a blue, finely dispersed O/W emulsion. Acidification with glacial acetic acid transforms this blue O/W emulsion into a microemulsion (ME) which, as a 30% concentrate, can be diluted with water to yield a clear solution prior to being used for textile finishing.

The number of possible liquid crystal phases in neutral and acidic environments is remarkable. If water is added to an acidic lamellar liquid crystal (L_α) a highly

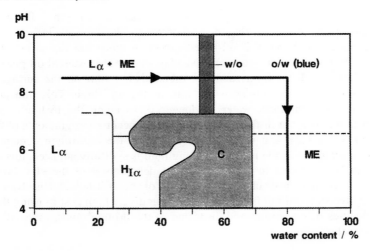

Figure 12.12 *pH–water content map of water/amino-functional silicone oil/$C_{12/14}E_4$ mixtures with $C_{12/14}E_4$/silicone oil =1:3 at 25 °C*

viscous hexagonal phase is formed first of all ($H_{I\alpha}$), followed by a solid-like cubic phase (C). The sequence in which the individual components are added is also of crucial importance for the microemulsification process. Although only low-viscous emulsion phases were passed through in the first sequence (water followed by acid), the reverse addition sequence resulted in the formation of highly-viscous, non-pumpable liquid crystal phases. Systematic examination of the pH/water-content cards enables optimal microemulsion processes to be developed for a variety of combinations of amino-functional silicone oils and emulsifiers.

12.5   Oilfields

Rock drilling, for the purpose of recovering oil, natural gas or water, makes a variety of demands on drilling fluids, depending on the type of drill hole and the local circumstances. In the simplest case, that of a land-based shallow drill hole, an aqueous system, thickened with layered silicates or polymers, suffices for removing the drill cuttings and cooling the drill bit. In the case of deeper holes or side-tracked drill holes, which are familiar in the offshore sector, good lubrication of the rotating parts is also required, and is provided by water-based O/W emulsions.[65] Another factor is that the hydrostatic pressure increases with the depth of the drill hole, and this pressure has to be compensated for by fillers, *e.g.* $BaSO_4$ particles.

The main problem associated with deep drilling, however, is that of stabilizing the wall of the drill hole. Depending on the geological properties of the rock formation, the wall of the hole may exhibit cracks. Water from the drilling fluid will then be forced into these cracks by the prevailing hydraulic pressure.[66] In rock formations with a large proportion of clay this water causes the clay minerals to swell, and as a consequence they lose their mechanical strength.[67] The rock may then break up, ultimately causing the wall of the drill hole to cave in.

Experience has shown that oil-based inversion fluids are the safest operating fluids for drilling through water-sensitive clay layers. The external oil phase of the W/O drilling emulsion forms a dense oil/emulsifier film on the surface of the drilled rock layers, and this seals the wall of the drill hole. This film acts as a semipermeable membrane: water no longer flows along the hydraulic pressure gradient in the rock but diffuses only along the concentration gradients of the salts dissolved in the rock and in the drilling fluid. If the electrolyte content of the internal water phase of the inverted drilling fluid is skillfully adjusted, this osmotic effect can be exploited to achieve a slight dehydration of the clay layers and therefore to strengthen mechanically the wall of the drill hole.[68] Disadvantages of the W/O inversion drilling fluids are the comparatively high cost of using them and the difficulties of processing and disposing of the drilling muds, especially in the offshore sector.

12.6   Washing and Cleaning Processes

12.6.1   Detergency Mechanisms

The cleaning action of aqueous cleaning solutions is based above all on the action of surfactants. By rolling up and lifting off the soil particles, then emulsifying and solubilizing them in the cleaning solution, they make the main contribution to removing soil from fabrics. A decisive factor in this context is the lowest possible interfacial tension between cleaning solution and soil or fabric.[69]

Of relevance is Young's equation, which relates the interfacial tension between the surfactant solution and oil (γ_{ow}), oil and solid substrate (γ_{os}) and surfactant solution and solid substrate (γ_{ws}) to the equilibrium contact angle θ:

$$\cos \theta = \frac{\gamma_{ws} - \gamma_{os}}{\gamma_{ow}} \qquad (12.1)$$

For quite hydrophilic surfaces like cotton, γ_{ws} is smaller than γ_{os}, and a contact angle greater than 90° is commonly achieved. In this case, the roll-up mechanism is operative: the water preferentially wets the fabric, causing the oily stains to be entirely lifted off the fibers into the washing solution. This behavior, shown schematically in Figure 12.13b for soil removal from a flat surface, is enhanced on cotton fabric due to swelling of the cotton fibers with water, which increases the hydrophilicity of the fabric surfaces.[70]

For low energy surfaces, *i.e.* hydrophobic materials such as polyester, a contact angle of less than 90° is usually observed, and small amounts of the oily soil may be removed by hydraulic forces at the soil–water interface, as shown in Figure 12.13a. If the fabric surface is completely covered by oily soil, no location is available for the surfactant solution to reach the fiber surface and undercut the soil. Observation of this emulsification mechanism has been made by many investigators for mineral oils and mineral oil/polar soil mixtures on hydrophobic flat films and

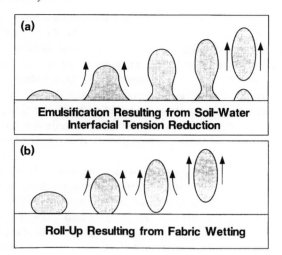

Figure 12.13 *Mechanisms of liquid soil removal: emulsification* (a) versus *roll-up* (b) *mechanism*
(Reproduced by permission of Elsevier from ref. 86)

fibers.[71,72] Removal in this manner is enhanced by low interfacial tension at the oil–water interface, which allows the oil film to be deformed easily to form small emulsion droplets.

Several factors have been studied with regard to their effect on the emulsification mechanism for the removal of mixtures of mineral oil and polar organic alcohols or acids from polyester.[71] The rate of emulsification of mineral oil/oleic acid mixtures from polyester films was found to change as the oleic acid content was varied. Other factors such as electrolyte concentration and temperature were also found to have large effects on the rate of soil removal by this mechanism.

Direct solubilization of oily soils into surfactant micelles can occur to a significant extent if a large excess of surfactant relative to oil is present and if the surfactant is above its critical micelle concentration (CMC). The solubilization of very small oil drops from polymer fibers has been visualized for a variety of nonpolar oils representative of liquid laundry soils.[73]

Another mechanism of oily soil removal involves the formation of liquid crystalline phases at the detergent solution/oil interface.[74–76] After formation the intermediate phase is broken off by agitation and emulsified into the aqueous solution, allowing fresh contact of the remaining soil with the detergent solution.

Soil solubilization rates are often enhanced when surfactant-rich phases, either isotropic or liquid crystalline in nature, are present in the washing solution. Such phases exist, for instance, when nonionic surfactants are above their cloud points. These phases can either solubilize oily soils directly or interact with soil to form intermediate surfactant-rich phases such as microemulsions containing large amounts of oil. Under favorable conditions, the intermediate phases can be emulsified into the washing bath.

Clear evidence exists that solubilization and emulsification are major factors in removal of oily soils from hydrophobic, synthetic fabrics.[77,78] Unlike roll-up, in which the interaction of the fabric with the oily soil and water is the most critical factor, the solubilization–emulsification mechanism occurs primarily at the oil/detergent solution interface and is therefore directly influenced by the phase behavior of the corresponding oil–water–surfactant system.

An important feature of the phase behavior of systems containing water, surfactants, and hydrocarbon soils is the existence of microemulsions (see Section 12.2), which are thermodynamically stable liquid phases containing substantial amounts of both water and oil.

The condition for which the hydrophilic and lipophilic properties are exactly balanced and the surfactant films have no spontaneous tendency to curve in either direction has been called the phase inversion temperature (PIT) or hydrophile–lipophile balance (HLB) temperature by Shinoda and Friberg[79] for the case of nonionic surfactants for which temperature is usually the variable of greatest interest (see Section 12.2). For ionic surfactants it is more common to speak of 'optimal' conditions, *e.g.* optimal salinity.[80]

Schambil and Schwuger compared the phase behavior of water/oil/surfactant systems and detergency.[81] When three-phase microemulsions are formed, extremely low interfacial tensions between two phases are observed. Figure 12.14 shows the change of the interfacial tension at equilibrium in the proximity of the three-phase range for the system water/decane/$C_{12}E_4$.[82] Because the interfacial tension is generally the restraining force with respect to the removal of liquid soil in the washing and cleaning process, it should be as low as possible for optimal soil

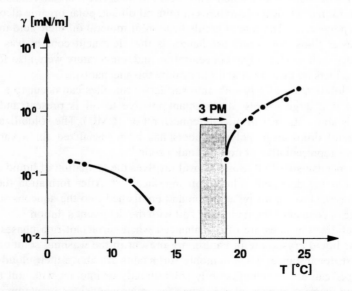

Figure 12.14 *Phase behavior and interfacial tension of water/decane/$C_{12}E_4$ systems; 3-PM = three-phase microemulsion*
(Reproduced by permission of Steinkopff from ref. 81)

removal. For instance, in mixtures of alkylbenzene sulfonate and alkyldiglycol ether sulfate, the minimum interfacial tension coincides with the optimum oil removal. Other quantities such as the wetting energy and the contact angle on polyester, as well as the emulsifying ability of olive oil, also show optima at the same mixture ratio at which the minimum interfacial tension is observed.[83]

Figure 12.15 (right) represents the three-phase temperature intervals for $C_{12}E_4$ and $C_{12}E_5$ *versus* the number (n) of carbon atoms in n-alkanes.[84] The left side of Figure 12.15 shows the results of detergency performance tests.[85] Comparison of both diagrams indicates that the maximum oil removal is in the three-phase interval of the soil used (n-hexadecane). This means that not only the solubilization capacity of the concentrated surfactant phase but also the minimum interfacial tension existing in the zone of the three-phase body are responsible for the maximum oil removal. Further details about the influence of the polarity of the oil, the type of surfactant and the addition of salt are summarized in the review of Miller and Raney.[86]

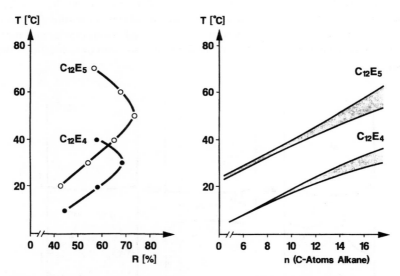

Figure 12.15 *Phase behavior of nonionic surfactants and detergency: three-phase temperatures* versus n *(right) and detergency (measured as reflectancy* R*) versus temperature (left)*
(Reprinted by permission of Steinkopff from ref. 81)

Studies of diffusional phenomena have direct relevance to detergency processes. Experiments are reported which investigate the effects of changes in temperature on the dynamic phenomena, which occur when aqueous solutions of pure nonionic surfactants contact hydrocarbons such as tetradecane and hexadecane.[77,87] These oils can be considered to be models of nonpolar soils such as lubricating oils. The dynamic contacting phenomena, at least immediately after contact, are representative of those which occur when a detergent solution contacts an oily soil on a synthetic fabric surface. The following is a summary of the observed behavior interpreted through the use of schematic diffusion paths. Detailed studies of the phase behavior in such systems were used in the construction of the diffusion paths.[88]

With $C_{12}E_5$ as the nonionic surfactant at a 1 wt% level in water, quite different phenomena were observed below, above, and well above the cloud point when tetradecane or hexadecane was carefully layered on top of the aqueous solution. Below the cloud point temperature of 31 °C, very slow solubilization of oil into the one-phase micellar solution occurred. At 35 °C, which is just above the cloud point, a much different behavior was observed. The surfactant-rich L_1 phase separated to the top of the aqueous phase prior to the addition of hexadecane. Upon addition of the oil, the L_1 phase rapidly solubilizes the hydrocarbon to form an oil-in-water microemulsion containing an appreciable amount of the nonpolar oil. After depletion of the larger surfactant-containing drops, a front developed as smaller drops were incorporated into the microemulsion phase. This behavior is shown schematically in Figure 12.16. Unlike the experiments carried out below the cloud point temperature, appreciable solubilization of oil was observed in the time frame of the study, as indicated by upward movement of the oil–microemulsion interface. Similar phenomena were observed with both tetradecane and hexadecane as the oil phases.

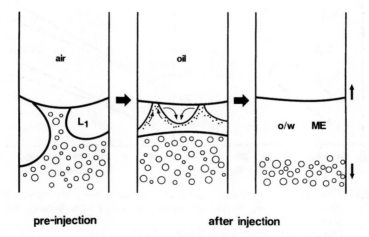

Figure 12.16 *Conversion of the L_1 phase into an oil-in-water microemulsion at temperatures above the cloud point*
(Reproduced by permission of Elsevier from ref. 86)

When the temperature of the system was raised to just below the phase inversion temperatures of the hydrocarbons with $C_{12}E_5$ (45 °C for tetradecane and 50 °C for hexadecane), two intermediate phases formed when the initial dispersion of L_1 drops in the water contacted the oil. One was the lamellar liquid crystalline phase

L_a (probably containing some dispersed water). Above it was a middle-phase microemulsion. In contrast to the studies below the cloud point temperature, there was appreciable solubilization of hydrocarbon into the two intermediate phases. A diagram of the phenomena observed is shown in Figure 12.17. A similar progression of phases was found at 35 °C using *n*-decane as the hydrocarbon. At this temperature, which is near the phase inversion temperature of the water–$C_{12}E_5$– decane system, the existence of a two-phase dispersion of L_a and water below the middle-phase microemulsion was clearly evident.

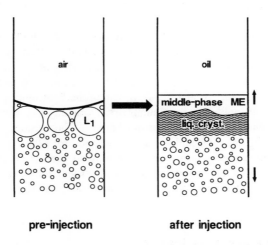

pre-injection   **after injection**

Figure 12.17 *Conversion of the L_1 phase into a middle-phase microemulsion and a liquid crystal dispersion at temperatures just below the PIT*
(Reproduced by permission of Elsevier from ref. 86)

These results can be utilized to optimize surfactant systems in detergents, and in particular to improve the removal of oily soils. The formation of microemulsions is also described in the context of the pretreatment of oil-stained textiles with a mixture of water, surfactants and cosurfactants.[89-91]

12.6.2 Dry Cleaning

Aqueous cleaning liquors represent by far the best solution for the majority of practical problems encountered in the field of cleaning technology. However, some stains are very difficult to remove from textiles with aqueous cleaning liquors, *e.g.* bad oil stains or high-molecular weight lipophilic compounds (resins, waxes, paints). Such soil components can only be extracted by the organic solvents used in dry cleaning.[92] In practice, pure organic solvents are only rarely used to clean textiles. Usually small amounts of surfactants, so-called cleaning boosters, are also present. These enable the water in the fabric to be dissolved in inverse micelles. Often small amounts of water are also added, to promote the removal of hydrophilic soil and impart a higher soil-carrying capacity to the cleaning solution. The cleaning media used in dry cleaning are therefore, strictly speaking, oil-rich W/O microemulsions.

In the mid-1970s the idea of uniting wet cleaning and dry cleaning in one process was put forward, with a view to exploiting the advantages of both techniques.[93,94] Two different processes were subsequently developed, referred to as dual and emulsion cleaning. Dual cleaning involves pretreating the textiles in an aqueous cleaning medium, in which most of the water-soluble soils and pigments are removed. The lipophilic components are then extracted in a series of solvent baths and the remaining water in the fabric is solubilized by cleaning boosters.

Emulsion cleaning combines both cleaning media in the form of an emulsion in a washing bath. The textiles are pretreated in a solvent bath, in which most of the oily soils and the pigments attached to them are lifted off. In a second step the textiles are treated with an emulsion consisting of organic solvents, a surfactant and water (usually between 6 and 50%), in which the hydrophilic soil components and the redeposited lipophilic components are extracted. The fabric is then rinsed in organic solvents in a series of rinsing baths.[95] The cleaning performance of emulsions consisting of water and tetrachloroethane proved to be much better than that of the aqueous cleaning liquid or the organic solvent tetrachloroethane alone. Emulsion cleaning also has the advantage that detergent additives such as builders, bleaches, optical brighteners and enzymes are just as active as in the aqueous cleaning medium.[92]

12.6.3 Microemulsion Cleaners

In contrast to the formation of microemulsions from aqueous surfactant systems and oily soils during the cleaning process, very little basic research has been carried out on microemulsions as a cleaning medium. Initial studies of textile cleaning with microemulsions by Solans *et al.*[96,97] were published in 1985. At washing temperatures between 296 and 307 K, homogeneous microemulsions obtained from the system water/$C_{12}E_4$/n-hexadecane and systems with technical nonionic surfactant mixtures removed 1.5 to 2 times more soil from wool, cotton and cotton–polyester blended fabrics stained with oily and particulate soils than a highly concentrated commercial liquid detergent (Figure 12.18). Soil removal by the microemulsions was increased by 20 to 25% by adding 0.05 M of the electrolytes sodium triphosphate and sodium citrate, which act as builders. The microemulsions also proved superior to the liquid detergent, in that they could be used seven times without losing any of their cleaning effectiveness.

Dörfler *et al.* systematically studied the phase behavior of quaternary systems, consisting of water, nonionic surfactants, a cosurfactant and a hydrocarbon, with regard to possible applications in the textile cleaning sector.[98] As an example, Figure 12.19 shows the influence of the cosurfactant on the phase behavior of the water–oil–surfactant system. In this case the phase inversion range decreases by an average of about 5 K per added mol% cosurfactant. The extent of the three-phase zone is scarcely affected.

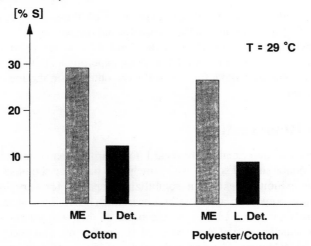

Figure 12.18 *Soil removal* (S) *by a surfactant phase microemulsion (ME) and by a 1%*
aqueous liquid detergent solution (L. Det.) from different fabrics
(Reproduced by permission of Dekker from ref. 96)

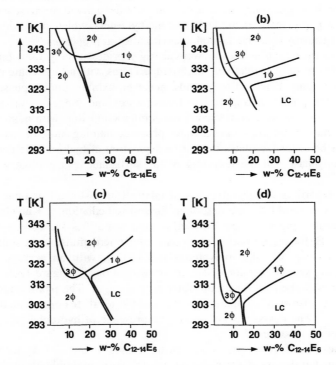

Figure 12.19 *Phase behavior of water/oil/$C_{12/14}E_6$ mixtures without cosurfactant* (a), *with*
2 wt% n-pentanol (b), *with 4 wt% n-pentanol* (c), *and with 6 wt% pentanol*
(d); *water/oil ratio = 1:1*
(Reproduced by permission of Hanser from ref. 98)

The influence of the degree of ethoxylation j of the nonionic surfactants. ($C_{12-14}E_j$; $j = 5, 4, 6, 7, 8$) and added cosurfactant (*n*-pentanol) on the phase inversion temperature was quantified in the form of a phase diagram.[98] When microemulsions were subjected to model washing experiments with oil-stained test fabrics, the best cleaning results for oily soil were obtained in the low-water W/O microemulsion range.

12.7   Soil Remediation

Microemulsions are described with regard to their use as a medium for the remediation of contaminated soils.[99] The principle is based on the observation that microemulsions exhibit an excellent solubilizing capacity for very hydrophobic pollutants such as polycyclic aromatic hydrocarbons (PAH). Moreover, the use of bicontinuous microemulsion phases on the basis of their temperature-dependent phase behavior enables partial separation of the pollutants and recovery of the surfactants to be achieved. The demands made on the system in practice are a slightly increased temperature for the extraction, minimum oil and surfactant content, maximum extraction performance, and good separation of the pollutants. A decisive criterion for the use of microemulsions for soil remediation is the choice of suitable biodegradable surfactants and of oils derived from renewable raw materials, *e.g.* triglycerides and fatty acid methyl esters, which are readily degraded by microorganisms and have favorable environmental properties.[100]

Owing to their high molecular weight, triglycerides usually form bicontinuous microemulsions only at relatively high temperatures with a high content of hydrophobic surfactants. Esters of fatty acids, however, exhibit suitable phase behavior, forming three phases at appropriate temperatures and moderate surfactant content.[101] At lower temperatures, bicontinuous microemulsions with nonionic surfactants split into an oil phase and a water phase containing most of the surfactant. This phase behavior facilitates the partial separation of hydrophobic contaminants with the oil phase and the recycling of the surfactant by a multistep extraction process.[99,102,103]

Studies regarding ternary mixtures of rape oil or castor oil or rape oil methyl ester with water and C_iE_j are described for soil remediation.[104] Knowledge of the effective chain length of rape oil methyl ester simplifies the choice of suitable technical C_iE_j for the formation of bicontinuous microemulsions at suitable temperatures. A series of alkyl polyethoxylates $C_{9/11}E_j$ with an average degree of ethoxylation j was chosen for systematic investigations.[104] Figure 12.20 shows the phase diagrams obtained for an oil/water ratio $\alpha = 0.5$. The phase boundaries are considerably distorted compared to the nearly ideal features found for pure surfactants.[105] The deviations are due to the presence of hydrophobic components in the technical mixtures.[106]

The series of surfactants shows the typical behavior found for C_iE_j with different j. The lower the degree of ethoxylation, *i.e.* the more hydrophobic the surfactant, the lower the temperature range for microemulsions. However, technical surfactants induce the formation of liquid crystalline (LC) phases to a much greater extent than pure surfactants covering parts of the one-phase microemulsion, thus restricting or

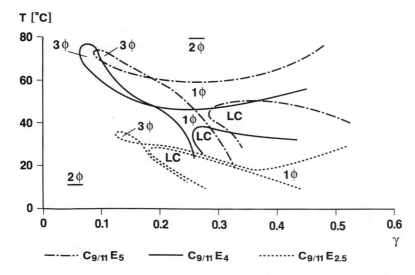

Figure 12.20 *Phase behavior of water/rape oil methyl ester/$C_{9/11}E_j$ mixtures with water/oil ratio = 1:1 as function of surfactant mass fraction*
(Reproduced by permission of Hanser from ref. 104)

even preventing the use of bicontinuous microemulsions in soil extraction. $C_{9/11}E_5$ forms microemulsions with rape oil methyl ester and water at temperatures between 39 and 60 °C. The crossing of the phase boundaries is found at 60 °C and 20% emulsifier. $C_{9/11}E_4$ makes even lower temperatures, between 30 and 47 °C, possible. For this ternary system, values of $T = 47$ °C and 20% emulsifier are obtained for the transition point from three-phase to one-phase microemulsions. $C_{9/11}E_{2.5}$, however, cannot be used owing to the extended LC phases.

The results shown in Table 12.4 demonstrate that for an extractant/soil mass ratio of six the extracted amounts of pyrene from the contaminated soil material is nearly the same for $\alpha = 0.5$ and $\alpha = 0.3$. Even for the lower oil content the pyrene concentration is far from the solubilization limit. It is worth noting that the results for the extraction with microemulsions at extraction temperatures much lower than the boiling point of toluene are higher than for the extraction with hot toluene.

Table 12.4 *Extracted amount of pyrene (Q) per mass of solid for microemulsions with different content (α) of rape oil methyl ester (RME) at temperatures T_{extr} compared to hot extraction with toluene (Q_{tol})*

System	α	γ	T_{extr}/ °C	Q/mg/kg soil	Q/Q_{tol}
Toluene (soxhlet)	–	–	111	400	1.00
$C_{9/11}E_4$–RME–water	0.5	22	43	452	1.13
$C_{9/11}E_4$– RME–water	0.3	17	39	432	1.08

Splitting the microemulsion by lowering the temperature results in partial separation of pyrene with the oil phase. The results obtained for model systems given in Table 12.5 show that the amount of pyrene which can be removed from the microemulsion is closely connected to the volume of oil which separates. Whereas the concentration in the oil phase is nearly independent of the separated volume, the concentration in the aqueous phase decreases considerably. The temperature dependence of the separated oil volume for different microemulsion compositions is shown in Figure 12.21. The separated volume increases up to a certain limit with

Table 12.5 *Separated oil volume* $(V(oil)/V_0(oil))$ *and distribution equilibrium of pyrene (Py) between the aqueous (aq) and the oil phase (oil) for the microemulsion system* $C_{9/11}E_4$/*rape oil methyl ester/water* (μE) *at different splitting temperatures* T_{sp} *and different oil content* (α) *as a function of surfactant mass fraction* (γ)

α	γ	$T_{sp}/\ °C$	$V(oil)/$ $V_0(oil)$	$c_{Py}(oil)/$ $c_{Py}(\mu E)$	$c_{Py}(aq)/$ $c_{Py}(\mu E)$	$m_{Py}(oil)/$ $m_{Py}(\mu E)$	$m_{Py}(aq)/$ $m_{Py}(\mu E)$
0.3	0.17	30	0.32	2.88	0.94	0.24	0.84
0.3	0.17	22	0.56	2.94	0.71	0.43	0.58
0.3	0.17	16	0.67	3.00	0.63	0.52	0.50
0.3	0.17	10	0.73	2.90	0.55	0.58	0.46
0.5	0.22	34	0.38	1.88	0.93	0.28	0.73
0.5	0.22	26	0.57	1.87	0.72	0.44	0.54
0.5	0.22	20	0.66	1.94	0.65	0.52	0.48
0.5	0.22	16	0.69	1.93	0.63	0.54	0.45

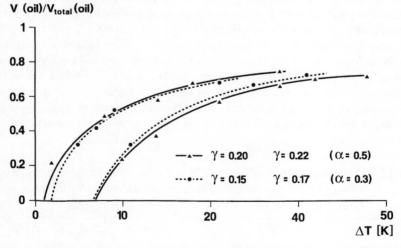

Figure 12.21 *Temperature dependence of the separated oil volume for different compositions of the microemulsion formed by water/rape oil methyl ester/$C_{9/11}E_4$ for different oil/water ratios* α *and surfactant content* γ
(Reproduced by permission of Hanser from ref. 104)

increasing temperature difference. A temperature difference of about 20–25 K is sufficient to reduce the contaminant concentration in the microemulsion to about half of its initial value. These data clearly demonstrate the possible applications of microemulsions for soil remediation.

12.8   References

1. J.H. Schulman, W. Stoeckenius and L.M. Prince, *J. Phys. Chem.*, 1959, **63**, 1677.
2. V. Gold, K. Loening, A.D. McNaught and P. Semi, 'Compendium of Chemical Terminology: IUPAC Recommendations', Blackwell Scientific, Oxford, 1987.
3. W. Umbach, 'Kosmetik', Thieme, Stuttgart, 1995.
4. U. Zeidler, *Fette Seifen Anstrichmittel*, 1985, **87**, 403.
5. A. Ansmann, R. Kawa, E. Prat and A. Wadle, *Seife Öle Fette Wachse*, 1994, **120**, 158.
6. Th.F. Tadros, in 'Emulsions—A Fundamental and Practical Approach', ed. J. Sjöblom, Kluwer, Dordrecht, 1992, p. 173.
7. Th. Förster, B. Jackwerth, W. Pittermann, W. von Rybinski and M. Schmitt, *Cosmet. Toiletries*, 1997, **112**, 73.
8. Th. Förster, F. Schambil and H. Tesmann, *Int. J. Cosmet. Sci.*, 1990, **12**, 217.
9. B.W. Barry and G.M. Saunders, *J. Colloid Interface Sci.*, 1972, **38**, 616.
10. J. Talman and E.M. Rowan, *J. Pharm. Pharmacol.*, 1970, **22**, 338.
11. G.M. Eccleston, *Pharm. Int.*, 1986, **63**, 70.
12. H. Junginger, C. Führer, J. Ziegenmeyer and S. Friberg, *J. Soc. Cosmet. Chem.*, 1979, **30**, 9.
13. S. Fukushima, M. Takahashi and M. Yamaguchi, *J. Colloid Interface Sci.*, 1976, **57**, 201
14. N. Krog and K. Larsson, *Chem. Phys. Lipids*, 1968, **2**, 129.
15. R. Pal, *J. Rheol.*, 1992, **36**, 1245.
16. Y. Otsubo and R.K. Prud'homme, *Rheol. Acta*, 1994, **33**, 29.
17. C.R. Chen and J.L. Zatz, *J. Soc. Cosmet. Chem.*, 1992, **43**, 1.
18. N. Gladwell, M.J. Grimson, R.R. Rahalkar and P. Richmond, *J. Chem. Soc., Faraday Trans.*, 1984, **81**, 643.
19. B.W. Barry and A.J. Grace, *J. Pharm. Sci.*, 1972, **61**, 335.
20. R. Hofmann, Th. Förster, W. von Rybinski and A. Wadle, *Prog. Colloid Polym. Sci.*, 1995, **98**, 106.
21. M. Kahlweit and R. Strey, *Angew. Chem.*, 1985, **97**, 665.
22. Th. Förster, in 'Surfactants in Cosmetics', ed. M.M. Rieger and L.D. Rhein, Surfactant Science Series, vol. 68, Dekker, New York, 1997, p. 105.
23. R. Aveyard, B.P. Binks and P.D.I. Fletcher, *Langmuir*, 1989, **5**, 1210.
24. Th. Förster, and H. Tesmann, *Cosmet. Toiletries*, 1991, **106**, 49.
25. Th. Förster, F. Schambil and W. von Rybinski, *J. Disp. Sci. Technol.*, 1992, **13**, 183.
26. Th. Förster, W. von Rybinski, H. Tesmann and A. Wadle, *Int. J. Cosmet. Sci.*, 1994, **16**, 84.
27. A. Wadle, H. Tesmann, M. Leonard and Th. Förster, in 'Surfactants in Cosmetics', ed. M. M. Rieger and L.D. Rhein, Surfactant Science Series, vol. 68, Dekker, New York, 1997, p. 207.
28. S.H. Gohla and J. Nielsen, *Seife Öle Fette Wachse*, 1995, **121**, 707.
29. M. Seiller, F. Puisieux and J.L. Grossiord in 'Surfactants in Cosmetics', ed. M.M. Rieger and L.D. Rhein, Surfactant Science Series, vol. 68, Dekker, New York, 1997, p. 139.

30. I. Terrisse, M. Seillier, A. Rabaron, A. Magnet, C. Le Hen-Ferrenbach and J.L. Grossiord, *Int. J. Cosmet. Sci.*,1993, **15**, 53.
31. Th. F. Tadros, C. Py, J. Rouviere, M. C. Taelman and P. Loll, *Seife Öle Fette Wachse*, 1995, **121**, 714.
32. G.H. Dahms and M. Tagawa, 'Proceedings of the 19th IFSCC Congress', Society of Cosmetic Chemists, Sydney, 1996, no. 21.
33. E. Friberg and M. Podzimek, *Colloid Polym. Sci.*, 1984, **262**, 252.
34. A. Zombeck and G.H. Dahms, 'Proceedings of the 19th IFSCC Congress', Society of Cosmetic Chemists, Sydney, 1996, no. 24.
35. K. Shinoda and H. Kunieda, in 'Encyclopedia of Emulsion Technology', ed. P. Becher, Dekker, New York, 1983, vol. 1, p. 337.
36. Th. Förster, W. von Rybinski and A. Wadle, *Adv. Colloid Interface Sci.*, 1995, **58**, 119.
37. E. Nürnberg and W. Pohler, *Prog. Colloid Polym. Sci.*, 1984, **69**, 48.
38. H. Sagitani, *J. Disp. Sci. Technol.*, 1988, **9**, 115.
39. F. Comelles, V. Megias, J. Sanchez, J.L. Parra, J. Coll, F. Balaguer and C. Pelejero, *Int. J. Cosmet. Sci.*, 1989, **11**, 5.
40. T. Suzuki, M. Nakamura, H. Sumida and A. Shigeta, *J. Soc. Cosmet. Chem.*, 1992, **43**, 21.
41. K. Hill, W. von Rybinski and G. Stoll, 'Alkyl Polyglycosides', VCH, Weinheim, 1997.
42. K. Fukuda, O. Söderman, B. Lindman and K. Shinoda, *Langmuir*, 1993, **9**, 2921.
43. Th. Förster, B. Guckenbiehl, H. Hensen and W. von Rybinski, *Prog. Colloid Polym. Sci.*, 1996, **101**, 105.
44. Th. Förster, B. Guckenbiehl, A. Ansmann and H. Hensen, *Seife Öle Fette Wachse*, 1996, **122**, 746.
45. *Ger. Pat.* 19 62 40 51.4 (Henkel) 1997.
46. B. Guckenbiehl, W. von Rybinski and H. Tesmann, 'Proceedings of the 20th IFSCC Congress', Hungarian Chemical Society, Budapest, 1997, p. 224.
47. M. Kietzmann, W. Löscher, D. Arens, P. Maaß and D. Lubach, *J. Pharmacol. Toxicol. Methods*, 1993, **30**, 75.
48. W. Pittermann, M. Kietzmann and B. Jackwerth, *ALTEX*, 1995, **12**, 196.
49. H.E. Bodde, T. De Vringer and H. E. Junginger, *Prog. Colloid Polymer Sci.*, 1986, **72**, 37.
50. S.E. Friberg and B. Langlois, *J. Disp. Sci. Technol.*, 1992, **13**, 223.
51. B. Langlois and S. E. Friberg, *J. Soc. Cosmet. Chem.*, 1993, **44**, 23.
52. S.E. Friberg, *J. Disp. Sci. Technol.*, 1994, **15**, 359.
53. S.E. Friberg and A. J. Brin, *J. Soc. Cosmet. Chem.*, 1995, **46**, 255.
54. D. Paesold, A. van Drunen, M. Höglinger, H. Niedermayr, R. Cornely and M. Köhler, *Stahl Eisen*, 1994, **114**, 75.
55. H. Turner and K. Oldridge, *Stahl Eisen*, 1994, **114**, 71.
56. C.A. Smits, *Manuf. Eng. Mater. Process.*, 1994, **41**, 99.
57. W.E. Lucke, *Lubr. Eng.*, 1996, **52**, 596.
58. R.M. Dick, *Manuf. Eng. Mater. Process.*, 1994, **41**, 339.
59. R.R. Mathis, in 'Polyester: 50 Years of Achievements', The Textile Institute, Manchester, 1993, p. 74.
60. *Ger. Pat.* 40 04 946 A1 (Wacker) 1991.
61. H.J. Lautenschläger, J. Bindl and K. G. Huhn, *Textil Praxis Int.*, 1992, **47**, 460.
62. *Ger. Pat.* 37 23 697 A1 (Pfersee) 1988.
63. *Eur. Pat.* 0 13 81 92 B1 (Dow Corning) 1988.
64. *Eur. Pat.* 44 20 98 A2 (Wacker) 1991.

65. G.R. Gray and H.C.H. Darley, 'Composition in Properties of Oil Well Drilling Fluids', Gulf Publishing Company, Houston, 1980/81.
66. T.J. Ballard, S. P. Beare and T. A. Lawless, 'SPE European Petroleum Conf.', Cannes, 1992, paper 24974.
67. L. Bailey, J.H. Denis and G.C. Maitland, in 'Chemicals in the Oil Industry: Developments and Applications', ed. P.H. Ogden, The Royal Society of Chemistry, Cambridge, 1991, p. 53.
68. F.K. Mody and A.H. Hale, 'SPE/IADC Drilling Conf.', Amsterdam, 1993, paper 25728.
69. H.-D. Dörfler, 'Grenzflächen- und Kolloidchemie', VCH, Weinheim, 1994.
70. K.W. Dillan, E.D. Goddard and D.A. McKenzie, *J. Am. Oil Chem. Soc.*, 1980, **57**, 230.
71. K.W. Dillan, E.D. Goddard and D.A. McKenzie, *J. Am. Oil Chem. Soc.*, 1979, **56**, 59.
72. K.H. Raney and H.C. Benson, *J. Am. Oil Chem. Soc.*, 1990, **67**, 722.
73. B.J. Carroll, *J. Colloid Interface Sci.*, 1981, **79**, 126.
74. H.S. Kielman and P.J.F. van Steen, in 'Surface Active Agents', Society of Chemical Industry, London, 1991, p. 191.
75. D.G. Stevenson, in 'Surface Activity and Detergency', ed. K. Durham, Macmillan, London, 1961, p. 6.
76. A.S.C. Lawrence, *Nature*, 1959, **183**, 1491.
77. K.H. Raney, W.J. Benton and C.A. Miller, *J. Colloid Interface Sci.*, 1987, **117**, 282.
78. C. Solans and N. Azemar, in 'Organized Solutions', ed. S. Friberg and B. Lindman, Surfactant Science Series, vol. 44, Dekker, New York 1992, p. 19.
79. K. Shinoda and S. Friberg, 'Emulsions and Solubilization', Wiley, New York, 1986.
80. R.L. Reed and R.N. Healy, in 'Improved Oil Recovery by Surfactant and Polymer Flooding', ed. D.O. Shah and R.S. Schechter, Academic Press, New York, 1977, p. 383.
81. F. Schambil and M.J. Schwuger, *Colloid Polymer Sci.*, 1987, **265**, 1009.
82. G.H. Findenegg and B. Föllner, unpublished data, 1986.
83. M.J. Schwuger, in 'Structure/Performance Relationships in Surfactants', ed. M.J. Rosen, ACS Symp. Ser. 253, Washington, 1984, p. 3.
84. M. Kahlweit and R. Strey, in 'Proceedings of Vth Int. Conf. on Surface and Colloid Sci.', Potsdam, ed. H. L. Rosano, New York, American Chemical Society, 1985.
85. H.L. Benson, K.R. Cox and J.E. Zweig, *HAPPI*, 1985, **22**, 50.
86. C.A. Miller and K.H. Raney, *Colloids Surf. A*, 1993, **74**, 169.
87. W.J. Benton, K.H. Raney and C.A. Miller, *J. Colloid Interface Sci.*, 1986, **110**, 363.
88. H. Kunieda and K. Shinoda, *J. Disp. Sci. Technol.*, 1982, **3**, 233.
89. C.-P. Kurzendörfer and W. Wichelhaus (Henkel) *Eur Pat.* 0288 858 1988.
90. H. Krüßmann and R. Bercovici, *J. Chem. Tech. Biotechnol.*, 1991, **50**, 399.
91. H. Krüßmann and R. Bercovici, *Tenside Surfactants Det.*, 1993, **30**, 99.
92. S.V. Vaeck, *Tenside Det.*, 1975, **12**, 107.
93. S.V. Vaeck, D. Toureille, J. Constant and W. Verlege, *Tenside Det.*, 1977, **14**, 15.
94. S.V. Vaeck, *Tenside Det.*, 1984, **21**, 74.
95. M. Richter, *Textilreinigung*, 1977, **24**, 159.
96. C. Solans, J. Garcia Dominguez and S.E. Friberg, *J. Disp. Sci. Technol.*, 1985, **6**, 523.
97. F. Comelles, C. Solans, N. Azemar, J. Sánchez Leal and J.L. Porra, *Tenside Det.*, 1985, **22**, 323.
98. H.-D. Dörfler, A. Große and H. Krüßmann, *Tenside Surfactants Det.*, 1995, **32**, 484.
99. W.D. Clemens, F. H. Haegel, K. Stickdorn, M.J. Schwuger and L. Webb, in 'Contaminated Soil '93', ed. F. Arendt, G.J. Annokée, R. Bosman and W.J. van den Brink, Kluwer, Dordrecht, 1993, p. 1315.
100. W. Fabig, K. Hund and K.J. Groß, *Fat. Sci. Technol.* 1989, **9**, 357.

101. K. Mönig, W.D. Clemens, F.H. Haegel and M.J. Schwuger, in 'Micelles, Micro-emulsions, and Monolayers: Science and Technology', ed. D.O. Shah, Dekker, New York, 1998, p. 215.
102. W.D. Clemens, F. H. Haegel, K. Stickdorn, M.J. Schwuger and L. Webb, *World Pat.* 94/04289, 1994.
103. K. Bonkhoff, F.H. Haegel, K. Mönig, M.J. Schwuger and G. Subklew, 'Contaminated Soil '95', ed. W.J. von den Brink, R. Bosman and F. Arendts, Kluwer, Dordrecht, 1995, p. 1157.
104. K. Mönig, F.H. Haegel and M.J. Schwuger, *Tenside Surfactants Det.*, 1996, **33**, 228.
105. K.V. Schubert, R. Strey and M. Kahlweit, *J. Colloid Interface Sci.*, 1987, **141**, 21.
106. H. Kunieda and M. Yamagata, *Langmuir*, 1993, **9**, 3345.

Subject Index